Computational Complexity and Feasibility of Data Processing and Interval Computations

Applied Optimization

Volume 10

Series Editors:

Panos M. Pardalos
University of Florida, U.S.A.

Donald Hearn
University of Florida, U.S.A.

The titles published in this series are listed at the end of this volume.

Computational Complexity and Feasibility of Data Processing and Interval Computations

by

Vladik Kreinovich

University of Texas at El Paso,
Texas, U.S.A.

Anatoly Lakeyev

Computing Center, Russian Academy of Sciences,
Irkutsk, Russia

Jiří Rohn

Charles University and Academy of Sciences,
Prague, Czech Republic

Patrick Kahl

IBM, Tucson, Arizona, U.S.A.

KLUWER ACADEMIC PUBLISHERS
DORDRECHT / BOSTON / LONDON

A C.I.P. Catalogue record for this book is available from the Library of Congress.

ISBN 0-7923-4865-6

Published by Kluwer Academic Publishers,
P.O. Box 17, 3300 AA Dordrecht, The Netherlands.

Sold and distributed in the U.S.A. and Canada
by Kluwer Academic Publishers,
101 Philip Drive, Norwell, MA 02061, U.S.A.

In all other countries, sold and distributed
by Kluwer Academic Publishers,
P.O. Box 322, 3300 AH Dordrecht, The Netherlands.

Printed on acid-free paper

Printed in the Netherlands

CONTENTS

PREFACE

Targeted audience

- *Specialists in numerical computations*, especially in *numerical optimization*, who are interested in designing algorithms with automatic result verification, and who would therefore be interested in knowing how general their algorithms can in principle be.
- *Mathematicians* and *computer scientists* who are interested in the *theory of computing* and computational complexity, especially computational complexity of numerical computations.
- *Students* in applied mathematics and computer science who are interested in computational complexity of different numerical methods and in learning general techniques for estimating this computational complexity. The book is written with all explanations and definitions added, so that it can be used as a *graduate level textbook*.

What this book is about

Data processing. In many real-life situations, we are interested in the value of a physical quantity y that is difficult (or even impossible) to measure directly. For example, it is impossible to directly measure the amount of oil in an oil field or a distance to a star. Since we cannot measure such quantities *directly*, we measure them *indirectly*, by measuring some other quantities x_i and using the known relation between y and x_i's to reconstruct y. The algorithm that transforms the results $\tilde{x}_i$ of measuring x_i into an estimate $\tilde{y}$ for y is called *data processing*.

Error estimation for data processing: interval computations. The input data for *data processing* algorithms come from measurements and are, therefore, not precise: the measured value $\tilde{x}_i$ is, in general, different from the (unknown) actual value x_i of the measured quantity. In some cases, we know the ***probabilities*** of different measurement errors $\Delta x_i = \tilde{x}_i - x_i$, but in many other practical situations, we only know the ***upper bound*** Δ_i for this error. In such situations, the only information that we have about the (unknown) actual value x_i is that $|x_i - \tilde{x}_i| \leq \Delta_i$, i.e., that the value x_i belongs to the ***interval*** $\mathbf{x}_i = [\tilde{x}_i - \Delta_i, \tilde{x}_i + \Delta_i]$.

Since the ***input*** data of the data processing algorithm are not precise, the ***result*** $\tilde{y}$ of data processing is also not precise: it may differ from the (unknown) actual value y of the estimated quantity. It is, therefore, necessary that the data processing algorithm return not only the numerical ***estimate*** $\tilde{y}$ itself, but also the ***bound*** Δ of the possible ***inaccuracy*** of this estimate, so that we will be able to deduce the ***interval*** $\mathbf{y} = [\tilde{y} - \Delta, \tilde{y} + \Delta]$ of possible values of the desired quantity y.

> For example, if our estimate of an oil deposit is $\tilde{y} = 100$ mln. tons, it may mean 100 ± 10, in which case it is reasonable to start exploiting this well, or100 ± 100, in which case additional measurements are in order.

In other words, we need data processing methods that produce not just ***approximate*** results but results with ***automatic result verification***. Such interval-processing algorithms are called ***interval computations***.

Interval computations: from Archimedes to success. Interval computations can be traced to Archimedes who produced two-sided estimates for π, i.e., an ***interval*** that is guaranteed to contain π. In this century, basic interval methods were pioneered as early as 1914 by Norbert Wiener, the future father of cybernetics, in his analysis of measurement accuracy. The real boom started with the space age. In 1959, Ramon E. Moore, a young Stanford student working for Lockheed Missiles and Space Co., published a technical report in which he developed a new technique called ***interval computations*** and applied this technique to the problem of computing the trajectory of an Earth-Moon spaceship (a problem for which errors can indeed be disastrous). After Moore's 1966 pioneer book, interval computations started to actively flourish. Methods with automatic results verification were developed for numerous computational problems, and these methods were applied to areas ranging from manufacturing to economy to quantum physics to space exploration.

Interval computations: problems. In spite of numerous success stories, there were also many problems in which all known algorithms either produced unrealistically overestimated bounds, or required unrealistically long computations. In 1981, A. A. Gaganov, a student from St. Petersburg (then, Leningrad) University, proved that some of these problems are indeed computationally intractable (in some precise sense). Since then, many other intractability results appeared. It turned out (somewhat unexpectedly) that even for the *simplest* data processing algorithms (e.g., quadratic or piecewise-linear), the above-described problem of estimating accuracy of the results of data processing (i.e., interval computations) is, in general, *computationally intractable.*

It is important to know what problems are feasible. Since these intractability results occur even for simple algorithms, it is *extremely important* for algorithm developers to know which classes of problems are solvable, and which are intractable, so that they will be able to concentrate on the classes of problems for which *feasible* general *algorithms exist.*

A gap. Unfortunately, there has been, so far, no book specifically devoted to computational complexity (feasibility and intractability) of interval computations. There are several good books that describe interval computations techniques, and applications of interval computations. These books mention some computational complexity results; however, in general, these results are mainly *scattered* in various journal papers and monographs, and *finding* out whether a given problem is known to be feasible or intractable is often very *difficult.*

The main goal of this book is to fill this gap.

Our main goal

This book describes classes of problems for which interval computations (i.e., data processing with automatic results verification) are feasible, and when they are intractable.

This book is not only a comprehensive survey of known results; many of the results published in this book have never been published before.

Thanks

This book is based on the authors' invited talks at several *international conferences* and workshops. The authors are thankful to all the participants of the International Symposium on Scientific Computing, Computer Arithmetic, and Validated Numerics (SCAN'95), of post-STOC'97 (ACM Symposium on Theory of Computing) workshop on computational complexity and interval computations, and of several other conferences for the attention to this work and for valuable discussions.

We are especially thankful to Gerhard Heindl, Hoon Hong, R. Baker Kearfott, Werner Krandick, Luc Longpré, Wolfram Luther, Arnold Neumaier, Dietmar Ratz, Sergey Shary, Yuri Shokin, Bill Walster, and Jürgen Wolff von Gudenberg. Our special thanks to Tran Cao Son for proof-reading the text.

We want to thank the *granting agencies* that made this book possible:

- the Charles University Grant Agency through grant GAUK 195/96;
- the Czech Republic Grant Agency through grant GACR 201/95/1484;
- Future Aerospace Science and Technology Program (FAST) Center for Structural Integrity of Aerospace Systems, supported by the US Air Force Office of Scientific Research (AFOSR), Air Force Materiel Command, under grant number F49620-95-1-0518;
- the German Science Foundation;
- Laforia, Laboratory of Forms and Artificial Intelligence of the University of Paris VI;
- National Aeronautic and Space Administration (NASA) through grants No. NAG 9-757 and NCCW-0089;
- NASA Pan-American Center for Earth and Environmental Studies (PACES); and
- National Science Foundation (NSF) through grants Nos. CDA-9015006, EEC-9322370, and DUE-9750858.

And, last but not the least, we want to thank John Martindale, Arlene Apone, Sharon Donovan, and all the wonderful staff of Kluwer Academic Publishers for their inspiration, patience, and help.

Vladik Kreinovich, Anatoly Lakeyev, Jiří Rohn, and Patrick Kahl
El Paso, Irkutsk, Prague, and Tucson, Summer 1997

1

INFORMAL INTRODUCTION: DATA PROCESSING, INTERVAL COMPUTATIONS, AND COMPUTATIONAL COMPLEXITY

This introduction starts with material aimed mainly at those readers who are not well familiar with interval computations *and/or with the computational complexity aspects of data processing and interval computations. It provides the motivation for the basic mathematical and computational problems that we will be analyzing in this book. Readers who* are *well familiar with these problems can skip the bulk of this chapter and go straight to the last section that briefly outlines the structure of the book.*

In brief, this chapter's analysis starts with the the following problem (that is considered one of the basic problems of interval computations): given a function $f(x_1, ..., x_n)$ of n real variables, and n intervals $\mathbf{x}_i = [\underline{x}_i, \overline{x}_i]$, compute the range

$$\mathbf{y} = f(\mathbf{x}_1, ..., \mathbf{x}_n) = \{f(x_1, ..., x_n) | x_1 \in \mathbf{x}_1, ..., x_n \in \mathbf{x}_n\}.$$

A typical application of this problem is: from the measurements, we know the approximate values $\tilde{x}_i$ of physical quantities x_i, and we know the guaranteed accuracy Δ_i of each measurement. As a result, we know that x_i belongs to the interval $\mathbf{x}_i = [\tilde{x}_i - \Delta_i, \tilde{x}_i + \Delta_i]$. We also know the algorithm $f(x_1, \ldots, x_n)$ that transforms the values x_i into the value of the desired quantity y. We want to know the set of possible values of y. For a continuous function $f(x_1, \ldots, x_n)$, this set is an interval (we will denote it by $\mathbf{y} = [\underline{y}, \overline{y}]$). So, the question is: can we compute the endpoints $\underline{y}$ and $\overline{y}$ of this interval $\mathbf{y}$ in reasonable time?

1.1. Data processing: what is it and why we need it

In many real-life situations, we are interested in the value of a physical quantity y that is difficult (or even impossible) to measure directly.

For *example*, it is difficult to directly measure the amount of oil in a well, the distance to a quasar, or the mass of an elementary particle.

Since we cannot measure these quantities *directly*, we have to measure them *indirectly*; in other words, instead of directly measuring y, we do the following:

- First, we measure some *other* (easier-to-measure) quantities $x_1, \ldots, x_n$ that are related with y by a known relation $y = f(x_1, \ldots, x_n)$.
- Then, we use the *results* $\tilde{x}_1, \ldots, \tilde{x}_n$ of measuring x_i and the known *function* $f(x_1, \ldots, x_n)$ to compute the *estimate* $\tilde{y} = f(\tilde{x}_1, \ldots, \tilde{x}_n)$ of the desired quantity y.

  For *example*, to estimate the amount of oil in a well, we perform several measurements of density, conductivity, and sound velocity, and use the results of these measurements to reconstruct y.

The two-stage procedure that we have just described is called *indirect measurement* (see, e.g., Rabinovich [333]):

- The first stage of this procedure, on which we actually measure x_i, is called *direct measurements*. The results $\tilde{x}_i$ of these direct measurements will be used for computations and are therefore called *data*.
- The second stage, on which we use the measurement data $\tilde{x}_1, \ldots, \tilde{x}_n$ to compute the desired estimate $\tilde{y}$, is called *data processing*.

1.2. Error estimation for indirect measurements: an important practical problem

Measurements are never absolutely accurate; as a result, the actual value x_i of the measured quantity may differ from the measured value $\tilde{x}_i$. The resulting measurement errors $\Delta x_i = \tilde{x}_i - x_i$ cause our estimate $\tilde{y} = f(\tilde{x}_1, \ldots, \tilde{x}_n)$ to differ from the actual value $y = f(x_1, \ldots, x_n)$ of the desired quantity.

In many cases, it is extremely important to know how different the actual value y can be from our estimate. For *example*, if our estimate for the amount of oil in a well is 100 million ton, then our actions will be depend on how *accurate* this estimate is:

- If the estimate is reasonably *accurate*, e.g., if the actual value is 100 ± 10 million ton, then we should probably start commercially exploiting this well.
- On the other hand, if the estimate is *not accurate*, and the error $\Delta y = \tilde{y} - y$ can be as large as ± 100, then, in spite of the optimistic estimate $\tilde{y}$, it is quite possible that the actual amount of oil y is close to 0, i.e., that there is no commercial amount of oil at all. This means that the measurements that we have performed so far do not give us enough information to decide on what to do, and so, we must perform further measurements.

Hence, it is not sufficient to have the *estimate* $\tilde{y}$; we also need to know the *accuracy* of this estimate, i.e., in other words, we need to know what values of the error $\Delta y = \tilde{y} - y$ are possible.

1.3. Interval computations: what they are and why we need them

The problem of error estimation of the results of indirect measurements and data processing occurs in many different areas of science and engineering. Traditional engineering methods of solving this problem are based on the assumption that we know the *probabilities* of different possible values of measurement errors $\Delta x_i = \tilde{x}_i - x_i$.

These probabilities can be determined, e.g., by a *calibration* of the corresponding measuring instrument, in which:

- First, we measure, in a laboratory, several different quantities $x^{(1)}, \ldots, x^{(K)}$ by *two* measuring instruments:

- the instrument that we want to calibrate, and
- a more precise instrument (e.g., an instrument using a regional or national standard for the corresponding unit).

Since the second measuring instrument is much more accurate than the first instrument (which we are calibrating), we can *neglect* the measurement errors of this second instrument, and take the results of measuring the values $x^{(k)}$ by this instrument as the *actual* values of the measured quantity. In this case, for each "double" measurement, the difference $\Delta x^{(k)} = \tilde{x}^{(k)} - x^{(k)}$ between the two measurement results becomes a good estimate for the measurement error of the calibrated instrument.

- As a result of several tests, we get a sample of values $\Delta x^{(1)}, \ldots, \Delta x^{(K)}$. We can now use the standard methods of mathematical statistics to reconstruct the probability distribution for the errors Δx.

First probability-based error estimation methods were proposed by Gauss in the very beginning of the 19th century. During the following two centuries, many probability-based algorithms have been designed, perfected, and thoroughly analyzed (see, e.g., Wadsworth [422], Rabinovich [333]). Of course, there still are open problems, but in most practical cases in which we *know the probabilities* of measurement errors, the existing methods work reasonably well.

The problem is that in many practical situations, we *do not know the probabilities.* This happens for two different reasons:

- In *fundamental physics* and in other types of cutting-edge research, the measurements that we are dealing with are the most accurate that we can get. There is simply no more accurate measuring instrument that we could use for calibration, and therefore, we *cannot* determine the probabilities.
- In many *manufacturing* situations, calibration is, in principle, possible, but the cost of calibrating every sensor makes it economically unrealistic. Most manufacturing processes use mass-produced and therefore (relatively) cheap sensors, and individual calibration would prohibitively skyrocket their cost.

In both types of situations, we do not know the probabilities of different values of the measurement error Δx_i. At best, we know the *upper bound* Δ_i on the possible values of measurement error (i.e., we know that $|\Delta x_i| \le \Delta_i$). This *guaranteed upper bound* is usually supplied by the manufacturer of the measuring instrument.

If no bound is known for the error, this means that the actual value x_i of the measured quantity can be arbitrarily different from the measurement result $\tilde{x}_i$, and therefore, even after this "measurement", arbitrarily large and arbitrarily small values of x_i are possible. In other words, such "measurements" give us no information about x_i and are, therefore, of no use.

In such non-probabilistic situations, after we have measured the quantity x_i and obtained the measured result $\tilde{x}_i$, the only information that we get about the *actual* (unknown) value x_i of the measured quantity is that this value belongs to the *interval* $\mathbf{x}_i = [\tilde{x}_i - \Delta_i, \tilde{x}_i + \Delta_i]$. (Strictly speaking, we have a closed non-empty bounded interval. In this book, when we say "interval", we will always mean an interval of this type.)

In this setting, the problem of estimating errors of data processing takes the following form:

- From the measurements, we know the *approximate* values $\tilde{x}_i$ of the physical quantities x_i.
- We also know the guaranteed *accuracy* Δ_i of each measurement.

  As a result, we know that the actual value of x_i belongs to the interval $\mathbf{x}_i = [\tilde{x}_i - \Delta_i, \tilde{x}_i + \Delta_i]$.
- We know the *algorithm* $f(x_1, \ldots, x_n)$ that transforms the values x_i into the value of the desired quantity y.

We want to know the set of possible values of y.

This range of possible values of $y = f(x_1, \ldots, x_n)$ is equal to

$$\mathbf{y} = f(\mathbf{x}_1, \ldots, \mathbf{x}_n) = \{f(x_1, \ldots, x_n) | x_1 \in \mathbf{x}_1, \ldots, x_n \in \mathbf{x}_n\}.$$

For continuous functions $f(x_1, \ldots, x_n)$, this range is also an interval. Because of this, we must *compute an interval* $\mathbf{y}$ based on the intervals $\mathbf{x}_i$. Computing intervals based on interval data is usually called *interval computations*; so, the above range estimation problem is one of the problems of *interval computations.* (Due to the practical importance of this problem, it is considered one of *basic* problems of interval computations.)

1.4. Interval computations: approximate methods, traditional optimization techniques, naive interval computations, and more sophisticated interval methods (in brief)

Approximate methods. In many engineering applications, we do not necessarily need to know the *exact* endpoints $\underline{y}$ and $\overline{y}$ of the interval $\mathbf{y} = [\underline{y}, \overline{y}]$. It is quite sufficient to get *approximate* values for these endpoints. One way to get such approximate estimates is to *linearize* the function $f(x_1, \ldots, x_n)$, i.e., represent x_i as $x_i = \tilde{x}_i - \Delta x_i$, expand the resulting expression

$$f(x_1, \ldots, x_n) = f(\tilde{x}_1 - \Delta x_1, \ldots, \tilde{x}_n - \Delta x_n)$$

in Taylor series in Δx_i, and neglect quadratic and higher order terms in this expansion. As a result, we get an approximate formula

$$f(x_1, \ldots, x_n) \approx a_0 + a_1 \cdot \Delta x_1 + \ldots + a_n \cdot \Delta x_n,$$

where $a_0 = f(\tilde{x}_1, \ldots, \tilde{x}_n) = \tilde{y}$ and

$$a_i = -\frac{\partial f}{\partial x_i}(\tilde{x}_1, \ldots, \tilde{x}_n).$$

For the resulting (approximate) linear function, the desired range $\mathbf{y} = [\underline{y}, \overline{y}]$ (over all possible values Δx_i for which $|\Delta x_i| \le \Delta_i$) is easy to compute:

$$\underline{y} = \tilde{y} - |a_1| \cdot \Delta_1 - \ldots - |a_n| \cdot \Delta_n;$$

$$\overline{y} = \tilde{y} + |a_1| \cdot \Delta_1 + \ldots + |a_n| \cdot \Delta_n.$$

Numerous engineering and physical situations, in which this linearization techniques have been successfully used, are described, e.g., in Rabinovich [333].

In many practical situations, approximate methods do not work: we need guaranteed estimates. The linearization method works well if two conditions are met:

1) first, quadratic and higher order terms that we neglected are really small; and,

2) second, no big harm is done if we slightly overestimate or underestimate the values $\underline{y}$ and $\overline{y}$.

The first condition, i.e., the possibility to neglect the terms

$$\frac{\partial^2 f}{\partial x_i \partial x_j} \cdot \Delta x_i \cdot \Delta x_j,$$

means, in its turn, that:

1a) first, the measurement errors Δx_i are (relatively) small;

1b) second, that the function $f(x_1, \ldots, x_n)$ is smooth enough so that the second (and higher order) derivatives are not too large;

1c) third, that there are not too many quadratic (and higher order) terms (i.e., not too many variables), because otherwise, while each of these terms is small and quite negligible, their sum may add up to a large and non-negligible value.

Alas, in many practical problems, these conditions are not always met:

1a) Measurement errors are sometimes relatively large, so that their squares cannot be safely neglected.

> In applications to *fundamental physics*, this low accuracy may be the best we can *achieve*, while in *manufacturing* applications, the low accuracy may be the best we can *afford*.

1b) The function $f(x_1, \ldots, x_n)$, that describes the relation between the directly measured quantities x_i and the desired quantity y, may be non-smooth.

> For *example*, many detectors of elementary particles (bubble cameras, etc.) use *phase transitions* to amplify (and thus, detect) small signals. The very fact that phase transition processes drastically amplify small differences means exactly that the derivatives of the corresponding data processing function $f(x_1, \ldots, x_n)$ are huge and therefore, even for small measurement errors Δx_i, the resulting quadratic terms may not be negligible.

1c) Many data processing algorithms process lots of values, e.g., when they process the values measured at different moments of time.

> For *example*, to determine the parameters of a quasar, we must analyze the radio signal coming from this quasar. On a typical centimeter wavelength, the signal frequency is in Giga Hertz (billions of cycles per second), and so, even a short (e.g., second-long) observation means that we get *billions* of values to process.

For $n \approx 10^9$, there are so many quadratic terms that their sum is no longer negligible.

2) Finally, there are many problems in which we need a ***guaranteed*** estimate, because mis-estimation can be disastrous.

For *example*, when we plan a mission to another planet, it is not enough to prove that we will ***approximately*** get into this planet: we must ***guarantee*** that we get there and not miss it. When we control a nuclear power plant, it is not sufficient to say that, according to approximate computations, we seem to be in the safe zone: we must *guarantee* that we are safe and that the reactor will not explode.

In all such situations, approximate methods are not sufficient, we must have *guaranteed* estimates for the range $\mathbf{y}$.

Traditional optimization techniques and why they are not always applicable. From the mathematical viewpoint, the problem of finding the endpoints for the range is a typical optimization problem: the lower endpoint $\underline{y}$ is the solution of the *minimization* problem

$$f(x_1, \ldots, x_n) \to \min$$

under the conditions

$$\underline{x}_i \le x_i \le \overline{x}_i, \ 1 \le i \le n$$

(where $\underline{x}_i = \tilde{x}_i - \Delta_i$ and $\overline{x}_i = \tilde{x}_i + \Delta_i$), and the upper endpoint is the solution to the *maximization* problem

$$f(x_1, \ldots, x_n) \to \max$$

under the conditions

$$\underline{x}_i \le x_i \le \overline{x}_i, \ 1 \le i \le n.$$

For a smooth function $f(x_1, \ldots, x_n)$, these are simple optimization problems of the type solved in the first calculus courses. Let us recall the corresponding method and explain why it is not always practically useful. (We urge the reader to read through this explanation, because we *do not simply repeat* well known things from calculus, but we repeat them in such a way that leads us straight *into* the *computational complexity* aspects of data processing problems.)

Let us start with the simplest case of one input ($n = 1$). In this case, the maximum (or minimum) of a function of *one* variable $f(x_1)$ on an interval

$\mathbf{x}_1 = [\underline{x}_1, \overline{x}_1]$ is attained either at one of the *endpoints* $\underline{x}_1$ and $\overline{x}_1$, or *inside* this interval. If the maximum is attained at some point x_1 inside the interval, then the derivative df/dx_1 must be equal to 0 at this point. Therefore, to find the maximum, it is sufficient:

- to find the point at which the derivative is equal to 0 (or all such points if there are several such points on the interval $\mathbf{x}_1$),
- to compute the value $f(x_1)$ for all such "candidate" points and for all the endpoints, and
- to find the *largest* of the values of $f(x_1)$ for all these points; this largest value is the desired maximum (correspondingly, the smallest of these values $f(x_1)$ is the desired minimum).

Of course, if the function $f(x_1)$ is very complicated, then the equation $df/dx_1 = 0$ is also very complicated and therefore, difficult to solve, but for reasonably simple functions $f(x_1)$, this method is very efficient. So, whether this method is useful for our problems, depends on what functions $f(x_1)$ we are interested in.

The main reason why we started looking for new methods is that *linearization*, that is based on approximating an arbitrary function by *linear* terms from its Taylor series, does not always work. If a linear approximation does not work, then the next approximation is *quadratic*. For quadratic functions, the derivative is linear and therefore, the equation $df/dx_1 = 0$ is easy to solve. So, at least for quadratic functions $f(x_1)$, this method (borrowed from calculus textbooks) works pretty well.

A similar textbook method can be used to find a maximum of a function $f(x_1, \ldots, x_n)$ of several variables $x_1, \ldots, x_n$ on a *box* (parallelepiped) $\mathbf{x}_1 \times \ldots \times \mathbf{x}_n$. This method is based on the following idea: When the function $f(x_1, \ldots, x_n)$ attains its maximum at some point $\vec{x}^{\,\text{opt}} = (x_1^{\text{opt}}, \ldots, x_n^{\text{opt}})$, this means that if we change the values of *some* (or all) of the variables, we get a smaller (or equal) value of $f(x_1, \ldots, x_n)$. In particular, if we change the value of only *one* variable x_i, we get a smaller (or equal) value of $f(x_1, \ldots, x_n)$. Thus, each of the functions $f_i(x_i) = f(x_1^{\text{opt}}, \ldots, x_{i-1}^{\text{opt}}, x_i, x_{i+1}^{\text{opt}}, \ldots, x_n^{\text{opt}})$ attains its maximum at $x_i = x_i^{\text{opt}}$. Therefore, either this maximum is attained at one of the *endpoints* $\underline{x}_i$ or $\overline{x}_i$ of the interval $\mathbf{x}_i = [\underline{x}_i, \overline{x}_i]$, or it is attained at the *internal* point, in which case, the corresponding partial derivative much be equal to 0: $\partial f/\partial x_i = 0$. Therefore, at each optimum point $\vec{x}^{\,\text{opt}}$, each of n variables x_i falls into one of the three groups:

- variables for which $x_i^{\text{opt}} = \underline{x}_i$;
- variables for which $x_i^{\text{opt}} = \overline{x}_i$;
- variables for which $\partial f/\partial x_i = 0$.

Hence, we can find the maximum of a function $f(x_1, \ldots, x_n)$ on a given box as follows:

- We consider all possible partitions $\mathcal{P}$ of the set $\{1, \ldots, n\}$ into three subsets $\{1, \ldots, n\} = L \cup R \cup I$.
- For each partition $\mathcal{P} = \langle L, R, I \rangle$, we:
  - take $x_i^{\mathcal{P}} = \underline{x}_i$ for all $i \in L$,
  - take $x_i^{\mathcal{P}} = \overline{x}_i$ for all $i \in R$, and
  - solve the system of equations

    $$\frac{\partial f}{\partial x_i}(x_1, \ldots, x_n) = 0, \text{ for all } i \in I$$

    (with the selected values $x_i = x_i^{\mathcal{P}}$ for $i \notin I$) to find the remaining values $x_i^{\mathcal{P}}$;

  and then compute $f^{\mathcal{P}} = f(x_1^{\mathcal{P}}, \ldots, x_n^{\mathcal{P}})$.
- Finally, we find the largest of the values of $f^{\mathcal{P}}$ and produce this largest value as the desired maximum (correspondingly, the smallest of these values $f^{\mathcal{P}}$ is the desired minimum).

For *quadratic* functions $f(x_1, \ldots, x_n)$, each partial derivative is a *linear* expression and therefore, the corresponding system of *linear* equations is easy to solve. Thus, for small n, it is a reasonable and efficient method.

We said "for small n", because we have to analyze *all* possible partitions, and the number of partitions (3^n) grows very fast with n. Of course, we cannot use this method for $n = 10^9$, but even for a reasonable number of inputs $n \approx 300$, the resulting computations require at least 3^{300} computational steps. How large is this number? Even if assume that each step takes as short a time as possible, i.e., each step takes the time that is necessary for light to pass through the smallest elementary particle, the resulting computation time will exceed the lifetime of the Universe. This is clearly *not* computationally *feasible*.

Of course, the above method is just the *simplest* possible, there exist numerous *faster* optimization methods (see, e.g., Pardalos [324], Horst *et al.* [157], Kearfott [175]), but still, all the methods that *guarantee* the result, require (in some cases) *exponential time*.

Naive interval computations: why and what. Traditional optimization techniques work very well for small n, and only for large n, these methods become non-feasible. Before the computer age, when data were processed by hand, it was extremely difficult even to *process* large amounts of data; estimating errors was an unrealistic dream. For example, practically no error estimation was done for extensive computations used in the design of the first atomic bombs. With the advent of (reasonably) fast computers in the 1950s, it became possible not only to *process* huge amounts of data, but also to try to *estimate errors* of the resulting data processing. At first, mainly linearized error estimation methods were used, but at the end of the 50s, there appeared an important class of real-life computation-intense problems for which linearized methods were not applicable: the problems of *space exploration*. Indeed, for these problems, not just *one* of the conditions (for applicability of linearization) is not met, but *all* of them:

1a) First of all, since we are sending a mission into the *unknown*, we have only very crude estimates of the values of many relevant quantities. For example, when planning a mission to the Moon, we had to rely on Earth-based (and therefore, inaccurate) measurements of the Moon terrain. Since measurement errors are large, quadratic terms cannot be safely neglected.

1b) Second, many processes that we try to control are highly unstable. A minor change in an angle with which the spaceship enters the atmosphere can mean a difference between a safe return and a disaster. Thus, the data processing algorithms $f(x_1, \ldots, x_n)$ are *very sensitive* to the errors in the input data.

1c) The success of a space flight depends on lots of factors, and many of these factors are dynamic (i.e., rapidly changing); in terms of computations, this means that we have to process huge amount of measurements (mainly, sent by telemetry).

2) Finally, the consequences of an error can be truly disastrous, so, we need *guaranteed* estimates.

For these problems, there was a clear need for new error estimation methods, and these methods were invented in 1959 by Ramon E. Moore and published in two technical reports [288, 289] of Lockheed Missiles and Space Co. (Moore also coined the very terms "interval analysis", "interval computations", etc.)

Moore's innovative idea was very natural: since it is difficult to design an algorithm that solves the error estimation problem for *all* functions $f(x_1, \ldots, x_n)$, let us use an *incremental* approach: first solve this problem for *simple* functions $f(x_1, \ldots, x_n)$, and then modify the resulting algorithm so that it will be applicable to *more and more complicated* functions $f(x_1, \ldots, x_n)$. The very possibility to use an incremental approach comes from the fact that inside a computer, every algorithm is performed as a sequence of *elementary operations* (usually, hardware supported) such as *arithmetic operations* addition x_1+x_2, subtraction x_1-x_2, multiplication $x_1 \cdot x_2$, and division x_1/x_2, and (sometimes) applications of elementary functions such as $\exp(x)$, $\log(x)$, $\sin(x)$, $\cos(x)$, $\tan(x)$, $\cot(x)$, $\arcsin(x)$, etc. So, it is reasonable to start with computing the range for these *simple* functions, and then extend the resulting error estimation methods to arbitrarily *complex* algorithms (consisting of multiple elementary operations).

For the *simplest* case when $f(x_1, x_2)$ coincides with one of the *arithmetic* operations, Moore immediately obtained explicit formulas for the resulting interval $\mathbf{y} = [\underline{y}, \overline{y}] = f(\mathbf{x}_1, \mathbf{x}_2) = f([\underline{x}_1, \overline{x}_1], [\underline{x}_2, \overline{x}_2])$:

$$[\underline{x}_1, \overline{x}_1] + [\underline{x}_2, \overline{x}_2] = [\underline{x}_1 + \underline{x}_2, \overline{x}_1 + \overline{x}_2];$$

$$[\underline{x}_1, \overline{x}_1] - [\underline{x}_2, \overline{x}_2] = [\underline{x}_1 - \overline{x}_2, \overline{x}_1 - \underline{x}_2];$$

$$[\underline{x}_1, \overline{x}_1] \cdot [\underline{x}_2, \overline{x}_2] =$$

$$[\min(\underline{x}_1 \cdot \underline{x}_2, \underline{x}_1 \cdot \overline{x}_2, \overline{x}_1 \cdot \underline{x}_2, \overline{x}_1 \cdot \overline{x}_2), \max(\underline{x}_1 \cdot \underline{x}_2, \underline{x}_1 \cdot \overline{x}_2, \overline{x}_1 \cdot \underline{x}_2, \overline{x}_1 \cdot \overline{x}_2)];$$

$$1/[\underline{x}_1, \overline{x}_1] = [1/\overline{x}_1, 1/\underline{x}_1] \text{ if } 0 \notin [\underline{x}_1, \overline{x}_1];$$

$$[\underline{x}_1, \overline{x}_1]/[\underline{x}_2, \overline{x}_2] = [\underline{x}_1, \overline{x}_1] \cdot (1/[\underline{x}_2, \overline{x}_2]).$$

These interval computation formulas for arithmetic operations are called formulas of *interval arithmetic.* Similar formulas exist for elementary functions as well: e.g., since $\exp(x)$ and $\log(x)$ are monotonically increasing functions, we have

$$\exp([\underline{x}, \overline{x}]) = [\exp(\underline{x}), \exp(\overline{x})];$$

$$\log([\underline{x}, \overline{x}]) = [\log(\underline{x}), \log(\overline{x})].$$

Sometimes the square function x^2 is also hardware supported. For such cases, we can compute the range $f(\mathbf{x})$ of the function $f(x) = x^2$ on an interval $\mathbf{x} = [\underline{x}, \overline{x}]$ as follows:

- $[\underline{x}, \overline{x}]^2 = [\underline{x}^2, \overline{x}^2]$ if $0 \le \underline{x}$;
- $[\underline{x}, \overline{x}]^2 = [\overline{x}^2, \underline{x}^2]$ if $\overline{x} \le 0$; and
- $[\underline{x}, \overline{x}]^2 = [0, \max(\underline{x}^2, \overline{x}^2)]$ if $\underline{x} < 0 < \overline{x}$.

These (and similar) formulas can be combined to compute the range of an arbitrarily complex algorithmic function $f(x_1, \ldots, x_n)$: Indeed, e.g., the formula for the *sum* means that if $x_1 \in \mathbf{x}_1 = [\underline{x}_1, \overline{x}_1]$ and $x_2 \in \mathbf{x}_2 = [\underline{x}_2, \overline{x}_2]$, then $x_1 + x_2 \in \mathbf{x}_1 + \mathbf{x}_2 = [\underline{x}_1 + \underline{x}_2, \overline{x}_2 + \overline{x}_2]$. Therefore, if we, e.g., apply multiplication to this result (i.e., take $f(x_1, x_2, x_3) = (x_1 + x_2) \cdot x_3$), then from the fact that $x_3 \in \mathbf{x}_3 = [\underline{x}_3, \overline{x}_3]$ and $x_1 + x_2 \in \mathbf{x}_1 + \mathbf{x}_2$, we can conclude that $(x_1 + x_2) \cdot x_3 \in (\mathbf{x}_1 + \mathbf{x}_2) \cdot \mathbf{x}_3$.

In general, if we have an *arbitrary* data processing algorithm $f(x_1, \ldots, x_n)$, and we want to find a range $\mathbf{y} = f(\mathbf{x}_1, \ldots, \mathbf{x}_n)$ for given intervals $\mathbf{x}_1, \ldots, \mathbf{x}_n$, then we can do the following:

- *represent* the algorithm $f(x_1, \ldots, x_n)$ as a sequence of elementary steps;

  This representation is automatically produced by a compiler from any "high-level" programming language, i.e., programming language that allows arithmetic expressions.

- *replace* each numerical operation by an interval one, and
- *apply* this new interval algorithm to the inputs $\mathbf{x}_1, \ldots, \mathbf{x}_n$.

As a result, we get an interval $\mathbf{Y}$ that is *guaranteed* to contain the desired value y, i.e., an interval that is guaranteed to contain ("enclose") the range $f(\mathbf{x}_1, \ldots, \mathbf{x}_n)$ of the function $f(x_1, \ldots, x_n)$. Such an interval is called an *enclosure*. This enclosure $\mathbf{Y}$ *contains* the interval $\mathbf{y}$, but *does not* necessarily *coincide* with $\mathbf{y}$.

For *example*, suppose that we are interested in the range $\mathbf{y} = f(\mathbf{x}_1)$ of the function $f(x_1) = x_1 - x_1^2$ on the interval $\mathbf{x}_1 = [0, 1]$. In the computer, a function $f(x_1) = x_1 - x_1^2$ is computed as follows:

- first, we compute the intermediate result $r_1 = x_1 \cdot x_1$;
- then, we compute $y = x_1 - r_1$.

If we replace each arithmetic operation with numbers by a corresponding operation with intervals, we get the following:

- first, we compute the interval

$$\mathbf{r}_1 = \mathbf{x}_1 \cdot \mathbf{x}_1 = [0,1] \cdot [0,1] =$$

$$[\min(0 \cdot 0, 0 \cdot 1, 1 \cdot 0, 1 \cdot 1), \max(0 \cdot 0, 0 \cdot 1, 1 \cdot 0, 1 \cdot 1)] = [0,1]$$

  that is guaranteed to contain r_1;

- then, we compute the interval

$$\mathbf{Y} = \mathbf{x}_1 - \mathbf{r}_1 = [0,1] - [0,1] = [0-1, 1-0] = [-1,1]$$

  that is guaranteed to contain y.

Using the textbook calculus method, we can easily find out that the actual range $\mathbf{y}$ is equal to $[0, 0.25]$. Thus, the enclosure $\mathbf{Y} = [-1,1]$ indeed *encloses* the range $\mathbf{y}$, but it *does not coincide* with the range.

More sophisticated interval methods. The main problem of *naive* interval computations is that this method often drastically overestimates. To get better estimates, we need *more sophisticated* interval techniques.

One possibility to improve the results of naive interval computations comes from the fact that for one and the same function, there are usually several different algorithms for computing it. So, instead of simply applying naive interval computations to *an* algorithm, we may try to:

- first, find a *better* algorithm for computing the same function (better in the sense that it leads to narrower interval estimates), and then
- apply naive interval computations to this better algorithm.

Let us give a simple *example*. The function $f(x_1, x_2) = x_1^2 - x_2^2$ can be computed in two different ways:

- A standard compiler will translate this expression into a natural sequence of computations: first, we first compute $r_1 = x_1 \cdot x_1$, then $r_2 = x_2 \cdot x_2$, and finally, $y = r_1 - r_2$.
- On the other hand, an *optimizing* compiler may try to rearrange the function in the form $y = (x_1 - x_2) \cdot (x_1 + x_2)$. In this case, we first compute $r_1 = x_1 - x_2$, then $r_2 = x_1 + x_2$, and finally, $y = r_1 \cdot r_2$.

In a computer, the first computation requires two multiplications and one subtractions, while the second one requires one subtraction, one addition, and one multiplication. Both algorithms require the same total number of arithmetic operations (three), with the only difference that where the first algorithm contained an extra multiplication, the second algorithm contains an addition. Since addition is much faster than multiplication (actually, multiplication of binary numbers is performed as a sequence of additions), an optimizing compiler would definitely prefer the *second* algorithm. However, from the *interval* viewpoint, the first algorithm is often better; for example, for $\mathbf{x}_1 = \mathbf{x}_2 = [0, 1]$, we get the following:

- If we use the first algorithm, we get $\mathbf{Y} = \mathbf{x}_1 \cdot \mathbf{x}_1 - \mathbf{x}_2 \cdot \mathbf{x}_2 = [0, 1] \cdot [0, 1] - [0, 1] \cdot [0, 1] = [0, 1] - [0, 1] = [-1, 1]$; this enclosure happens to coincide with the exact range.
- If we use the second algorithm, we get $\mathbf{Y} = (\mathbf{x}_1 - \mathbf{x}_2) \cdot (\mathbf{x}_1 + \mathbf{x}_2) = ([0, 1] - [0, 1]) \cdot [0, 1] + [0, 1]) = [-1, 1] \cdot [0, 2] = [-2, 2]$. This enclosure is a drastic overestimation of range $\mathbf{y} = [-1, 1]$.

Many other ideas of improving interval computations have been proposed, such as a *centered form* that leads to *asymptotically optimal* enclosures Ratschek *et al.* [336]. Some of these more sophisticated methods were described already in Moore's 1966 pioneer monograph [290], other methods are described, e.g., in Moore [291], Alefeld *et al.* [10], Hammer [140], and Kearfott [175].

Interval computations have many successful applications. Sophisticated interval methods have been applied to various areas ranging from manufacturing to control to robotics to geophysics to economics etc.; see, e.g., Kearfott *et al.* [209, 176].

1.5. Computational complexity and feasibility of interval computations: the problem and first results

Problems. In spite of all the success stories, there were still problems with interval computations, the main problem being that the existing methods often return an enclosure $\mathbf{Y}$ that *overestimates* the desired range $\mathbf{y}$: $\mathbf{y} \subseteq \mathbf{Y}$ and $\mathbf{y} \neq \mathbf{Y}$. Why is overestimation a serious problem?

For *example*, suppose that in oil exploration, the estimate is $\tilde{y} = 100$ mln. ton, and the actual range $\mathbf{y}$ of the corresponding data processing function is $\mathbf{y} = [90, 110]$.

- If we knew the *exact* range, we would be able to make a reasonable decision: to start exploiting the well.
- However, in reality, instead of the *actual* range, we get an *enclosure* $\mathbf{Y} \supset \mathbf{y}$. It is not un-typical for the enclosure to be ten times wider than the range itself, so we can have an enclosure $\mathbf{Y} = [0, 200]$. Based on this enclosure, we are not even sure whether there is any commercially viable amount of oil at all. Thus, based on this enclosure, we would recommend that further measurements are done.

These very costly and time-consuming measurements could have been avoided if we had a better enclosure in the first place.

This example illustrates a typical drawback of an *overestimation*: it leads to a *waste of resources* on further measurements. These measurements are usually very *costly* and *time-consuming* (if they were cheap and fast, we would have done them in the first place). Thus, if we can spend *a little more time* and money and compute *a better enclosure*, we would be able to *save a lot of time* and money by not performing unnecessary further measurements.

In view of this desirability, researchers have been trying to develop algorithms that lead to the smallest possible overestimation, ideally, to the *optimal* enclosure, i.e., to the enclosure that coincides with the desired range.

The original optimism soon turned into a doubt. Originally, some researchers believed that it will be possible to develop algorithms that produce optimal enclosures in reasonable computation time. This optimism was justified by the fact that algorithms did become faster and faster and the resulting enclosures did become better and better. However, even for the simplest (polynomial) functions $f(x_1, \ldots, x_n)$, all algorithms that have been proposed were:

- either computing a *non-optimal* enclosure,
- or computing the optimal enclosure, but (sometimes) in un-realistic *exponential time* (like the above textbook algorithm).

So the natural question appeared: can we have an "ideal" (fast and optimal) algorithm at all?

First negative result: in general, computing the exact range is computationally intractable. The question of whether an "ideal" algorithm is possible was formulated, among others, by Yuri Matiyasevich, known for his (negative) solution of Hilbert's 10th Problem [276] (see also Matiyasevich [277] and Davis *et al.* [86]).

- Matiyasevich's solution of Hilbert's problem (one of the 23 problems that the 19th century mathematics presented as a challenge to the 20th century, Hilbert [152]) was that *no algorithm* can check whether a given polynomial $f(x_1, \ldots, x_n)$ with integer coefficients has integer roots, i.e., check whether 0 belongs to the range $f(Z, \ldots, Z)$ of the polynomial $f(x_1, \ldots, x_n)$ on *integer* inputs x_i (i.e., inputs from the set Z of all integers).
- Matiyasevich, therefore, conjectured that, similarly, no feasible algorithm can find the range of a polynomial on *interval* inputs.

This conjecture was confirmed by Matiyasevich's student A. A. Gaganov who proved, in 1981, that the problem of computing the range of a given polynomial on given intervals is indeed *computationally intractable* in some precise sense (called *NP-hard*) [115, 116].

Follow-up research. Crudely speaking, Gaganov's result means that no feasible algorithm can solve *all* instances of this problem. On the other hand, there exist feasible algorithms that compute the optimal enclosures for *some* problems. A natural question is: *How general can these feasible algorithms be?*

This question is very important for algorithm designers, because an answer to this question enables them to *avoid* useless attempts of trying to find a general solution for a class for which no general algorithm is possible, and to concentrate on classes of problems for which such an algorithm has been proven to exist.

Lots of effort went into answering this question, and the answers did not come easily: e.g., for quadratic polynomials, NP-hardness was proven ten years after the original Gaganov's result (in 1991), and NP-hardness of the problem of solving a system of interval linear equations was proven even later: only in 1993.

State-of-the-art and the main goal of this book. As of now (1997), although some open problems still remain, these open problems are more of a technical than fundamental type. The *big picture* is now clear: we have a pretty good understanding which classes of problems are *feasible* and which are *intractable*. There are a few surveys (see, e.g., Rohn *et al.* [352, 353, 209, 224, 358]), but these surveys are short, somewhat sketchy, and far from being comprehensive. It is therefore time to present the big picture to the interested readers. This is what this book intends to do.

1.6. The structure of the book

The main goal of this book is to present the results about feasibility and intractability of different problems of interval computations. To be able to prove results about feasibility and intractability, we must use the formal definitions of these notions. Since not all potential readers of this books are well familiar with these definitions, we start our book with a short tutorial in which these definitions are given and explained. This tutorial forms *Chapter 2*.

Chapter 3 starts with the basic problem: *given* a function $f(x_1, ..., x_n)$ of n real variables, and n intervals $\mathbf{x}_i$, *compute* the range

$$\mathbf{y} = f(\mathbf{x}_1, ..., \mathbf{x}_n) = \{f(x_1, ..., x_n) | x_1 \in \mathbf{x}_1, ..., x_n \in \mathbf{x}_n\}.$$

Usually, these intervals $\mathbf{x}_i$ have rational endpoints. The question is: *Is this basic problem of interval, computations feasible or intractable?*

If the function $f(x_1, \ldots, x_n)$ is itself difficult to compute, then the answer to this question is easy: it is difficult to compute the endpoints of the interval $\mathbf{y}$ even for *degenerate* input intervals $\mathbf{x}_i = [x_i, x_i]$. To avoid this trivial situation, in this book we will mainly restrict ourselves to the simplest possible functions: functions that can be obtained by finitely many applications of the basic arithmetic operations (addition +, subtraction −, multiplication ·, and division /), i.e., to ***rational*** functions with rational coefficients.

> Let us describe this restriction in more precise terms. We have mentioned that in a computer, for every set of inputs, every algorithm is implemented as a sequence of arithmetic operations and applications of elementary functions.
>
> - Most of the problems that we will analyze turn out to be *NP-hard* already for rational functions (even for polynomials). For such problems, there is ***no need to analyze*** the computational consequences of ***adding*** other ***elementary functions*** (or if–then statements): the resulting more general problem will, of course, still be intractable.
> - In a few cases in which the computational problem for rational functions turns out to be *feasible*, we ***analyze*** what will happen if we ***allow elementary functions*** in addition to arithmetic operations.

Chapter 3 contains Gaganov's 1981 result, that the main problem is computationally intractable even for polynomials $f(x_1, \ldots, x_n)$.

Gaganov's result means that if we allow polynomials with arbitrarily many variables, of arbitrary degree, with arbitrary rational numbers for coefficients, and arbitrary intervals $\mathbf{x}_i$, then the problem is computationally intractable. A natural question is: what if we restrict some or all of these "arbitrary" parameters? E.g., what will happen if we only consider polynomials with bounded number of variables? of bounded degree? with bounded coefficients? with bounded (or even fixed) data intervals? These restrictions are explicitly described and analyzed, one-by-one, in *Chapters 4–9*.

Of all analyzed classes of polynomials, only for *linear* functions the basic problem has a simple and feasible algorithm. Since we cannot extend this result to more general polynomials, it is natural to extend it for different functions. For *fractionally* linear functions, the problem is still feasible (*Chapter 10*); for a more general class of functions described by systems of linear equations, the problem is again NP-hard (*Chapters 11–12*). This NP-hardness can be traced on ***practically meaningful*** problems of prediction in physics and signal processing (*Chapters 13–14*).

At first glance, it seems that all the above NP-hardness results paint a gloomy picture of computational intractability of data processing. The reality is, hopefully, not so gloomy: in *Chapter 15*, we show that these results can actually help to apply interval heuristics to other hard problems (see also *Appendix F*), and in *Chapter 16*, we show that in spite of the *worst-case* complexity, some good heuristic algorithms are feasible in *almost all* cases.

In addition to the above-mentioned problems in which input is known with *interval* uncertainty, interval computations are also used to get guaranteed interval estimates for problems with purely *numerical* inputs, such as optimization, solving systems of equations, approximation, etc. The computational complexity and feasibility of such problems is overviewed in *Chapters 17–20*.

In Chapters 3–20, we analyze the complexity of the problems in which the main goal is to compute a number (or an interval). In many practical situations, however, we are not interested in the exact *value* of this number; all we need to know is whether a certain *property* is true or not: e.g., whether a given control is stable, etc. The computational complexity and feasibility of such *checking* problems is analyzed in *Chapters 21–22*.

Finally, in *Chapters 23–26*, we analyze the computational complexity and feasibility of the problems with *non-interval* uncertainty (as described by ellipsoids, multi-intervals, probabilities, etc.).

Several related computational complexity and feasibility results are presented as *Appendices*. In *Appendix A*, we describe the computational complexity of the problem of finding the *simplest* representative of an interval. In *Appendix B*, we consider the case when the function $f(x_1, \ldots, x_n)$ is known only approximately. In this case, the error estimation problem becomes exponentially hard even for linear functions $f(x_1, \ldots, x_n)$. In *Appendix C*, we consider another natural generalization of interval computations: to *modal mathematics*.

One of the main objectives of *theoretical* computational complexity is to explain the *empirically* observed difference in computer time required for solving different problems. If one of the problems is feasible and the other is NP-hard, then, in general, solving the first problem will require less time than solving the second one. But what if both compared problems are NP-hard? In *Appendix D*, we consider such an example when we have to go *beyond* NP-hardness, and show that in some cases, we can successfully rank different NP-hard problems by their complexity. In the process of ranking these problems, we get a philosophical by-product: foundations for *optimism*.

This optimism may be even more justified if some new physical device would enable us to solve intractable (NP-hard) problems in reasonable time. Such possibilities are briefly mentioned in *Appendix E*.

We conclude the book with a special *Appendix G* in which we challenge the reader to explain why the feasibility results for data processing are often counter-intuitive.

Proofs. For reader's convenience, in each chapter, all the proofs are concentrated in the special (last) *Proofs* section, so that a reader who is only interested in the results will be able to easily skip the corresponding proofs.

We tried to present as many proofs as possible, and we present proofs for all *major* results. However, we also wanted this book to be as comprehensive as possible. There are many computational complexity and feasibility results about interval computations and related data processing problems, and if we included the proofs of all these results, we would not fit into a single book. Therefore, for some *auxiliary* results whose proofs are published in easily accessible journals, we had to restrict ourselves to presenting *references* to the proofs and not the proofs themselves.

Notations: boldface letters, possibly with indices, like $\mathbf{a}$, $\mathbf{b}_i$, will indicate *intervals*; their endpoints will be denoted by *underline* and *overline*: e.g., $\underline{a}$ and $\overline{a}$ are endpoints of the interval $\mathbf{a}$; $\underline{b}_i$ and $\overline{b}_i$ are endpoints of the interval $\mathbf{b}_i$, etc.

2

THE NOTIONS OF FEASIBILITY AND NP-HARDNESS: BRIEF INTRODUCTION

The main goal of this book is to analyze computational complexity and feasibility of data processing and interval computations. In Chapter 1, we defined the basic problems of data processing and interval computations. In this chapter, we give a brief introduction to the notions related to feasibility and computational complexity.

2.1. When is an Algorithm Feasible?

2.1.1. *What Does "Feasible" Mean? The Main Idea*

Some algorithms are not feasible. In theory of computation, it is well known that not all algorithms are feasible (see, e.g., Garey *et al.* [121], Lewis *et al.* [251], Martin [274]): whether an algorithm is feasible or not depends on how many computational steps it needs.

For example, if for some input x of length $\text{len}(x) = n$, an algorithm requires 2^n computational steps, then for an input of a reasonable length $n \approx 300$, we would need 2^{300} computational steps. Even if we use a hypothetical computer for which each step takes the smallest physically possible time (the time during which light passes through the smallest known elementary particle), we would still need more computational steps than can be performed during the (approximately 20 billion years) lifetime of our Universe.

A similar estimate can be obtained for an arbitrary algorithm whose running time $t(n)$ on inputs of length n grows at least as an exponential function, i.e., for which, for some $c > 0$, $t(n) \geq \exp(c \cdot n)$ for all n. As a result, such algorithms (called *exponential-time*) are usually considered *not feasible.*

Comment. The fact that an algorithm is not feasible, does not mean that it can never be applied: it simply means that there are cases when its running time will be too large for this algorithm to be practical; for other inputs, this algorithm can be quite useful.

Some algorithms are feasible. On the other hand, if the running time grows only as a polynomial of n (i.e., if an algorithm is *polynomial-time*), then the algorithm is usually quite feasible.

Existing definition of feasibility: the main idea. As a result of the above two examples, we arrive at the following idea: An algorithm $\mathcal{U}$ is called *feasible* if and only if it is *polynomial-time*, i.e., if and only if there exists a polynomial $P(n)$ such that for every input x of length len(x), the computational time $t_{\mathcal{U}}(x)$ of the algorithm $\mathcal{U}$ on the input x is bounded by $P(\text{len}(x))$: $t_{\mathcal{U}}(x) \leq P(\text{len}(x))$.

In most cases, this idea works. In most practical cases, this idea *adequately* describes our intuitive notion of feasibility: *polynomial-time* algorithms are usually *feasible*, and *non-polynomial-time* algorithms are usually *not feasible.*

This idea is not perfect, but it is the best we can do. Although in *most* cases, the above idea adequately describes the intuitive notion of feasibility, the reader should be warned that this idea is *not perfect*: in some (very rare) cases, it does not work (see, e.g., Garey *et al.* [121], Lewis *et al.* [251], Martin [274]):

- Some algorithms are polynomial-time but not feasible: e.g., if the running time of an algorithm is $10^{300} \cdot n$, this algorithm is polynomial-time, but, clearly, not feasible. (Other examples of such algorithms, examples that are directly related to data processing and interval computations, will be given in Chapter 4.)
- Vice versa, there exist algorithms whose computation time grows, say, as $\exp(0.000\ldots01 \cdot \text{len}(x))$. Legally speaking, such algorithms are exponential time and thus, not feasible, but for all practical purposes, they are quite feasible.

It is therefore desirable to look for a *better* formalization of feasibility, but as of now, "polynomial-time" is the best known description of feasibility.

In view of this, in the following text, we will use the terms "feasible" and "polynomial-time" interchangingly, and we will specifically mention the rare cases when these two notions differ.

2.1.2. *How Can We Formalize the Main Idea?*

How can we formalize the main idea? The above idea of polynomial-time algorithms is based on the following two fundamental notions: the *length* of the input and the *computation time*. So, to formalize the notion of feasible algorithms, we must formalize the notions of *input length* and *computation time*.

First try: computer-specific measures of input length and computation time. At first glance, these seems to be no problem with formalizing the notions of input length and computation time: if we fix a computer and if we fix a programming language, then we can define the *length* of the input, e.g., as the number of symbols on it, and the *computation time* as the actual (physical) time from the start of the program to the moment when it produces the results.

To make computer-independent conclusions about feasibility of algorithms, we need computer-independent measures of input length and computation time. The above-described "computer-specific" approach to defining input length and computation time is not perfect:

- Different languages use different symbols to describe the same data; as a result, the *length* of the input may drastically change if we change the language.
- Similarly, different computers require different *times* to perform different operations. Even on one and the same computer, the computation time for one and the same operation may change from time to time, depending, e.g., on whether the registers are currently available, whether any periodic automatic maintenance operations are currently being performed, etc.

As a result, the dependence of the computation time on the input length may drastically change from one computer to another. Hence, if we use the computer-specific measures of input length and computation time, we *will* be able to show that some algorithms *are* polynomial-time, but it will be very *difficult* to prove that some problems are *not* polynomial-time: Indeed, we can only show that the algorithm is not polynomial-time on a *given* computer, but it is very difficult to argue that it will *not* become polynomial-time on some other computer.

If we want *computer-independent* results, we must, therefore, be able to describe *computer-independent* measures of input length and time complexity.

2.1.3. *First Step Towards Formalization of Feasibility: Computer-Independent Notion of the Input Length (Number of Bits)*

General idea. It is easy to come up with a computer-independent notion of the input length. Indeed, although computers use different hardware, most of them use the same way of representing all their data: each element of data is represented by a sequence of 0's and 1's. Thus, we can always measure the length of each input by the number of *binary units* (also called *bits*), i.e., by the number of 0's or 1's, that are needed to represent it.

Integers. In particular, inside the computer, *integers* are usually represented in their binary form. Therefore, the input length of an integer can be naturally defined as the number of bits in its binary expansion: e.g., the number $8_{10} = 1000_2$ requires 4 bits, while $26_{10} = 11010_2$ requires 5 bits.

Binary-rational numbers. Similarly, *binary-rational* numbers, i.e., numbers of the type $p/2^q$, are usually represented in their fixed-point binary form: e.g., $3/8_{10} = 0.011_2$ takes 4 bits, while $19/16_{10} = 1.0011_2$ requires 5 bits. So, e.g., if we say that "there is a polynomial-time algorithm that, for every binary-rational number ε, computes ...", we mean, in particular, that for the values $\varepsilon_n = 0.0\ldots01_2 = 2^{-n}$ of length n, the running time of this algorithm is bounded by a polynomial of n.

General rational numbers. A general *rational* number m/n, where m and n are integers, can be naturally represented as a *pair* of integers m and n; therefore, we take the total length of the binary representations of m and n as the input length: e.g., $5/7_{10} = 101/111_2$ requires 6 bits to describe.

Fixed-point and floating point numbers: a comment. In most computers, there is no special *rational* data type, there is a type *real* which actually describes binary rational numbers. In most computers, there are two *different* representations of these "real" numbers: in addition to the above-described *fixed-point* real numbers, there are also *floating-point* real numbers, in which a binary rational number is represented as $m \cdot 2^e$, where m is a fixed-point real number (usually, with only zero before the binary point) and e is an integer.

In this book, we describe the input length in terms of the *fixed point* representation. Most of our results about computational complexity and feasibility of data processing and interval computations, both positive (that some problems can be solved by feasible algorithms) and negative (that for some other problems no feasible algorithm is possible) are true for floating point numbers as well: e.g., since every fixed point number is at the same time a floating point number (with $e = 0$), *negative* results about fixed point inputs are automatically transformed into *negative* results about the floating point inputs.

However, to avoid potential confusion, we decided, in this book, not to mention the floating point representation versions of our results at all. The reason why such a confusion can occur is that with floating point representation, there are some negative results that are caused not by a complexity of the problem, but by a *peculiarity* of this representation. Let us give a simple example of this peculiarity:

- For *fixed-point* binary-rational numbers x and y, *exactly* computing their *sum* $x + y$ is easy: the standard bit-after-bit addition requires linear time.
- However, for *floating point* binary-rational numbers, this same addition problem requires *exponential* time: e.g., if we take $x = 1$ and $y = 1 \cdot 2^e$ with $e = -2^n = -10\ldots0_2$ (n zeros), with the total length $\approx n$, then the exact description of $x + y = 1.0\ldots01$ (2^n zeros) requires *exponentially many* bits, so generating these bits would require *exponentially long* time.

This example does *not* mean that floating point representation is in any way inferior and bad: it is known to be very useful, and the above difficulty with addition can be easily overcome if we require that the result be known with a given *accuracy*. However, this example clearly shows that floating-point formulations require extra accuracy and complexity that appears even for such simple operations as addition. In other words, floating-point formulations contain extra complexity that is unrelated to the complexity of data processing and interval computations; thus, if we try to reformulate our results in terms of floating point numbers, the resulting combination of two complexities would make the corresponding results very un-intuitive.

2.1.4. *Second Step Towards Formalization of Feasibility: Computer-Independent Notion of Computation Time*

General idea. On each computer, every algorithm, every program consists of a sequence of *elementary steps*, i.e., hardware-supported simple operations such as operations with bits, or addition, etc. We can estimate the computation time by simply counting the number of these elementary steps, and multiplying this number by the average time of each step. (We may also want to take into consideration that different elementary steps take different times.) Then, if we have a similar computer, with similar (but faster) elementary steps, we can estimate the running time on a new computer if we already know the number of steps needed to run the algorithm.

This is the general idea behind different computer-independent estimates.

The situation is not so easy as it may seem. The actual estimates are somewhat more difficult to get because different computers may have different elementary steps. Let us give two examples:

- First computers only had hardware-supported *fixed-point* operations, so a floating-point operation had to be implemented as several fixed-point ones, and thus, was counted as *several* steps. In most modern computers, *floating-point* operations are also directly hardware supported and thus, each such operation can be counted as *one* step.
- In many modern computers, computing the element-wise sum $c_i \leftarrow a_i + b_i$ of two vectors $(a_1, \ldots, a_n)$ and $(b_1, \ldots, b_n)$ means n additions, i.e., n elementary steps. On the other hand, in computers with a math co-processor, the entire addition is directly hardware supported and therefore, can be counted as a single elementary step.

To get a computer-independent definition, we have to consider different types of computers and corresponding definitions.

Turing machines, and a standard complexity measure. Most textbooks on complexity theory and theory of computation start with the simplest possible "computer", that was designed by Turing long before the actual computers. This toy computer (called *Turing machine*) consists of a potentially infinite *tape* divided into sequentially located *cells*, and a *head* that moves along this tape, from one cell to another (the reader can see that Turing machine probably resembles a tape recorder more than it resembles a modern computer).

To specify a Turing machine, we must describe three things:

- The *first* thing we must describe is the (finite) set of all possible symbols that it can place in each cell. This set (called ***alphabet***) can simply consist of two possible symbols 0 and 1, in which case, each cell can be in one of the three states: nothing is written in this cell, 0 is written, and 1 is written. We can also consider Turing machines with a larger alphabet: e.g., cells can simulate ***bytes*** (i.e., 8-bit sequences), in which case, the alphabet consists of $2^8 = 256$ possible symbols: 0000 0000, 0000 0001, ..., 1111 1111.
- *Second*, we must describe the possible state of the head. In more precise terms, we must describe another finite set, whose elements will be called ***states*** (or ***states of the head***). There must be two special states: ***starting*** state (in which we start computations) and the ***halting*** state, on reaching which the computer stops.
- Finally, we must describe the *action* of the computer. Namely, for every state s of the head and for every symbol σ in the cell to which the head is currently pointing, we must describe what this computer will do. It can do one of the three things:
  - It can *overwrite* the symbol σ with some other symbol σ' that, in general, depends on s and σ ($\sigma' = \sigma'(s, \sigma)$), and at the same time ***change*** its state into the some other state $s' = s'(s, \sigma)$;
  - It can, instead, move one step to the ***right***.
  - It can also move one step to the *left*.

We start in the ***starting*** state, with the input written on this tape, and then ***apply*** the computer's ***action*** again and again until we reach the ***halting*** state. When this happens, the tape should contain the result of the computations.

Let us give a very simple *example* of a Turing machine, that adds 1 to a binary number. Since we are dealing with binary numbers, we only need the alphabet $\{0, 1\}$. The machine starts at the end of the input binary number, and proceeds as follows:

- If it is in the starting state s_0, and it sees 0, it changes it to 1 and gets into the halting state h (and stops).
- If it is in the starting state s_0, and it sees 1, it changes it to 0 and gets into the state s_1 ("ready to move, carrying a carry").
- If it is in the state s_1, then, no matter what it sees, it moves one step to the left and gets into the state s_2 ("I have a carry").
- If it is in the state s_2, and it sees 0, then it changes 0 to 1 and halts.
- If it is in the state s_2 and it sees 1, it changes 1 to 0 and goes into the state s_1.
- Finally, if it is in the state s_2 and sees an empty cell, it changes the contents of this cell to 1 and halts.

For Turing machines, we have clearly defined steps, so we can define computational complexity $t_{\mathcal{U}}(x)$ of an algorithm $\mathcal{U}$ on an input x as the number of the corresponding steps. This is the *standard definition* of computational complexity in theory of computing.

Turing machine is a very primitive computer, on which a simple operation that is hardware supported on a normal computer can take a very long time. So, the Turing machine-based complexity is a rather poor estimate for the *actual* computation time of an algorithm on a real computer. If we use more sophisticated computer models, we can get much *better* estimates.

What makes Turing machine-based complexity *standard* and widely used is the fact that although the *actual* computation time changes from one type of computer to another, but whether the algorithm is *polynomial-time* or not does not depend on our choice of the computer. Thus, whatever reasonable class of computers we consider, a problem can be solved in polynomial-time on computers of this class if and only if we can solve it in polynomial time on a Turing machine (for details, see, e.g., Emde Boas [100]). Thus, if all we are interested in is whether an algorithm is feasible or not, Turing machines are quite sufficient.

Since we are also interested in more realistic estimates of computation time, we will use more realistic computer models.

RAM and bit complexity. In Turing machines, if we are currently facing cell 1, and we want to use the contents of cell n, we actually have to move step-by-step, and spend n computation steps just to reach this new cell. This is definitely not realistic. In real computers, if we know the number of the cell, we immediately go there. In other words, we can go to an arbitrarily ("randomly") chosen cell in a single step. If we add this ability to the Turing machine, we get the so-called *RAM* (Random Access Memory) computers. For the particular case when cells can only contain 0 and 1, the number of computational steps on RAM is sometimes called *bit complexity*.

Algebraic complexity. RAM is slightly more realistic than a Turing machine, but it is still not very realistic. In real-life computers, in addition to operations with bits, we have a hardware support for elementary *arithmetic* operations such as addition and multiplication of two integers. It is therefore reasonable, given an input of length n, to assume that we can perform addition and multiplication of integers of length $C \cdot n$ (for some reasonable C) in a single step. The number of computational steps on such machine is called *algebraic complexity*.

Algebraic complexity is the closest to the actual computation time and therefore, we will use this complexity measure in the book. To be more precise:

- When we claim that something is *polynomial-time*, this claim (as we have already mentioned) will be independent on the choice of complexity measure. But:
- If we claim something more specific, like *linear time* (i.e., $t_U(n) \leq C \cdot n$) or *quadratic time* (i.e., $t_U(n) \leq C \cdot n^2$), we will mean exactly linear time (correspondingly, quadratic time) in the sense of algebraic complexity.

Comment. The relation between algebraic and bit complexity is analyzed, e.g., in Pan [322].

A remark about BSS complexity. To go from bit complexity to algebraic complexity, we counted each arithmetic operation with numbers of *fixed* length as a single computation step. We can go further and count each operation with binary-rational numbers of *arbitrary* length (or even with arbitrary *real* numbers, not necessarily rational) as a single computation step. This definition was, in effect, proposed by Blum, Shub, and Smale [50] and is called BSS complexity (see also Smale [398], Cucker *et al.* [82, 81], Meer [281], Grädel *et al.* [133], Lickteig *et al.* [255]).

For our problems, BSS complexity is very useful in proving *negative* results: If a problem is *difficult* according to this BSS complexity, then it will be even more difficult if we only allow a narrower class of operations, i.e., it will be difficult according to algebraic complexity as well.

On the other hand, if a problem is *easy* in the sense of BSS complexity, it often means that this problem is actually easy, but sometimes, it can become difficult if we only allow operations with bounded numbers; examples of such *difference* between BSS and algebraic complexity are given, e.g., in Meer [280, 282]. Because of this difference, in this book, we will not use BSS complexity.

2.1.5. *Formal Definition of Feasibility*

Now, we are ready for the precise definition:

Definition 2.1. *An algorithm U is called feasible if there exists a polynomial $P(n)$ such that for every input x, the running time $t_U(x)$ of this algorithm does not exceed $P(\text{len}(x))$, where by $\text{len}(x)$, we denoted the length of the input x (i.e., the number of bits that form this input).*

2.2. When is a Problem Tractable?

2.2.1. *Ideal Solution is not yet Possible*

What would be an ideal solution. At first glance, now, that we have a definition of a feasible algorithm, we can describe which problems are tractable and which problems are intractable: If there exists a polynomial-time algorithm that solves all instances of a problem, this problem is tractable, otherwise, it is intractable.

Sometimes, this ideal solution is possible. In some cases, this ideal solution is possible, and we either have an explicit polynomial-time algorithm, or we have a proof that no polynomial-time algorithm is possible.

Alas, for many problems, we do not know. Unfortunately, in many cases, we do not know whether a polynomial-time algorithm exists or not. This does not mean, however, that the situation is hopeless: instead of the missing *ideal* information about intractability, we have another information that is almost as good:

What we have instead of the ideal solution. Namely, for some cases, we do not know whether the problem can be solved in polynomial time or not, but we do know that this problem is as hard as practical problems can get: if we can solve *this* problem easily, then we would have an algorithm that solves *all* problems easily, and the existence of such universal solves-everything-fast algorithm is very doubtful. We can, therefore, call such "hard" problems *intractable*.

In order to formulate this notion in precise terms, we must describe what we mean by a problem, and what we mean by the ability to *reduce* other problems to this one.

2.2.2. *How Can We Define a General Practical Problem?*

What is a practical problem: informal idea. What is a practical problem? When we say that there is a practical problem, we usually mean that:

- we have some information (we will denote its computer representation by x), and
- we know the relationship $R(x, y)$ between the known information x and the desired object y.

In the computer, everything is represented by a binary sequence (i.e., sequence of 0's and 1's), so we will assume that x and y are binary sequences.

Two examples of problems. In this section, we will trace all the ideas on two examples, one taken from mathematics and one taken from physics. Readers who do not feel comfortable with one of the example (say, with a physical one) are free to simply skip it.

- (Example from *mathematics*) We are given a mathematical statement x. The desired object y is either a proof of x, or a "disproof" of x (i.e., a proof of "not x"). Here, $R(x, y)$ means that y is a proof either of x, or of "not x".

- (Example from *physics*) x is the results of the experiments, and the desired y is the formula that fits all these data. Imagine that we have a series of measurements of voltage and current: e.g., x consists of the following pairs $(x_1^{(k)}, x_2^{(k)})$, $1 \le k \le 10$: $(1.0, 2.0), (2.0, 4.0), \ldots, (10.0, 20.0)$; we want to find a formula that is consistent with these experiments (e.g., y is the formula $x_2 = 2 \cdot x_1$).

Solution must be checkable. For a problem to be practically meaningful, we must have a way to check whether the proposed solution is correct. In other words, we must assume that there exists a feasible algorithm that checks $R(x, y)$ (given x and y). If no such feasible algorithm exists, then there is no criterion to decide whether we achieved a solution or not.

Solution must not be too long. Another requirement for a real-life problem is that in such problems, we usually know an *upper bound* for the length $\text{len}(y)$ of the description of y. In the above examples:

- In the *mathematical* problem, a proof must be not too huge, else it is impossible to check whether it is a proof or not.
- In the *physical* problem, it makes no sense to have a formula $x_2 = f(x_1, C_1, \ldots, C_{40})$ with, say, 40 parameters to describe the results $(x_1^{(1)}, x_2^{(1)}), \ldots, (x_1^{(10)}, x_2^{(10)})$ of 10 experiments, for two reasons:
  - First, one of the goals of physics is to discover the laws of nature. If the number of parameters exceeds the number of experimental data, then no matter what dependency $f(x_1, C_1, \ldots)$ we choose, in order to determine C_i, we have, say, 10 equations with 40 unknowns. Such under-determined system usually has a solution, so the fact that, say, a linear formula with many parameters fits all the experimental data does not mean that the dependency is proven to be linear: a quadratic or cubic formula with as many parameters will fit the same data as well.
  - Second, another goal of physics (definitely related to the first one) is to find a way to *compress* the data, so that we will not need to store all billions of experimental results in order to make predictions. A dependency y that requires more storage space than the original data x is clearly not satisfying this goal.

In all cases, it is necessary for a user to be able to read the desired solution symbol-after-symbol, and the time required for that reading must be feasible.

In the previous section, we have formalized "feasible time" as a time that is bounded by some polynomial of $\text{len}(x)$. The reading time is proportional to the length $\text{len}(y)$ of the answer y. Therefore, the fact the reading time is bounded by a polynomial of $\text{len}(x)$ means that the length of the output y is also bounded by some polynomial of $\text{len}(x)$, i.e., that $\text{len}(y) \leq P_L(\text{len}(x))$ for some polynomial P_L.

So, we arrive at the following formulation of a problem:

Definition 2.2. *By a* ***general practical problem*** *(or simply a* ***problem****, for short), we mean a pair* $\langle R, P_L \rangle$*, where* $R(x, y)$ *is a feasible algorithm that transforms two binary sequences into a Boolean value ("true" or "false"), and* P_L *is a polynomial.*

Definition 2.3. *By an* ***instance*** *of a (general) problem* $\langle R, P_L \rangle$*, we mean the following problem:*

GIVEN: *a binary sequence* x.

GENERATE

- *either* y *such that* $R(x, y)$ *is true and* $\text{len}(y) \leq P_L(\text{len}(x))$,
- *or, if such a* y *does not exist, a message saying that there are no solutions.*

For example, for the general mathematical problem described above, an instance would be: given a statement, find its proof or disproof.

Comments. What we called "general practical problems" is usually described as "problems from the class NP" (to separate them from more complicated problems in which the solution may not be easily verifiable). Problems for which there is a feasible algorithm that solves all instances are called *tractable*, *easily solvable*, or "problems from the class P" (P from *P*olynomial). It is widely believed that not all (general practical) problems are easily solvable (i.e., that NP$\neq$P), but it has never been proved.

One way to solve an NP problem is to check $R(x, y)$ for all binary sequences y with $\text{len}(y) \leq P_L(\text{len}(x))$. This algorithm (called *British Museum* algorithm) requires $2^{P_L(\text{len}(x))}$ checks. This algorithm takes exponential time and is therefore, not feasible.

2.2.3. *Reducing a Problem to Another One*

Example. Let us start with an example. Suppose that we can have an algorithm that checks whether a given system of linear inequalities

$$a_{i1} \cdot x_1 + \ldots + a_{im} \cdot x_m \geq b_i, \quad 1 \leq i \leq n,$$

with known a_{ij} and b_i, has a solution. A problem of checking whether a given system of inequalities *and equalities* $c_{k1} \cdot x_1 + \ldots + c_{km} \cdot x_m = d_k$ is consistent can be *reduced* to the problem of checking inequalities if we replace each equality by two inequalities: $c_{k1} \cdot x_1 + \ldots + c_{km} \cdot x_m \geq d_k$ and $(-c_{k1}) \cdot x_1 + \ldots + (-c_{km}) \cdot x_m \geq -d_j$ (the latter being equivalent to $c_{k1} \cdot x_1 + \ldots + c_{km} \cdot x_m \leq d_k$).

General definition. In general, we can say that a problem $\mathcal{P} = \langle R, P_L \rangle$ can be *reduced* to a problem $\mathcal{P}' = \langle R', P'_L \rangle$ if there exist three feasible algorithms U_1, U_2, and U_3 with the following properties:

- The (feasible) algorithm U_1 transforms each input x of the first problem into an input of the second problem.
- The (feasible) algorithm U_2 transforms each solution y of the first problem into the solution of the corresponding case of the second problem: i.e., if $R(x, y)$ is true, then $R'(U_1(x), U_2(y))$ is also true.
- The (feasible) algorithm U_3 transforms each solution y' of the corresponding instance of the second problem into the solution of the first problem: i.e., if $R'(U_1(x), y')$ is true, then $R(x, U_3(y'))$ is also true.

(In the above example, U_1 transforms each equality into two inequalities, and U_2 and U_3 simply do not change the values x_i at all.)

If there exists a reduction, then an instance x of the first problem is solvable if and only if the corresponding instance $U_1(x)$ of the second problem is solvable. Moreover, if we can actually solve the second instance (and find a solution y'), we will then be able to find a solution to the original instance x of the first problem (as $U_3(y')$). Thus, if we have a *feasible* algorithm for solving the second problem, we would thus design a *feasible* algorithm for solving the first problem as well.

Comment. We only described the simplest way of reducing one problem to another one: when a *single* instance of the first problem is reduced to a *single* instance of the second problem. In some cases, we cannot reduce to a *single* case, but we can reduce to *several* cases, solving which helps us solve the original instance of the first problem.

2.2.4. *When is a Problem Tractable, and When is It Intractable?*

Definition 2.4.

- *A problem (not necessarily from the class NP) is called **NP-hard** if every problem from the class NP can be reduced to it.*
- *If a problem from the class NP is NP-hard, it is called **NP-complete**.*

If a problem $\mathcal{P}$ is NP-hard, then every feasible algorithm for solving *this* problem $\mathcal{P}$ would lead to feasible algorithms for solving *all* problems from the class NP, and this is generally believed to be hardly possible.

- For example, mathematicians believe that not only there is *no algorithm* for checking whether a give statement is provable or not (the famous Gödel's theorem has proven that), but also they believe that there is *no feasible way* to find a proof of a given statement even if we restrict the lengths of possible proofs. (In other words, mathematicians believe that computers cannot completely replace them.)
- Similarly, physicists believe that what they are doing cannot be completely replaced by computers.

In view of this belief, NP-hard problems are also called *intractable*.

Comment. It should be noted that although most scientists *believe* that intractable problems are not feasible, we still *cannot prove* (or disprove) this fact. If a NP-hard problem *can* be solved by a feasible algorithm, then (by definition of NP-hardness) *all* problems from the class NP will be solvable by feasible algorithms and thus, P=NP. Vice versa, if P=NP, then all problems from the class NP (including all NP-complete problems) can be solved by polynomial-time (feasible) algorithms.

So, if P$\neq$NP (which is a common belief), then the fact that the problem is NP-hard means that *no matter what algorithm we use, there will always be some cases for which the running time grows faster than any polynomial.* Therefore, for these cases, the problem is truly intractable.

2.3. Three Examples of NP-Hard Problems

Examples are needed. In this book, we will prove that some problems of data processing and interval computations are NP-hard. A typical way of proving that a certain problem $\mathcal{P}'$ is NP-hard is to reduce some problem $\mathcal{P}$, that is already known to be NP-hard, to $\mathcal{P}'$.

Why the existence of such a reduction proves NP-hardness of $\mathcal{P}'$ is easy to explain: Since we already know that $\mathcal{P}$ is NP-hard, this means that *any* NP problem $\mathcal{P}''$ can be reduced to $\mathcal{P}$. Since $\mathcal{P}$, in its turn, is reducible to $\mathcal{P}'$, thus, $\mathcal{P}''$ can be reduced to $\mathcal{P}'$ as well. Thus, every problem from NP can be reduced to $\mathcal{P}'$ and hence, by definition, $\mathcal{P}'$ is NP-hard.

In view of this, it is very important to have *examples* of problems that are already known to be NP-hard, problems that we will use in our proofs. Many examples of such problems can be found in Garey *et al.* [121]. In this book, we will mainly use the following three NP-complete problems:

First example: Satisfiability. Historically the NP-complete problem proved to be NP-complete was the so-called *propositional satisfiability* (*3-SAT*) problem for 3−CNF formulas.

This problem consists of the following: Suppose that an integer v is fixed, and a formula F of the type $F_1\&F_2\&\ldots\&F_k$ is given, where each of the expressions F_j has the form $a \vee b$ or $a \vee b \vee c$, and a, b, c are either the variables $z_1, \ldots, z_v$, or their negations $\neg z_1, \ldots, \neg z_v$ (these $a, b, c, \ldots$ are called *literals*).

For *example*, we can take a formula $(z_1 \vee \neg z_2)\&(\neg z_1 \vee z_2 \vee \neg z_3)$.

If we assign arbitrary Boolean values ("true" or "false") to v variables $z_1, \ldots, z_v$, then, applying the standard logical rules, we get the truth value of F. We say that a formula F is *satisfiable* if there exist truth values $z_1, \ldots, z_v$ for which the truth value of the expression F is "true". The problem is: given F, check whether it is satisfiable.

Second example: Partition. In the *PARTITION* problem, given n integers $s_1, \ldots, s_n$, we must check whether there exist values $x_1, \ldots, x_n \in \{-1, 1\}$ for which $s_1 \cdot x_1 + \ldots + s_n \cdot x_n = 0$.

Third example: Max-cut problem. (*MAX-CUT*) For every graph $(\mathcal{V}, \mathcal{E})$ with vertices (nodes) $\mathcal{V}$ and edges $\mathcal{E}$, and for every subset S of the set of all vertices, the *cut* $c(S)$ is defined as the number of edges from $\mathcal{E}$ that connect vertices from the set S with vertices that are outside the set S. The problem is: given a graph $(\mathcal{V}, \mathcal{E})$ and a positive integer L, check whether there exists a set $S \subseteq \mathcal{V}$ with the cut $c(S) \geq L$.

3

IN THE GENERAL CASE, THE BASIC PROBLEM OF INTERVAL COMPUTATIONS IS INTRACTABLE

In this chapter, we describe the first negative result: that even for polynomials $f(x_1, \ldots, x_n)$, the basic problem of interval computations – the problem of computing the range $f(\mathbf{x}_1, \ldots, \mathbf{x}_n)$ for given intervals $\mathbf{x}_i$ – is computationally intractable (NP-hard).

3.1. Is Our Problem Feasible?

There are many reasonable algorithms that compute an enclosure $\mathbf{Y}$ for the range $\mathbf{y} = [\underline{y}, \overline{y}] = f(\mathbf{x}_1, \ldots, \mathbf{x}_n)$ of a given continuous function $f(x_1, \ldots, x_n)$ on given intervals $\mathbf{x}_1, \ldots, \mathbf{x}_n$. Ideally, we would like to *always* compute the *exact* range (i.e., its endpoints) in *feasible* time. Alas, known algorithm are not yet ideal:

- Some algorithms are *always feasible* (i.e., polynomial-time), but they *sometimes overestimate*, i.e., result in a non-optimal enclosure $\mathbf{Y} \supset \mathbf{y}$, $\mathbf{Y} \neq \mathbf{y}$.
- Other algorithms *always* compute the *exact* range, but these algorithms *require, in some cases, exponential time.*

A natural question is: *Is it possible to have an algorithm that always computes the exact range* (i.e., its endpoints) *in feasible time?*

To describe this problem in precise terms, first, we must *explain* what we mean by "*computing the endpoints*":

- If the endpoints are ***rational numbers***, then we may want to compute the explicit values for these points.
- ***In general***, we may be interested, e.g., in computing ***rational approximations*** to the desired endpoints with a given accuracy $\varepsilon > 0$.

Second, we must *fix the class of functions f*. If the function f itself is difficult to compute, then it is difficult to compute the endpoints of the interval $\mathbf{y}$ even for degenerate input intervals $\mathbf{x}_i = [x_i, x_i]$. To avoid this situation, in this book, we will restrict ourselves to the simplest possible functions: functions that can be obtained by finitely many applications of arithmetic operations $+$, $-$, $*$, and $/$, i.e., to ***rational*** functions with rational coefficients.

Comment. For rational functions, the problem of computing the range is, in principle, ***algorithmically solvable***: namely, we can apply the so-called Tarski's algorithm (for details, see the Proofs section)[408]. However, this algorithm takes too long [85]: it sometimes takes time $\approx 2^{2^n}$ for an input of size n. As a result, even for small n, it may take billions of years. This is not a practical solution.

3.2. Definitions and the Main Result

In 1981, Gaganov proved that even for polynomial functions f, the basic problem is computationally intractable (NP-hard); i.e. (unless P=NP), no feasible algorithm can solve ***all*** instances of this problem [115]). This result was first published in 1985 [116]. Let us describe the result in precise terms.

Definition 3.1. *By the basic problem of interval computations, we mean the following problem:*

GIVEN:

- *n rational intervals $\mathbf{x}_i$ (i.e., intervals with rational endpoints), and*
- *a computable continuous function f that transforms n real numbers $x_1, \ldots, x_n$ into a real number $y = f(x_1, \ldots, x_n)$.*

COMPUTE: the interval of possible values of y:

$$\mathbf{y} = [\underline{y}, \overline{y}] = f(\mathbf{x}_1, \ldots, \mathbf{x}_n) =$$
$$\{y | y = f(x_1, \ldots, x_n) \text{ for some } x_1 \in \mathbf{x}_1, \ldots, x_n \in \mathbf{x}_n\}.$$

Comments.

- By a *computable function* $f(x_1, \ldots, x_n)$, we mean an algorithm that, for arbitrary rational numbers $x_1, \ldots, x_n$, and $\delta > 0$, computes a rational number that is δ-close to $f(x_1, \ldots, x_n)$.
- Since we only consider *continuous* functions, the set of all possible values of y is indeed an interval.
- By *computing the interval* $\mathbf{y} = [\underline{y}, \overline{y}]$, we mean *computing its endpoints* $\underline{y}$ and $\overline{y}$. If these endpoints are not rational numbers, then computing these endpoints means being able to compute them with any given rational accuracy $\varepsilon > 0$, i.e., computing the rational numbers $\widetilde{\underline{y}}$ and $\widetilde{\overline{y}}$ for which $|\widetilde{\underline{y}} - \underline{y}| \leq \varepsilon$ and $|\widetilde{\overline{y}} - \overline{y}| \leq \varepsilon$. Thus, we arrive at the following definition:

Definition 3.2. *By the $\varepsilon-$approximate basic problem of interval computations, we mean the following problem:*

GIVEN:

- *n rational intervals $\mathbf{x}_i$, and*
- *a computable continuous function f that transforms n real numbers $x_1, \ldots, x_n$ into a real number $y = f(x_1, \ldots, x_n)$;*
- *a rational number $\varepsilon > 0$.*

COMPUTE: rational numbers $\widetilde{\underline{y}}$ and $\widetilde{\overline{y}}$ that are ε-close to the range's endpoints, i.e., for which $|\widetilde{\underline{y}} - \underline{y}| \leq \varepsilon$ and $|\widetilde{\overline{y}} - \overline{y}| \leq \varepsilon$, where:

$$\mathbf{y} = [\underline{y}, \overline{y}] = f(\mathbf{x}_1, \ldots, \mathbf{x}_n) =$$

$$\{y | y = f(x_1, \ldots, x_n) \text{ for some } x_1 \in \mathbf{x}_1, \ldots, x_n \in \mathbf{x}_n\}$$

Theorem 3.1. (Gaganov [115, 116]) *For every $\varepsilon > 0$, the $\varepsilon-$approximate basic problem of interval computations is NP-hard even for polynomial functions $f(x_1, \ldots, x_n)$ with rational coefficients.*

Technical comment. We will see from the proof that the problem is NP-hard even when we only consider such inputs for which the output interval $\mathbf{y}$ has rational endpoints.

3.3. The Basic Problem of Interval Computations is NP-Hard for Narrow Input Intervals

Motivation. In the previous section, we allowed input intervals $\mathbf{x}_i = [\underline{x}_i, \overline{x}_i]$ to be arbitrarily wide. These input intervals usually come from measurement, so that $\mathbf{x}_i = [\underline{x}_i, \overline{x}_i] = [\tilde{x}_i - \Delta_i, \tilde{x}_i + \Delta_i]$. In terms of measurements, wide intervals correspond to low accuracy.

If we have *perfect* accuracy $\Delta_i = 0$, then all the input intervals are degenerate $\mathbf{x}_i = [\tilde{x}_i, \tilde{x}_i]$, and the desired range $\mathbf{y}$ consists of a single easily computable point $f(\tilde{x}_1, \ldots, \tilde{x}_n)$. It is therefore natural to expect that for the *high* accuracy, when the values Δ_i are close to 0, and the corresponding input intervals are *narrow*, the basic interval computation problem is easier than in the general case. Alas, we will see that for *narrow* intervals, the problem remains NP-hard.

To describe this result, we must recall that in measurement theory (see, e.g., Rabinovich [333]), there are *two* definitions of accuracy: ***absolute*** accuracy Δ_i and *relative* accuracy $\Delta_i/\tilde{x}_i$. We will show that interval computations remain NP-hard even if we restrict ourselves to input intervals that are narrow both in the sense of absolute and relative accuracy.

Definition 3.3. *Let $\mathbf{x} = [\underline{x}, \overline{x}]$ be an interval. The value $\tilde{x} = (\underline{x} + \overline{x})/2$ is called the center of the interval $\mathbf{x}$, and the value $\Delta = (\overline{x} - \underline{x})/2$ is called the radius of the interval $\mathbf{x}$.*

Comment. One can easily see that for thus defined center and radius, $\mathbf{x} = [\tilde{x} - \Delta, \tilde{x} + \Delta]$.

Definition 3.4. *Let $\delta > 0$ be a rational number.*

- *We say that an interval $\mathbf{x} = [\tilde{x} - \Delta, \tilde{x} + \Delta]$ is **absolutely δ-narrow** if $\Delta \le \delta$.*
- *We say that an interval $\mathbf{x} = [\tilde{x} - \Delta, \tilde{x} + \Delta]$ is **relatively δ-narrow** if $\Delta \le \delta \cdot |\tilde{x}|$.*

Theorem 3.2. (Kahl [167]) ***For every $\varepsilon > 0$ and $\delta > 0$, the $\varepsilon-$approximate basic problem of interval computations for polynomial functions $f(x_1, \ldots, x_n)$ with rational coefficients and for intervals $\mathbf{x}_i$ that are both absolutely and relatively δ-narrow is NP-hard.***

3.4. What Do These Results Mean?

The situation is not as gloomy as it may seem: the problem is intractable only in some cases, but still feasible "on average". The fact that the problem is NP-hard means that (unless P=NP) every algorithm that computes the *exact* enclosure requires, in *some cases, exponential time.* This does not mean that the problem is inherently hard: we will see in the following chapters that, in spite of this *worst-case* complexity, *on average*, the range computation problem is feasible. In other words, *every* feasible algorithm has cases when it does not return the *exact* ("optimal") enclosure **y**, but for *good* algorithms, such cases are exceptionally *rare*.

One more reason why situation is not so gloomy: NP-hardness simply means that progress in data processing will never stop. The fact that the problem is NP-hard means that we cannot design an algorithm that solves *all* the problems and then rest: no matter how good our current algorithm is, there will always be problems to which this algorithm is not applicable, problems that require *creative thinking* (this idea is described convincingly, e.g., in Ratschek *et al.* [338], p. 796). When formulated in these terms, NP-hardness becomes not so much a pessimistic result, but rather a formal description of a typical situation in numerical mathematics and data processing. Crudely speaking, NP-hardness is a guarantee that new challenges will never stop and the progress in numerical methods and data processing will never end.

What if we are in a (rare) bad case: Necessity and possibility of interruptible algorithms. Since we cannot solve *all* the instances of our problem in polynomial time, for every algorithm that computes the *exact* enclosures, there are instances for which this algorithm requires unrealistically long time. If we have run into such an instance, then we have no other choice but to *interrupt* this algorithm. It is therefore desirable to design algorithms in such a way that after interrupting them, we still get a reasonable enclosure. The desirability of such *interruptible* algorithms was first considered in Artificial Intelligence (under the name of *anytime* algorithms; see, e.g., Pittarelli [326] and references therein), and it was explicitly formulated for interval computations and data processing in Shary [391, 392] (see also Shokin [396]). In Beltran *et al.* [30], it is shown that it's possible to make *every algorithm interruptible*: namely, it is proven that an arbitrary algorithm can be reformulated in interruptible form without a big increase in asymptotic computation time.

A search for feasible subclasses. The fact that a problem is NP-hard means, crudely speaking, that no feasible algorithm can solve *all* instances of this problem, i.e., that can solve this problem for polynomials of arbitrarily large number of variables, of arbitrarily large degree, with arbitrarily large coefficients, and with arbitrary intervals. A natural question is: what will happen if we limit the polynomials and/or intervals? Will the problem still be NP-hard, or will it become tractable (i.e., solvable by a feasible algorithm)? This question will be analyzed in the following few chapters. Namely, we will consider the following natural limitations:

- In Chapter 4, we consider polynomials of fixed *number of variables.*
- In Chapter 5, we consider polynomials of bounded *degree.*
- In Chapter 6, we consider polynomials with bounded *coefficients.*
- In Chapter 7, we show that the problem remains NP-hard even if we *fix* a sequence of *polynomials* $f_n(x_1, \ldots, x_n)$ and allow *arbitrary* (narrow) intervals.
- Finally, in Chapter 8, we will show that the main problem remains NP-hard even if we *fix* some narrow *intervals* and allow *arbitrary polynomials.*

Proofs

Proof of the Comment in Section 3.1. The algorithm proposed by Tarski (and later modified by Seidenberg and others; see Chapter 4 for more details) takes as input an arbitrary formula F that can be obtained from the polynomial equalities $f_i(x_1, \ldots, x_n) = 0$ and inequalities $f_j(x_1, \ldots, x_n) = 0$ (in which $f_i(x_1, \ldots, x_n)$ and $f_j(x_1, \ldots, x_n)$ are polynomials with integer or rational coefficients) by adding logical connectives ("and", "or", "not", "implies"), and quantifiers $\forall x$ and $\exists x$ that run over all real numbers. In particular, the upper endpoint $\overline{y}$ of the range $f(\mathbf{x}_1, \ldots, \mathbf{x}_n)$ of a polynomial $f(x_1, \ldots, x_n)$ over intervals $[\underline{x}_i, \overline{x}_i]$ with rational endpoints can be described by a formula

$$\forall x_1 \ldots \forall x_n (((\underline{x}_1 \le x_1 \le \overline{x}_1) \& \ldots \& (\underline{x}_n \le x_n \le \overline{x}_n)) \to f(x_1, \ldots, x_n) \le \overline{y}) \&$$

$$\exists x_1 \ldots \exists x_n ((\underline{x}_1 \le x_1 \le \overline{x}_1) \& \ldots (\underline{x}_n \le x_n \le \overline{x}_n) \& f(x_1, \ldots, x_n) = \overline{y}) .$$

If the resulting formula F contains *no free variables* (i.e., if each of its variables is bound by some quantifier), then this formula is either true or false, and

Tarski's algorithm determines, after a finite number of computational steps, whether this formula is true or false.

If the input formula F *contains free variables*, then this algorithm returns a quantifier-free formula F' that is equivalent to F, i.e., a finite sequence of equations and inequalities that is equivalent to the original formula F.

In particular, if the formula F has *exactly one* free variable, i.e., if it has the form $F(x)$ for a real-valued variable x, and if there exists a unique real number x_0 for which this formula is true, then Tarski-Seidenberg's algorithm returns a non-zero polynomial $P(x)$ with integer coefficients for which $P(x_0) = 0$, plus finitely many polynomial inequalities that uniquely determine x_0 out of all possible roots of the polynomial $P(x)$.

There exists algorithms that, given a polynomial $P(x)$ with integer coefficients and an arbitrary rational number $\varepsilon > 0$, compute rational numbers that are ε-close to the roots of $P(x)$, and these algorithms can be easily modified to take inequalities into consideration too. So, as a result, for every given $\varepsilon > 0$, we get an algorithm that computes an ε-approximation to the desired real number x_0. In particular, this means that we *can compute* the upper endpoint of the polynomial's range (and similarly, we can compute the range's lower endpoint).

Similarly, Tarski's algorithm can be used to compute the range of an arbitrary *rational* function with rational coefficients.

Proof of Theorem 3.1. Instead of the original Gaganov's proof, we will use a simpler proof (inspired by Vavasis [418]). First, we will show that the *exact* basic problem of interval computations is NP-hard.

To prove this result, we will show that if we are able to compute the desired interval $\mathbf{y}$ for quadratic polynomials $f(x_1, \ldots, x_n)$, then we will be able to solve propositional satisfiability problem for 3-CNF formulas. Since satisfiability problem is known to be NP-hard, we will thus prove that our problem is also NP-hard.

Indeed, let $F = F_1 \& \ldots \& F_k$ be a 3-CNF formula with the (Boolean) variables $z_1, \ldots, z_v$. (In the computer, usually, "true" is represented as 1, and "false" as 0.) The corresponding quadratic problem will have $v + k$ variables $x_1, \ldots, x_v, p_1, \ldots, p_k$. The construction of f is as follows:

- To each propositional variable z_i, we put into correspondence a real-number variable $f[z_i] = x_i$.
- To each negative literal $\neg z_i$, we put into correspondence a linear expression $f[\neg z_i] = 1 - x_i$.
- To each expression F_j of the type $a \vee b$, we put into correspondence the expression $f[F_j] = (f[a] + f[b] + p_j - 2)^2$. Since $f[a]$ and $f[b]$ are linear in the variables x_i, the resulting expression is quadratic in x_i and p_j.
- To each expression F_j of the type $a \vee b \vee c$, we put into correspondence the expression $f[F_j] = (f[a] + f[b] + f[c] + 2p_j - 3)^2$. The resulting expression is quadratic in x_i and p_j.
- To the formula F, we put into correspondence the quadratic function

$$f(x_1, \ldots, x_v, p_1, \ldots, p_k) = \sum_{i=1}^{v} x_i(1 - x_i) + \sum_{j=1}^{k} f[F_j].$$

We will choose $\mathbf{x}_i = \mathbf{p}_j = [0, 1]$, and try to compute the lower bound $\underline{y}$ of the interval

$$f(\mathbf{x}_1, \ldots, \mathbf{x}_v, \mathbf{p}_1, \ldots, \mathbf{p}_k).$$

Example. Let us take $F = (z_1 \vee z_2 \vee z_3) \& (z_1 \& \neg z_2)$. For this formula, $v = 2$, $k = 2$, so, we need $v + k = 4$ real-number variables x_1, x_2, p_1, and p_2. Here:

- $f[\neg z_2] = 1 - x_2$.
- $f[F_1] = (x_1 + x_2 + x_3 + 2p_1 - 3)^2$.
- $f[F_2] = (x_1 + (1 - x_2) + p_2 - 2)^2$.
- $f(x_{1,2}, p_1, p_2) = x_1(1 - x_1) + x_2(1 - x_2) + f[F_1] + f[F_2]$.

Before we start estimating $\underline{y}$, let us notice that $f[F_j]$ is defined as a square, and therefore, $f[F_j] \geq 0$. Also, if $x_i \in [0, 1]$, then $x_i(1 - x_i) \geq 0$. Therefore, the function f is a sum of non-negative numbers and is, therefore, non-negative. Hence, $\underline{y} \geq 0$.

Let us show that $\underline{y} = 0$ if and only if the formula F is satisfiable.

- If the formula F is satisfiable, i.e., it is true for some propositional vector $z_1, \ldots, z_v$, then we take $x_i = z_i$ (i.e., $x_i = 1$ if $z_i =$"true" and $x_i = 0$ if $z_i =$"false"). The values of p_j are chosen as follows:

  - If $F_j = a \vee b$, and both a and b are true for z_i, then we take $p_j = 0$.
  - If $F_j = a \vee b$, and only one of the literals a and b is true for a given choice of z_i, then we take $p_j = 1$.
  - If $F_j = a \vee b \vee c$, and all three literals are true, then $p_j = 0$.
  - If $F_j = a \vee b \vee c$, and two out of three literals are true, then $p_j = 0.5$.
  - If $F_j = a \vee b \vee c$, and only one of the three literals is true, then $p_j = 1$.

  In all five cases, $f[F_j] = 0$ for all j. Therefore, for these x_i and p_j, $f(x_1, \ldots, x_n, p_1, \ldots, p_k) = 0$; therefore, $\underline{y} = \min f \leq 0$. Since we know that $\underline{y} \geq 0$, we conclude that $\underline{y} = 0$.

- Vice versa, let us assume that $\underline{y} = \min f = 0$. The minimum of a continuous function of a compact $[0,1]^{v+2k}$ is always attained; therefore, there exist values x_i and p_j, for which $f = 0$. We have mentioned that f is a sum of several non-negative expressions. The sum of non-negative expressions is equal to 0 if and only if all these expressions are equal to 0, i.e., $x_i(1 - x_i) = f[F_j] = 0$ for all i and j. From $x_i(1 - x_i) = 0$, we conclude that $x_i = 0$ or $x_i = 1$. Let us show that the values $z_i = x_i$ (i.e., $z_i =$"true" if $x_i = 1$ and $z_i =$"false" when $x_i = 0$) make+ the formula F true. For that, we need to prove that each expression F_j is true.

  - For $F_j = a \vee b$, from $f[F_j] = 0$, it follows that $f[a] + f[b] + p_j = 2$. Since $p_j \leq 1$, we conclude that $f[a] + f[b] \geq 1$. Both values $f[a]$ and $f[b]$ are equal either to 0 or to 1. Since their sum is ≥ 1, they cannot be both equal to 0, so, one of them is equal to 1. Due to our choice of z_i, the corresponding literal a is then true, and therefore, F_j is true.
  - For $F_j = a \vee b \vee c$, from $f[F_j] = 0$, we conclude that $f[a] + f[b] + f[c] + 2p_j = 3$, and therefore, since $p_j \leq 1$, that $f[a] + f[b] + f[c] \geq 1$. Hence, one of these values is $= 1$. For this value, the corresponding literal is true, and therefore, F_j is true.

The result is proven.

Let us prove that for the function f constructed in the previous proof, if the formula F is not satisfiable, then $\underline{y} \geq 0.09$.

We will prove this statement by reduction to a contradiction: we will assume that $\underline{y} < 0.09$, and conclude that F is satisfiable. As in the previous proof, from the fact that $\underline{y} = \min f < 0.09$, it follows that there exist values x_i and p_j for which $f(x_1, \ldots, x_v, p_1, \ldots, p_k) = \underline{y} < 0.09$. Since f is the sum of non-negative terms, from this inequality, it follows that each term is < 0.09.

In particular, it follows that $x_i(1-x_i) < 0.09$. The function $x(1-x)$ is increasing for $x < 0.5$ and decreasing afterwards. So, from $x_i(1-x_i) < 0.09$ and from the fact that $0.1(1-0.1) = 0.9(1-0.9) = 0.09$, it follows that $x_i < 0.1$ or $x_i > 0.9$ for all i. Let us take $z_i =$"true" if $x_i > 0.9$, and $z_i =$"false" if $x_i < 0.1$, and let us show that these propositional values make the formula F true (i.e., they make all the expressions F_j true). Indeed:

- If $F_j = a \vee b$, then from $f[F_j] = (f[a] + f[b] + p_j - 2)^2 < 0.09$, it follows that $f[a] + f[b] + p_j - 2 > -0.3$, and $f[a] + f[b] > 1.7 - p_j$. Since $p_j \leq 1$, we conclude that $f[a] + f[b] > 0.7$. Therefore, the values $f[a]$ and $f[b]$ cannot be both < 0.1. Therefore, one of these two values is > 0.9. The corresponding literal is equal to "true", and hence, F_j is true.
- If $F_j = a \vee b \vee c$, then from $f[F_j] = (f[a] + f[b] + f[c] + 2p_j - 3)^2 < 0.09$, it follows that $f[a] + f[b] + f[c] + 2p_j - 3 > -0.3$, and $f[a] + f[b] + f[c] > 2.7 - 2p_j$. Since $p_j \leq 1$, we conclude that $f[a] + f[b] + f[c] > 0.7$. Therefore, the values $f[a]$, $f[b]$, and $f[c]$ cannot be all < 0.1. Therefore, one of these three values is > 0.9. The corresponding literal is equal to "true", and hence, F_j is true.

So, F is satisfiable. The statement is proven.

The value $\underline{y}$ is thus either $= 0$ (if the formula F is satisfiable), or ≥ 0.09 (if the formula F is not satisfiable). Therefore, if we had a polynomial-time algorithm that computes the desired interval with an accuracy 0.04, we would thus be able to distinguish between the two cases, and thus, tell whether a formula is satisfiable or not. So, for $\varepsilon = 0.04$, $\varepsilon-$accurate estimation of the interval $f(\mathbf{x}_1, \ldots)$ is NP-hard.

To prove NP-hardness for arbitrary ε, we must take into consideration that if we estimate, for an arbitrary quadratic function f, the interval for a function $g(x_1, \ldots,) = (\varepsilon/0.04) f(x_1, \ldots)$ with an accuracy ε, and divide this estimate by $\varepsilon/0.04$, then we have an $0.04-$accurate estimate for the original function f. So, since computing a $0.04-$accurate estimate is NP-hard, computing $\varepsilon-$accurate estimates is also NP-hard. The theorem is proven.

Proof of Theorem 3.2. Let us show that if we can solve the basic problem of interval computations for intervals that are both absolutely and relatively δ-*narrow*, then we will be able to solve it for *arbitrary* intervals $\mathbf{x}_i$. Indeed, assume that we can solve this problem for *narrow* intervals. How can we then compute the range of a polynomial $f(x_1, \ldots, x_n)$ over *arbitrary* rational intervals $\mathbf{x}_i = [\underline{x}_i, \overline{x}_i]$?

First of all, if one of the intervals $\mathbf{x}_i$ is *degenerate* (i.e., $\underline{x}_i = \overline{x}_i$ and $\mathbf{x}_i = \{\underline{x}_i\}$), then we can simply substitute this value $\underline{x}_i$ instead of the interval $\mathbf{x}_i$ into the polynomial $f(x_1, \ldots, x_n)$ and reduce our original problem to a range estimation problem for a polynomial with one fewer variable. We can thus easily "eliminate" all degenerate interval inputs.

Therefore, without losing generality, we can assume that all input intervals are non-degenerate: $\mathbf{x}_i = \tilde{x}_i - \Delta_i, \tilde{x}_i + \Delta_i]$ with $\Delta_i > 0$. Let us introduce new variables $y_i = a_i \cdot x_i + b_i$ in such a way that when x_i goes from $\underline{x}_i = \tilde{x}_i - \Delta_i$ to $\overline{x}_i = \tilde{x}_i + \Delta_i$, the new variable y_i goes from $\underline{y}_i = 1 - \delta$ to $\overline{y}_i = 1 + \delta$. The corresponding equations $a_i \cdot (\tilde{x}_i - \Delta_i) + b_i = 1 - \delta$ and $a_i \cdot (\tilde{x}_i + \Delta_i) + b_i = 1 + \delta$ lead to $a_i = \delta/\Delta_i$ and $b_i = 1 - \delta \cdot \tilde{x}_i/\Delta_i$. In terms of these new variables, $x_i = (y_i - b_i)/a_i$, $f(x_1, \ldots, x_n) = F(y_1, \ldots, y_n)$, where

$$F(y_1, \ldots, y_n) = f\left(\frac{y_1 - b_1}{a_1}, \ldots, \frac{y_n - b_n}{a_n}\right),$$

and the desired range $f(\mathbf{x}_1, \ldots, \mathbf{x}_n)$ coincides with the range $F(\mathbf{y}_1, \ldots, \mathbf{y}_n)$ of the new polynomial $F(y_1, \ldots, y_n)$ over the intervals $\mathbf{y}_i = [1 - \delta, 1 + \delta]$, each of which is both absolutely and relatively δ-narrow.

Thus, the basic problem of interval computations for arbitrary intervals (the problem that is already known to be NP-hard) is reduced to the basic problem for narrow intervals. Hence, the basic problem for narrow intervals is also NP-hard. The theorem is proven.

4

BASIC PROBLEM OF INTERVAL COMPUTATIONS FOR POLYNOMIALS OF A FIXED NUMBER OF VARIABLES

In the previous chapter, we proved that the problem of computing the range $f(\mathbf{x}_1, \ldots, \mathbf{x}_n)$ of a given polynomial $f(x_1, \ldots, x_n)$ over given intervals $\mathbf{x}_1, \ldots, \mathbf{x}_n$ is, in general, computationally intractable (NP-hard). Since this general problem is intractable, it is desirable to look for cases in which it is feasible. In this chapter, we analyze what happens when we restrict the number of variables n. Good news is that in this case, a polynomial-time algorithm is possible. Bad news is that the existing polynomial-time algorithms require too much computation time to be practical.

4.1. Good News

Theorem 4.1. *For every n, there exists a polynomial-time algorithm that, for any $\varepsilon > 0$, for any polynomial $f(x_1, \ldots, x_n)$, and for any n intervals $[\underline{x}_i, \overline{x}_i]$, computes ε-approximations to the endpoints of the range $f([\underline{x}_1, \overline{x}_1], \ldots, [\underline{x}_n, \overline{x}_n])$.*

The following table compares this result with the NP-hardness result from the previous chapter:

Polynomials of 1 variable	Polynomials of 2 variables	...	Polynomials of n variables (n fixed)	...	Polynomials of arbitrary number of variables
Polynomial time	Polynomial time	...	Polynomial time	...	NP-hard

Comment This result follows from an algorithm proposed by Grigor'ev *et al.* [134] (for checking whether a given system of polynomial inequalities has a solution).

A similar result is true not only for *polynomials*, but also for *algebraic* functions, i.e., functions $y = f(x_1, \ldots, x_n)$ that are uniquely determined by a system of polynomial equations $P_k(x_1, \ldots, x_n, y) = 0$ and polynomial inequalities $Q_k(x_1, \ldots, x_n, y) \geq 0$, where P_k and Q_k are polynomials with integer coefficients (e.g., $f(x) = \sqrt{x^2+1}$ is a solution of a system consisting of the equation $y^2 - (x^2+1) = 0$ and of the inequality $y \geq 0$).

In contrast to a *polynomial* which is always continuous (and therefore, always *bounded* on a box), a general *algebraic* function is *not necessarily bounded* (e.g., the function $f(x) = 1/x$ defined by the equation $x \cdot y - 1 = 0$ is not bounded on the interval $[0, 1]$). Therefore, to get a meaningful result, we will only consider *bounded* algebraic functions $f(x_1, \ldots, x_n)$, with a known bound Δ on $|f(x_1, \ldots, x_n)|$. For such functions, the following result is true:

Theorem 4.2. *For every n, there exists a polynomial-time algorithm that, for any $\varepsilon > 0$, for any bounded algebraic function $f(x_1, \ldots, x_n)$, and for any n intervals $[\underline{x}_i, \overline{x}_i]$, computes ε-approximations to the endpoints of the range $f([\underline{x}_1, \overline{x}_1], \ldots, [\underline{x}_n, \overline{x}_n])$.*

4.2. Not So Good News, and Hope

Not so good news: the existing algorithms are not yet practical. The algorithm described in Grigor'ev et al [134] requires time $M \cdot (k \cdot d)^{n^2}$, where 2^M is the upper bound for the values of the coefficients, k is the total number of inequalities, and d is the number of binary digits. For fixed n, this algorithm is polynomial, i.e., its computation time is bounded by a polynomial $P_n(L)$ of the bit length L of the input data, but the degrees of the polynomials P_n rapidly increase (as n^2) as n increases.

Heintz *et al.* [148] have shown that the algorithm of Grigor'ev *et al.* is an example of an algorithm that is polynomial-time but that is *not* yet *practical*: for polynomials of only four variables, it can take, for some instances, millions of years to compute, even on the fastest computers.

There is hope. Although the existing algorithms are not always practical, the fact that a *theoretically* feasible (i.e., polynomial-time) algorithm exists makes us hope that a *practically* feasible algorithm will appear. This hope is encouraged by the increasing amount of practical applications of similar algorithms; see, e.g., Hong *et al.* [155, 156], Loos *et al.* [259] Abdallah *et al.* [1], Dolzmann *et al.* [93, 94], Weispfenning *et al.* [425, 426, 427]. In particular, there exist applications to *transportation problems* Loos *et al.* [259], to *control system design* Abdallah *et al.* [1], to *geometric reasoning* Dolzmann *et al.* [94], to *stability analysis of partial differential equations* Hong *et al.* [156], and to *optimization* Weispfenning [425].

4.3. Can We Generalize This Feasibility Result?

We have just mentioned that the theoretical feasibility of range computations for polynomial (or algebraic functions) with fixed number of variables leads to the hope that some day, a practically feasible algorithm will appear. In view of this possibility, it is important to know whether theoretical feasibility can be proven for *more general* classes of interval computations problems. In this section, we consider three possible generalizations:

- First, a class that, strictly speaking, does not generalize the class of polynomials: In our results, we defined a polynomial as a sum of monomials. Some polynomials can be re-written in a more compact way: e.g., a binomial $x^3 + 3x^2 \cdot y + 3x \cdot y^2 + y^3$ can be re-written as $(x + y)^3$. A natural question is: if we allow such compact descriptions, will we still be able to have a feasible algorithm, i.e., an algorithm whose computation time is bounded by a polynomial of the length of the *compact* input?
- A natural *true* generalization of *polynomial* functions is the class of *piece-wise polynomial* functions (splines; see, e.g., Schumaker [385] and Eubank [103]). Is a polynomial-time algorithm possible for splines?
- Splines can be viewed as a very natural generalization of polynomials: Namely, polynomials can be defined as functions obtained from variables by using three arithmetic operations; if we add min and max to these operations, we get piece-wise polynomial functions, i.e., *splines.* What if we add *other operations*, such as sin, cos, exp?

We will show that all three generalizations make the problem intractable.

Polynomials with compact description

We will show that even allowing the iterated squaring makes the basic problem of interval computation intractable, even for polynomials of one variable:

Definition 4.1. *By a compact polynomial in variables $x_1, \ldots, x_n$, we mean an expression that is obtained from variables x_i by applying addition $+$, multiplication $\cdot$, and square operation x^2.*

As *examples* of compact polynomials, we can take $(x_1+x_2)^2 \cdot x_3$ or $x_1 \cdot x_1 \cdot (x_1^2)^2$.

Theorem 4.3. $(n = 1)$ *For every $\varepsilon > 0$, every algorithm that ε-accurately computes the range $f([\underline{x}, \overline{x}])$ of an arbitrary compact polynomial $f(x)$ over an arbitrary interval $[\underline{x}, \overline{x}]$, requires, in some instances, exponential time.*

	Polynomials $f(x_1, \ldots, x_n)$	Compact polynomials $f(x_1, \ldots, x_n)$
$n = 1$	Polynomial time	Exponential time (or worse)
$n = 2$	Polynomial time	Exponential time (or worse)
fixed n	Polynomial time	Exponential time (or worse)
arbitrary n	NP-hard	Exponential time (or worse)

Splines

For splines, we also get intractability even for functions of one variable and even for piece-wise linear functions (i.e., for linear splines):

A continuous function $f(x)$ defined on an interval $[\underline{x}, \overline{x}]$ is called a *piece-wise linear function*, or a *linear spline* if the interval can be divided into finitely many subintervals on each of which $f(x)$ is linear. A spline function is *feasible* if there exists a polynomial-time algorithm that, given rational numbers x and $\varepsilon > 0$, computes an ε-approximation to $f(x)$.

Theorem 4.4. $(n = 1)$ *For every $\varepsilon > 0$, the problem of ε-accurately computing the range $f([\underline{x}, \overline{x}])$ of a given feasible linear spline $f(x)$ on a given interval $[\underline{x}, \overline{x}]$ is NP-hard.*

Historical comment. This result was first announced (in a slightly different formulation) in Kreinovich [190, 194]; see also Chee [63, 64].

	Polynomials $f(x)$	Splines $f(x)$
Linear	Linear time	NP-hard
Quadratic	Polynomial time	NP-hard
Cubic and higher order	Polynomial time	NP-hard

Adding Non-Algebraic Functions to Polynomials

Definition 4.2. *Let $g(x)$ be a function. By a g-polynomial of n variables $x_1, \ldots, x_n$, we mean an arbitrary expression that can be obtained from the variables $x_1, \ldots, x_n$ and rational constants by applying arithmetic operations $+$, $-$, $\cdot$, and a function g.*

For g-polynomials, the range computation problem becomes intractable:

Theorem 4.5. (For $g(x) = \exp(x)$; $n = 1$) *For every $\varepsilon > 0$, every algorithm that ε-accurately computes the range $f([\underline{x}, \overline{x}])$ of an arbitrary* exp*-polynomial $f(x)$ of one variable over an arbitrary interval $\mathbf{x} = [\underline{x}, \overline{x}]$, requires, in some instances, exponential time.*

Comments.

- This result was first announced (in a slightly different formulation) in Kreinovich [190, 194].
- This result remains true even if we restrict ourselves to sub-intervals $\mathbf{x}$ of a fixed interval $\mathbf{a}$.

Theorem 4.6. (For $g(x) = \sin(x)$; $n = 1$) *For every $\varepsilon > 0$, the problem of ε-accurately computing the range $f([\underline{x}, \overline{x}])$ of an arbitrary* sin*-polynomial $f(x)$ over an arbitrary interval $\mathbf{x} = [\underline{x}, \overline{x}]$ is NP-hard.*

	Polynomials $f(x_1, \ldots, x_n)$	exp-polynomials $f(x_1, \ldots, x_n)$	sin-polynomials $f(x_1, \ldots, x_n)$
$n = 1$	Polynomial time	Exponential time (or worse)	NP-hard
$n \geq 2$	Polynomial time	Exponential time (or worse)	NP-hard

Adding Non-Algebraic Functions to Rational Functions

If, in addition to $+$, $-$, $\cdot$, and $g(x)$, we allow *division*, we get *g-rational* functions for which computing the range is harder than for g-polynomials:

Definition 4.3. *Let $g(x)$ be a function. By a g-rational function of n variables $x_1, \ldots, x_n$, we mean an arbitrary expression that can be obtained from the variables $x_1, \ldots, x_n$ and computable constants by applying arithmetic operations $+$, $-$, $\cdot$, $/$, and a function g.*

Theorem 4.7. ($n = 1$) *Suppose that $g(x)$ is a twice differentiable function for which $g'(x_0) = 0$ and $g''(x_0) \neq 0$ for some x_0. Then, for every $\varepsilon > 0$, every algorithm that ε-accurately computes the range $f([\underline{x}, \overline{x}])$ of an arbitrary g-rational function $f(x)$ of one variable over an arbitrary interval $\mathbf{x} = [\underline{x}, \overline{x}]$, requires, in some instances, exponential time.*

Comments.

- The class of such functions $g(x)$ includes $\sin(x)$ (with $x_0 = \pi/2$) and $\cos(x)$ (with $x_0 = 0$).
- A similar result is true for every $k \geq 2$ times differentiable function for which $g(x_0) = g'(x_0) = \ldots = g^{(k-1)}(x_0) = 0$ and $g^{(k)}(x_0) \neq 0$ for some x_0.
- Due to the previous remark, this exponential-time result is true for every computable analytical non-monotonic function $g(x)$ (for precise definition of computable functions, see, e.g., Bishop *et al.* [48, 49]). Indeed, since the function is not monotonic, it must have a local extremum, i.e., a point x_0 in which $g'(x_0) = 0$, and since it is analytical, at least one of the higher derivatives must be different from 0 at this point x_0. Computability of such values x_0 was proven in Orevkov [318].

	g-polynomials $f(x)$	g-rational functions $f(x)$
no $g(x)$	Polynomial time	Polynomial time
$g(x) = \exp(x)$	Exponential time (or worse)	Exponential time (or worse)
$g(x) = \sin(x)$	NP-hard	Exponential time (or worse)
non-monotonic analytical $g(x)$	?	Exponential time (or worse)

The problem also becomes intractable if we allow *complex numbers.*

Definition 4.4. *By a complex-rational function of n variables $x_1, \ldots, x_n$, we mean an arbitrary expression that can be obtained from the variables $x_1, \ldots, x_n$ and constants $0, 1$, and i, by applying arithmetic operations $+$, $-$, $\cdot$, $/$, and operations* Re *and* Im. *We say that a complex-rational function $f(x_1, \ldots, x_n)$ is real-valued if for all real values $x_1, \ldots, x_n$, the value $f(x_1, \ldots, x_n)$ is also real.*

Theorem 4.8. $(n = 1)$ *For every $\varepsilon > 0$, every algorithm that ε-accurately computes the range $f([\underline{x}, \overline{x}])$ of an arbitrary real-valued complex-rational function $f(x)$ of one variable over an arbitrary interval $\mathbf{x} = [\underline{x}, \overline{x}]$, requires, in some instances, exponential time.*

	Rational functions $f(x_1, \ldots, x_n)$	Complex-rational functions $f(x_1, \ldots, x_n)$
$n = 1$	Polynomial time	Exponential time (or worse)
$n \geq 2$	Polynomial time	Exponential time (or worse)

Are the Corresponding Range Estimation Problems Algorithmically Decidable At All?

The above results show that even for the simplest non-algebraic functions $g(x)$, crudely speaking, no *feasible algorithm* can estimate the range of the corresponding g-polynomials and g-rational functions. A natural next question is: Is there *any* algorithm at all, even non-feasible?

Even for the first non-algebraic function $\exp(x)$ that we considered, it is still not known whether the range-computation problem is algorithmically decidable. Macintyre *et al.* [265] (see also Marker [272] and Wilkie [428]) proved that a certain believed-to-be-true hypothesis about algebraic relations (due to Schanuel) implies the decidability of the first order theory of real numbers with arithmetic operations and exponentiation. So, if Schanuel's hypothesis is true, then for every rational number A, we can check, for every exp-polynomial $f(x_1, \ldots, x_n)$, whether the first order formula

$$\forall x_1 \ldots \forall x_n((\underline{x}_1 \le x_1 \le \overline{x}_1) \& \ldots \& (\underline{x}_n \le x_n \le \overline{x}_n) \rightarrow f(x_1, \ldots, x_n) \le A),$$

is true, and thus, check whether $\overline{y} \le A$; then, by using binary search, we will be able to compute $\overline{y}$ with an arbitrary accuracy ε ($\underline{y}$ can be computed in a similar manner).

Similarly, Schanuel's hypothesis would imply that the range problem is decidable not only for exp-*polynomials* and exp-*rational functions*, but also for exp-*algebraic functions*, i.e., for the functions $y = f(x_1, \ldots, x_n)$ that are uniquely determined by a system of exp-polynomial equations $P_k(x_1, \ldots, x_n, y) = 0$ and polynomial inequalities $Q_k(x_1, \ldots, x_n, y) \ge 0$, where P_k and Q_k are exp-polynomials. In particular, since the function $y = \log(x)$ is a unique solution of the exp-polynomial equation $\exp(y) - x = 0$, this function is exp-algebraic, and thus, Schanuel's hypothesis implies that the range computing problem is also algorithmically solvable for log-polynomials and log-rational functions.

The following table represents these results; in this table, "decidable*" means decidable if Schanuel's conjecture is true:

	g-polynomials $f(x_1, \ldots, x_n)$	g-rational functions $f(x_1, \ldots, x_n)$	g-algebraic functions $f(x_1, \ldots, x_n)$
no g	Algorithmically decidable	Algorithmically decidable	Algorithmically decidable
$g(x) = \exp(x)$	Algorithmically decidable*	Algorithmically decidable*	Algorithmically decidable*
$g(x) = \log(x)$	Algorithmically decidable*	Algorithmically decidable*	Algorithmically decidable*

Comment. Various other algorithms and negative results related to exp-polynomials and exp-rational functions can be also found in Gurevic [135, 136, 137, 138, 139].

Proofs

Proof of Theorem 4.1. Let us first describe the algorithm that finds the desired estimate. Since we are interested in *approximate* values of endpoints, it is sufficient to describe an algorithm that takes input intervals with *binary-rational* endpoints; the endpoints of a polynomial with *arbitrary rational* coefficients can be estimated, if we apply this algorithm to sufficiently accurate binary-rational approximations of the given rational values. Similarly, it is sufficient to consider polynomials whose coefficients are binary-rational numbers.

The algorithm restricted to binary-rational endpoints thus takes the following input:

- an integer n (describing the *number of direct measurements*);
- an integer d (describing the *computer precision*, or *the number of binary digits*);
- n *d-digit real numbers* $\widetilde{x}_i$, i.e., numbers of the type $d_k \ldots d_0.d_{-1} \ldots d_{-d}$ (with d digits after the binary point); these numbers correspond to *results of direct measurements*;
- n d-digit non-negative real numbers $\Delta_i > 0$ (these numbers describe the *accuracies of direct measurements*);
- a d-digit polynomial $f(x_1, \ldots, x_n)$, i.e., a sum of monomials $f_1(x_1, \ldots, x_n) + \ldots + f_m(x_1, \ldots, x_n)$, where each monomial $f_i(x_1, \ldots, x_n)$ is an expression of the type $a_{i_1 \ldots i_n} x_1^{i_1} \cdot \ldots \cdot x_n^{i_n}$, and each coefficient $a_{i_1 \ldots i_n}$ is a d-digit real number; (this polynomial $f(x_1, \ldots, x_n)$ describes a *data processing algorithm*);
- a rational number $\varepsilon > 0$ (that describes *precision with which we want the result.*)

Our objective is to compute the ε-approximations $\widetilde{\underline{y}}$ and $\widetilde{\overline{y}}$ to the endpoints $\underline{y}$ and $\overline{y}$ of the interval $\mathbf{y} = [\underline{y}, \overline{y}] = f(\mathbf{x}_1, \ldots, \mathbf{x}_n)$, where $\mathbf{x}_i = [\underline{x}_i, \overline{x}_i] = [\widetilde{x}_i - \Delta_i, \widetilde{x}_i + \Delta_i]$.

To compute these approximations, first, we use naive interval computations to compute the initial enclosures of the intervals $[\underline{F}, \overline{F}]$ and $[\underline{G}, \overline{G}]$ that contain $\underline{y}$ and $\overline{y}$:

- For each variable x_i, $1 \leq i \leq n$, we compute the upper bound X_i on its absolute values $|x_i|$ as $X_i = \max(|\underline{x}_i|, |\overline{x}_i|)$.
- For each monomial $f_j(x_1, \ldots, x_n) = a_{i_1 \ldots i_n} x_1^{i_1} \cdot \ldots \cdot x_n^{i_n}$, we compute the upper bounds F_j on its the absolute values $|f_j(x_1, \ldots, x_n)|$ of as $F_j = |a_{i_1 \ldots i_n}| \cdot X_1^{i_1} \cdot \ldots \cdot X_n^{i_n}$.
- We compute the upper bound F for the absolute value $|f(x_1, \ldots, x_n)|$ of the polynomial $f(x_1, \ldots, x_n)$ as $F = F_1 + \ldots + F_m$.

Finally, we take the smallest integer k for which $F \leq 2^k$, and take $\underline{F} = \underline{G} = -2^k$ and $\overline{F} = \overline{G} = 2^k$. (Since $-F \leq f \leq F$, and $F \leq 2^k$, we have $\underline{y} \in [\underline{F}, \overline{F}]$ and $\overline{y} \in [\underline{G}, \overline{G}]$.)

Then, we apply the following iterative bisection procedure for computing $\overline{y}$ (computing $\underline{y}$ is similar):

- At the beginning of each iteration, we have an interval $[\underline{G}, \overline{G}]$ that is known to contain $\overline{y}$. First, compute its midpoint $G = (1/2) \cdot (\underline{G} + \overline{G})$.
- Check whether $\overline{y} \leq G$, by applying Grigor'ev's algorithm to check the existence of a real solution of the following sequence of polynomial inequalities with integer coefficients: $2^D \cdot f(x_1, \ldots, x_n) - 2^D \cdot G \geq 0$, $2^D(x_i - \underline{x}_i) \geq 0, 1 \leq i \leq n$, and $2^D(\overline{x}_i - x_i) \geq 0$ for an appropriate D.
- If this set of inequalities has a solution, this means that the upper bound $\overline{y}$ for f is $\geq G$, so we take $[G, \overline{G}]$ as the new interval $[\underline{G}, \overline{G}]$; else, take $[\underline{G}, G]$.
- If $\overline{G} - \underline{G} \leq \varepsilon$, then return $\overline{G}$ as the desired estimate $\widetilde{\overline{y}}$; else, continue the iterations.

That this algorithm finds the desired estimate is clear. The only thing that needs to be proven is that it is feasible. Indeed:

- The initial step (computing F) is feasible.
- On each step of the bisection process, we are using the polynomial-time algorithm from Grigor'ev *et al.* [134]. How many times do we need to apply this algorithm? The initial size of the interval $[\underline{G}, \overline{G}] = [-2^k, 2^k]$ is 2^{k+1}. On each iteration step, we halve the size of the interval $[\underline{G}, \overline{G}]$. Therefore, after s steps, we get the interval of the size 2^{k+1-s}. To get a size $2^{-N} \leq \varepsilon$, we need $N + k + 1$ iterations. This number is polynomial in the length of the input data, and therefore, the total run-time for all these iterations is also polynomial.

So, both steps are polynomial, and therefore, the algorithm is feasible. The theorem is proven.

Proof of Theorem 4.2. One can easily verify that the following algorithm solves the problem:

The input to this algorithm consists of the same values n, d, $\tilde{x}_i$, Δ_i, and N as in Theorem 4.1; the only difference is that instead of a *single* polynomial $f(x_1, \ldots, x_n)$, we must now describe a finite *sequence* of (similarly defined) polynomials $f_1(x_1, \ldots, x_n, y), \ldots, f_k(x_1, \ldots, x_n, y)$ such that the given algebraic function $y = f(x_1, \ldots, x_n)$ is defined as a unique solution of the system $f_1(x_1, \ldots, x_n, y) = 0$, $f_2(x_1, \ldots, x_n, y) \geq 0$, ..., $f_k(x_1, \ldots, x_n, y) \geq 0$. We must also input the (known) upper bound Δ on $|f(x_1, \ldots, x_n)|$. Our goal is to compute the ε-approximations $\widetilde{\underline{y}}$ and $\widetilde{\overline{y}}$ to the endpoints $\underline{y}$ and $\overline{y}$ of the range $\mathbf{y} = [\underline{y}, \overline{y}] = f(\mathbf{x}_1, \ldots, \mathbf{x}_n)$.

Similarly to the proof of Theorem 4.1, we start with computing the initial value of the intervals $[\underline{F}, \overline{F}]$ and $[\underline{G}, \overline{G}]$ (that contain $\underline{y}$ and $\overline{y}$). For that, we simply take the smallest integer k for which $\Delta \leq 2^k$, and take $\underline{F} = \underline{G} = -2^k$ and $\overline{F} = \overline{G} = 2^k$.

Then, we apply the following iterative bisection procedure for computing $\overline{y}$ (computing $\underline{y}$ is similar):

- At the beginning of each iteration, we have an interval $[\underline{G}, \overline{G}]$ that is known to contain $\overline{y}$. First, compute its midpoint $G = (1/2) \cdot (\underline{G} + \overline{G})$.
- Then, we check whether $\overline{y} \geq G$, by applying Grigor'ev's algorithm to check the existence of a real solution of the following sequence of polynomial equalities and inequalities with integer coefficients: $2^d \cdot f_1(x_1, \ldots, x_n, y) = 0$, $2^d \cdot f_2(x_1, \ldots, x_n, y) \geq 0$, ..., $2^d \cdot f_k(x_1, \ldots, x_k, y) \geq 0$, $2^D(y - G) \geq 0$, $2^D(x_i - \underline{x}_i) \geq 0, 1 \leq i \leq n$, and $2^D(\overline{x}_i - x_i) \geq 0$ for an appropriate D.
- If this set of inequalities has a solution, this means that the upper bound $\overline{y}$ for f is $\geq G$, so we take $[G, \overline{G}]$ as the new interval $[\underline{G}, \overline{G}]$; else, take $[\underline{G}, G]$.
- Now, we can check whether we have achieved the desired solution, i.e., whether $\overline{G} - \underline{G} \leq \varepsilon$; if this is the case, then we simply return $\overline{G}$ as the desired estimate $\widetilde{\overline{y}}$; else, continue the iterations.

The proof that this algorithm is polynomial-time is similar to the proof of Theorem 4.1. The theorem is proven.

Proof of Theorem 4.3. Let us describe a sequence of compact polynomials for which the required time grows exponentially with the length of the polynomial's description: $f_0(x) = x$ and $f_{k+1}(x) = (f_k(x))^2$, i.e., $f_1(x) = x^2$, $f_2(x) = (x^2)^2$ $(= x^4)$, $f_3(x) = ((x^2)^2)^2$ $(= x^8)$, etc. We will take $\mathbf{x} = [0, 2]$.

The length of the description of $f_0(x)$ is 2: the symbol x itself and 2 indicating the squaring. To get $f_{k+1}(x)$ from $f_k(x)$, one must add three symbols to the description of f_k: "(", ")", and "2". So, the length L_k of the description of $f_k(x)$ is $3k + 2$.

One can easily prove by induction that $f_k(x) = x^{2^k}$. Hence, for $f_k(x)$, the desired interval $\mathbf{y}$ is equal to $[0, 2^{2^k}]$. For a fixed ε, for sufficiently large k, we will have $\widetilde{\overline{y}} \geq \overline{y} - \varepsilon \geq 2^{2^k - 1}$. Hence, to represent $\widetilde{\overline{y}}$ in the computer, we will need at least $2^k - 1$ binary digits. Generating one digit takes at least one computational step, so we need $\geq 2^k - 1$ computational steps. The theorem is proven.

Proof of Theorem 4.4. Let us assume that for some $\varepsilon > 0$, we can solve the ε-approximate basic problem of interval computations for feasible piece-wise linear functions $f(x)$ of one variable in polynomial time. Let us show that we will then be able to solve the propositional satisfiability problem in polynomial time and thus, our problem (ε-approximate basic problem of interval computations for feasible piece-wise linear functions of one variable) is NP-hard.

Indeed, suppose that we have a propositional formula F with v Boolean variables $z_1, \ldots, z_v$. In the following proof, we will design a feasible piece-wise linear function $f : [0,1] \to [0, 3\varepsilon]$ that satisfies the following two properties:

- If F is not satisfiable, then $f(x)$ is identically 0, so $\mathbf{y} = [0,0]$.
- If F is satisfiable, then $\mathbf{y} = [0, 3\varepsilon]$.

If we have such a function $f(x)$, then, if we are able to estimate $\overline{y}$ in polynomial time with a given accuracy ε, we will be able to distinguish between these two cases, and thus, decide in polynomial time whether a given formula F is satisfiable or not.

To achieve this goal, we want to use a piece-wise linear function $f(x)$ that is defined as follows:

- For binary-rational numbers $x = 0.z_1 \ldots z_v$, as a value of $f(x)$ we take 3ε times the truth value of a formula F for Boolean variables $z_1, \ldots, z_v$ (in computers, usually, "true" is represented as 1, and "false" is 0).
- We take $f(-0.0\ldots01) = f(0.0\ldots0)$ and $f(1.0\ldots0) = f(0.1\ldots1)$.
- In between these binary-rational numbers, the function $f(x)$ is linear.

This function $f(x)$ is piece-wise linear, and it satisfies the desired property. So, to complete our proof, we only need to prove that $f(x)$ is feasible, i.e., that there exists a polynomial-time algorithm that enables us to compute $f(x)$ for a given rational number x. This algorithm is easy to design:

- First, we ask for a 2^{-v}-approximation to x. As a result, we get a binary-rational number $\widetilde{x} = 0.z_1 \ldots z_v$ such that $|x - \widetilde{x}| \le 2^{-v}$. This inequality is equivalent to $\underline{x} \le x \le \overline{x}$, where $\underline{x} = \widetilde{x} - 2^{-v}$ and $\overline{x} = \widetilde{x} + 2^{-v}$ are also v-digit rational numbers.
- For each of the three v-digit numbers $\underline{x}$, $\widetilde{x}$, and $\overline{x}$, we compute the values of $f(x)$ by checking whether the formula F is satisfied by the propositional variables taken from the number's binary expansion, and multiplying the corresponding truth values by 3ε. As a result, we get the three values $f(\underline{x})$, $f(\widetilde{x})$, and $f(\overline{x})$. The application of F requires the time that is linear in the length of F.

- On the interval $[\underline{x}, \overline{x}]$ that contains x, our piece-wise linear function $f(x)$ consists of two linear parts:
  - For $x \in [\underline{x}, \widetilde{x}]$, we have
  $$f(x) = f(\underline{x}) + \frac{f(\widetilde{x}) - f(\underline{x})}{\widetilde{x} - \underline{x}} \cdot (x - \underline{x}).$$
  - For $x \in [\widetilde{x}, \overline{x}]$, we have
  $$f(x) = f(\widetilde{x}) + \frac{f(\overline{x}) - f(\widetilde{x})}{\overline{x} - \widetilde{x}} \cdot (x - \widetilde{x}).$$
- To complete the computation of $f(x)$, we must consider two possible cases:
  - if $f(\widetilde{x}) \le (1/2)\cdot(f(\underline{x})+f(\overline{x})))$, then on the interval $[\underline{x}, \overline{x}]$, the function $f(x)$ is convex, and therefore, we can compute $f(x)$ as

$$f(x) = \max\left\{ f(\underline{x}) + \frac{f(\widetilde{x}) - f(\underline{x})}{\widetilde{x} - \underline{x}} \cdot (x - \underline{x}), f(\widetilde{x}) + \frac{f(\overline{x}) - f(\widetilde{x})}{\overline{x} - \widetilde{x}} \cdot (x - \widetilde{x}) \right\}.$$

  - if $f(\widetilde{x}) \ge (1/2)(f(\underline{x})+f(\overline{x})))$, then on the interval $[\underline{x}, \overline{x}]$, the function f is concave, and therefore, we can compute $f(x)$ as

$$f(x) = \min\left\{ f(\underline{x}) + \frac{f(\widetilde{x}) - f(\underline{x})}{\widetilde{x} - \underline{x}} \cdot (x - \underline{x}), f(\widetilde{x}) + \frac{f(\overline{x}) - f(\widetilde{x})}{\overline{x} - \widetilde{x}} \cdot (x - \widetilde{x}) \right\}.$$

This algorithm takes polynomial time; hence, $f(x)$ is feasible. This completes the proof of the theorem.

Proof of Theorem 4.5. This proof is similar to the proof of Theorem 4.3. Namely, we will give an example of a sequence of exp-polynomial functions for which the required time grows at least exponentially with the length of the function's description: we take $f_0(x) = x$ and $f_{k+1}(x) = \exp(f_k(x))$ (i.e., $f_k(x) = \exp(\ldots \exp(x) \ldots)$ (k times), and $\mathbf{x}_1 = [0, 2]$.

The length of the description of $f_k(x)$ grows linearly with k. The value $\overline{y} = f_k(2)$, however, grows with k faster than 2^{2^k}. Hence, to represent $\widetilde{y}$ in the computer, we will need at least 2^k binary digits. Generating one digit takes at least one computational step, so we need $\ge 2^k - 1$ computational steps. The theorem is proven.

Comment. If we do not restrict ourselves to a single interval $\mathbf{x}$, but allow arbitrary intervals, then we do not even bother to consider a sequence of functions $f_k(x)$: we can simply take a function $\exp(x)$ for $\mathbf{x} = [n, n]$. Binary representation of the input requires $L \approx \log_2(n)$ bits. Since $\exp(n) > 2^n$, the binary representation of the result requires at least $n \approx 2^L$ bits. To generate each bit, we need at least one computational step. So, the number of computational steps does grow exponentially.

Proof of Theorem 4.6. Before we prove the NP-hardness of computing the range of sin-polynomials of *one* variable, let us first prove NP-hardness of sin-polynomials of *two* variables. To prove NP-hardness of the range estimation problem, we will reduce yet another known NP-hard problem to our problem: namely, the problem of *solving quadratic Diophantine equations.* This problem was first considered by Manders *et al.* [269] (see also Garey *et al.* [121], p. 250); it is formulated as follows: given positive integers a, b, and c, check whether there exist positive integers x_1 and x_2 for which $a \cdot x_1^2 + b \cdot x_2 = c$.

For each instance (a, b, c) of the above problem, we can construct a sin-polynomial $f(x_1, x_2)$ whose minimum $\underline{y}$ on $\mathbf{x}_1 = \mathbf{x}_2 = [1, c]$ is equal to 0 if and only if the given instance has a solution: namely, we will take $f(x_1, x_2) = \sin^2(\pi \cdot x_1) + \sin^2(\pi \cdot x_2) + (a \cdot x_1^2 + b \cdot x_2 - c)^2$. This function $f(x_1, x_2)$ is always non-negative (as a sum of non-negative squares), and it is equal to 0 if and only if all the squares, from which it is constructed, are equal to 0. From $\sin(\pi \cdot x_1) = 0$, we conclude that x_1 is an integer; since $x_1 \in [1, c]$, we conclude that it is a positive integer. Similarly, from $\sin(\pi \cdot x_2) = 0$ and $x_2 \in [1, c]$, we conclude that x_2 is a positive integer; finally, from the equality between the third term and 0 we conclude that $a \cdot x_1^2 + b \cdot x_2 = c$; therefore, x_1 and x_2 form a solution to the original instance. Vice versa, if (x_1, x_2) is a solution to the original instance, we have $f(x_1, x_2) = 0$, and therefore, $\underline{y} = 0$.

Thus, the problem of *exactly* computing the range of sin-polynomials is NP-hard. One can also show (similarly to the proofs from Chapter 3) that not only the original instance has a solution if $\underline{y} = 0$, but also that this instance has a solution if $\underline{y} \le \varepsilon_0$ for some small ε_0: Indeed, in this case, x_1 is close to a positive integer $\widetilde{x}_1$, x_2 is close to a positive integer $\widetilde{x}_2$, and from the closeness of $a \cdot x_1^2 + b \cdot x_2 - c$ to 0 we conclude that the value $a \cdot \widetilde{x}_1^2 + b \cdot \widetilde{x}_2 - c$ is also close to 0; since this value is an integer, it has to be *exactly* equal to 0. (The corresponding value ε_0 may actually depend on a, b, and c.) Thus, either $\underline{y} = 0$ and the original instance has a solution, or $\underline{y} > \varepsilon_0$, and the original instance does not have a solution. Therefore, if we can compute $\underline{y}$ with an accuracy $\varepsilon_0/2$, we will be able to tell whether the original instance has a solution or not. Hence, for a given $\varepsilon > 0$, we can consider a *modified* sin-polynomial $\widetilde{f}(x_1, x_2) = (2\varepsilon/\varepsilon_0) \cdot f(x_1, x_2)$.

Computing the lower bound $\widetilde{\underline{y}}$ for this new function with the accuracy ε is equivalent to computing the lower bound for the original function with the accuracy $\varepsilon_0/2$, and is, thus, equivalent to solvability of the original instance.

Strictly speaking, there is one minor point left in our proof: legally, the above function $f(x_1, x_2)$ is *not* a sin-polynomial because it uses an *irrational* constant π. So, instead of this function, we must consider a close function $f_r(x_1, x_2) = \sin^2(\pi_r \cdot x_1) + \sin^2(\pi_r \cdot x_2) + (a \cdot x_1^2 + b \cdot x_2 - c)^2$, where π_r is a *rational* approximation to π (the accuracy of this approximation may depend on a, b, and c). If π_r is close enough to π, then for solvable instances, the minimum $\underline{y_r}$ of the function $f_r(x_1, x_2)$ will be close to 0, while for non-solvable instances, it will be greater than or equal to some positive number that is close to ε_0. Thus, if we compute $\underline{y_r}$ with an accuracy $\varepsilon_r/2$ close to $\varepsilon_0/2$, we will then be able to distinguish between solvable and non-solvable instances. Therefore, for a function $\widetilde{f_r}(x_1, x_2) = (2\varepsilon/\varepsilon_r) \cdot f(x_1, x_2)$, computing the lower endpoint $\underline{y}$ of its range with accuracy ε is NP-hard.

To complete the proof, we must show that for sin-polynomials of *one* variable, range computation is also NP-hard. To prove this result, we will "combine" both values x_1 and x_2 into a single real-valued variable x. Namely, we will consider $x = x_1 + \alpha \cdot x_2$, where $\alpha > 0$ is some sufficiently small rational number. If we know that both x_1 and x_2 are integers, and if α is small enough ($\alpha \ll 1/c$), then all c^2 real numbers that correspond to different $x_1 = 1, \ldots, c$ and $x_2 = 1, \ldots, c$ are different, so, in principle, it is possible to reconstruct both integers x_1 and x_2 from this combination x.

Let us show that we can use sin-polynomials for this reconstruction. Indeed, since x_1 is an integer and α is small, we have $\sin(\pi \cdot x) = \sin(\pi \cdot x_1 + \pi \cdot \alpha \cdot x_2) = \pm \sin(\pi \cdot \alpha \cdot x_2) \approx \pm \pi \cdot \alpha \cdot x_2$. Therefore, $x_2 \approx \pm x_3 = (1/(\pi \cdot \alpha)) \cdot \sin(\pi \cdot x)$. So, we can represent the requirement that x_2 is an integer (i.e., that $\sin(\pi \cdot x_2) = 0$) in terms of x, as $\sin(\pi \cdot x_3) = 0$.

The above argument prompts us to consider the following sin-polynomial: $f(x) = \sin^2(\pi \cdot x) + \sin^2(\pi \cdot x_3) + (a \cdot x^2 + b \cdot x_3 - c)^2$, where we denoted $x_3 = (1/(\pi \cdot \alpha)) \cdot \sin(\pi \cdot x)$.

If the original instance of the quadratic Diophantine problem has a solution x, y, then for $x = x_1 + \alpha \cdot x_2$, we have $\sin^2(\pi \cdot x) \approx \pi^2 \cdot \alpha^2 \cdot x_2^2 \le \pi^2 \cdot \alpha^2 \cdot c^2$; the second term is approximately 0, and the third terms is approximately equal to $2a^2 \cdot \alpha \cdot x_1 \cdot x_2 \le 2a^2 \cdot \alpha \cdot c^2$. Thus, if α is small enough so that both bounds $\pi^2 \cdot \alpha^2 \cdot c^2$ and $2a^2 \cdot \alpha \cdot c^2$ are much smaller than ε_0, we conclude that for our sin-polynomial, $\underline{y} \ll \varepsilon_0$.

Vice versa, if for this polynomial, $\underline{y} \ll \varepsilon_0$, this means that for some x, all three squares in the sum that defines this polynomial $f(x)$ must be small. From the condition that $\sin^2(\pi \cdot x)$ is small, i.e., that $\sin(\pi \cdot x) \approx 0$, we conclude that x is close to some integer. We will denote this integer (closest to x) by x_1. Since x is close to an integer, i.e., $x \approx x_1$, we conclude that $\sin(\pi \cdot x) \approx \pi \cdot (x - x_1)$, and therefore, $x_3 = (1/(\pi \cdot \alpha)) \cdot \sin(\pi \cdot x)) \approx (x - x_1)/\alpha$. Thus, $x \approx x_1 + \alpha \cdot x_3$.

From the condition that $\sin^2(\pi \cdot x_3)$ is small, we conclude that $x_3 \approx (x - x_1)/\alpha$ is close to some integer x_2. From $x_3 \approx x_2$ and from the above approximate expression $x \approx x_1 + \alpha \cdot x_3$, we conclude that $x \approx x_1 + \alpha \cdot x_2$.

Finally, from the condition that $(a \cdot x^2 + b \cdot x_3 - c)^2$ is small, we conclude that $a \cdot x^2 + b \cdot x_3 - c \approx 0$, and therefore (since $x \approx x_1$ and $x_3 \approx x_2$), that $a \cdot x_1^2 + b \cdot x_2 - c \approx 0$. Since a, b, c, x_1, and x_2 are integers, for sufficiently small α, we will be able to conclude that $a \cdot x_1^2 + b \cdot x_2 - c = 0$, and therefore, that x_1 and x_2 form a solution to the original instance of the quadratic Diophantine problem.

Thus, if this instance has a solution, we have $\underline{y} \ll \varepsilon_0$, and if this instance does not have a solution, then we can similarly conclude that this lower bound $\underline{y}$ cannot be that small. So, if we are able to compute the lower bound $\underline{y}$ with a sufficiently good accuracy ε_1, then we will be able to solve the corresponding instance of the quadratic Diophantine problem.

Similarly to the case of two variables, to prove NP-hardness of range estimation for an arbitrary $\varepsilon > 0$, we consider a *modified* sin-polynomial $\widetilde{f}(x) = (\varepsilon/\varepsilon_1) \cdot f(x)$.

To finalize the proof, we must replace the *irrational* number π by a close *rational* number π_r. The theorem is proven.

Proof of Theorem 4.7. Since the function $g(x)$ is twice differentiable, we have $g(x_0+h) = g(x_0)+g'(x_0)\cdot h+(1/2)\cdot g''(x_0)\cdot h^2+o(h^2)$. Since $g'(x_0) = 0$ and $g''(x_0) \neq 0$, we can conclude that $g(x_0+h) - g(x_0) = (1/2) \cdot g''(x_0) \cdot h^2 + o(h^2)$, and therefore, that $h^2 = (2/g''(x_0)) \cdot (g(x_0+h) - g(x_0)) + o(h^2)$. So, if we define

$$g_1(h) = \frac{2}{g''(x_0)} \cdot (g(x_0 + h) - g(x_0)),$$

then for $h \to 0$, we get $\lim(g_1(h)/h^2) = 1$.

Length-wise, for every expression H, to get the expression for $g_1(H)$, we must add constantly many symbols to the description of H. So, if we define a sequence $g_2(h) = g_1(g_1(h))$, ..., $g_{k+1}(h) = g_1(g_k(h))$, ..., then each g-rational function in this sequence is obtained by adding constantly many symbols to the previous one, and therefore, the length of the description of the function $g_k(h)$ grows linearly with k.

Asymptotically, $g_1(h) \sim h^2$, so $g_2(h) = g_1(g_1(h)) \sim (h^2)^2 = h^4$, $g_3(h) = g_1(g_2(h)) \sim (h^4)^2 = h^8$, and in general, $g_k(h) \sim h^{2^k}$.

We can now define $f_k(x) = 1/g_k(1/x)$, and choose $\mathbf{x} = [X, 2X]$ for some $X > 0$. Then, as $x \to \infty$, we get $f_k(x) \sim x^{2^k}$, so for sufficiently large X, we get $\overline{y} = \sup f_k(x) \approx (2X)^{2^k}$. Hence, to represent $\widetilde{\overline{y}}$ in the computer, we will need at least 2^k binary digits. Generating one digit takes at least one computational step, so we need $\geq 2^k - 1$ computational steps. This number is exponential in k, and (since the length of the description of $f_k(x)$ grows linearly with k), this number is exponential in the length of the input as well. The theorem is proven.

Proof of Theorem 4.8. To prove this theorem, we will consider the following real-valued complex-rational function

$$g_1(h) = 1 - \mathrm{Re}\left(\frac{1}{1+ih}\right).$$

Since

$$\frac{1}{1+ih} = \frac{1-ih}{1+h^2},$$

we have $g_1(h) = 1 - 1/(1+h^2) \sim h^2$ as $h \to 0$. For every expression H, the expression for $g_1(H)$ is obtained by adding constantly many symbols to the description of H. So, we can construct the sequence $g_k(h)$ as $g_{k+1}(h) = g_1(g_k(h))$ and conclude the proof as in Theorem 4.7. The theorem is proven.

5

BASIC PROBLEM OF INTERVAL COMPUTATIONS FOR POLYNOMIALS OF FIXED ORDER

In Chapter 3, we showed that the basic problem of interval computations is NP-hard for polynomials $f(x_1, \ldots, x_n)$. In this chapter, we describe what happens if we restrict the order of these polynomials.

5.1. Results

Main result: intractable for quadratic functions

When the order is 1, we get *linear* functions. For linear functions $f(x_1, \ldots, x_n) = a_0 + \sum a_i \cdot x_i$, the problem is feasible: the desired range $\mathbf{y} = f([\underline{x}_1, \overline{x}_1], \ldots, [\underline{x}_n, \overline{x}_n])$ can be computed by naive interval computations, and it is equal to $\mathbf{y} = [\tilde{y} - \Delta, \tilde{y} + \Delta]$, where $\tilde{y} = f(\tilde{x}_1, \ldots, \tilde{x}_n)$ and $\Delta = \sum |a_i| \cdot \Delta_i$.

The next step is *quadratic* functions. Alas, as the proof of Theorem 3.1 shows, for quadratic functions, the problem is already NP-hard:

Theorem 5.1. *For every $\varepsilon > 0$, the problem of computing the endpoints $\underline{y}$ and $\overline{y}$ of the range $f([\underline{x}_1, \overline{x}_1], \ldots, [\underline{x}_n, \overline{x}_n])$ for a given quadratic polynomial $f(x_1, \ldots, x_n) = a_0 + \sum a_i \cdot x_i + \sum a_{ij} \cdot x_i \cdot x_j$ and for given intervals $[\underline{x}_i, \overline{x}_i]$ with accuracy ε is NP-hard.*

Comment. Quadratic polynomials are a particular case of cubic, quartic, etc., therefore, for cubic, etc. polynomials, the basic problem of interval computations is also NP-hard:

Linear $f(x_1,\ldots,x_n)$	Quadratic $f(x_1,\ldots,x_n)$	Cubic (and higher order) $f(x_1,\ldots,x_n)$
Linear time	NP-hard	NP-hard

Simple case: feasible for diagonal matrices and for linear functions

We cannot have a feasible algorithm for *all* quadratic polynomials, but maybe, we can have an algorithm for *some* of them? There are two trivial (or almost trivial) cases when this problem is feasible for quadratic functions $f(x_1,\ldots,x_n)$: one case is when $f(x_1,\ldots,x_n)$ is actually *linear*, and another is when the matrix a_{ij} is *diagonal*:

Theorem 5.2. *There exists a linear-time algorithm that computes the endpoints $\underline{y}$ and $\overline{y}$ of the range $f([\underline{x}_1,\overline{x}_1],\ldots,[\underline{x}_n,\overline{x}_n])$ for a given quadratic polynomial $f(x_1,\ldots,x_n)=a_0+\sum a_i\cdot x_i+\sum a_{ii}\cdot x_i^2$ and for given intervals $[\underline{x}_i,\overline{x}_i]$.*

Are there other feasible cases?

Both cases are very specific, but maybe, if we take *close* cases, we will still get a class for which feasible algorithms are possible? Unfortunately, the answer is negative: for all non-trivial generalizations of the above degenerate classes that we tried, the resulting problem is NP-hard:

First try: band matrices

A diagonal matrix is a matrix a_{ij} for which $a_{ij}=0$ for $i\neq j$, i.e., for $|i-j|\geq 1$. The first natural generalization of a diagonal matrix is a *w−band* matrix, for which $a_{ij}=0$ for $|i-j|\geq w$ for some $w\geq 1$ (diagonal matrices correspond to $w=1$):

Theorem 5.3. *For $w\geq 3$, and for every $\varepsilon>0$, the problem of ε-accurately computing the endpoints $\underline{y}$ and $\overline{y}$ of the range $f([\underline{x}_1,\overline{x}_1],\ldots,[\underline{x}_n,\overline{x}_n])$ for a given quadratic polynomial $f(x_1,\ldots,x_n)=a_0+\sum a_i\cdot x_i+\sum a_{ij}\cdot x_i\cdot x_j$ with a w-band matrix a_{ij} and for given intervals $[\underline{x}_i,\overline{x}_i]$ is NP-hard.*

Comments.

- w-band matrices naturally appear in many real-life problems (see, e.g., Schendel [377]). For example, they appear when we estimate the value of a functional of the type $f = \int \left(a_0 \cdot (x(t))^2 + a_1 \cdot (\dot{x}(t))^2\right) dt$, where $\dot{x}(t)$ denotes time derivative. This functional is important in ***signal processing*** and in ***image processing***, where it characterizes the smoothness of a signal (see, e.g., Tikhonov *et al.* [410], Inverse Problems [160, 161, 162], Glasko [127], Lavrentiev *et al.* [248]), and in *control*, where it characterizes the smoothness of a resulting trajectory (for a general justification of this functional, see, e.g., [231]). Usually, we are only given the values $x(t_i)$ of the signal for $t_i = t_0 + i \cdot h$. From these data, we can estimate the integral defining the functional by its integral sum $\sum \left[a_0 \cdot (x(t_i))^2 + a_1 \cdot (\dot{x}(t_i))^2\right] \cdot h$, and the derivative $\dot{x}(t_i)$ as $(x(t_{i+1}) - x(t_i))/h$. As a result, we get a ***quadratic*** form in which the elements a_{ij} are different from 0 only for $|i-j| \le 1$, i.e., we get a 2-band matrix. If we take second derivative into consideration, we get a 3-band matrix, etc.

- For 1-band (i.e., diagonal) matrices, we have a feasible (even linear-time) algorithm. We do not know whether a feasible algorithm is possible for 2-band matrices:

Diagonal (1-band) matrices	2-band matrices	3-band matrices
Linear time	?	NP-hard

Second try: almost scalar matrices

Another possibility is to generalize one particular case of a diagonal matrix: a *scalar* matrix, for which $a_{ii} = \text{const}$. These matrices can be characterized by the condition that all their eigenvalues λ_i are equal: $\lambda_1 = \ldots = \lambda_n = \lambda$ for some number λ. A natural generalization is the notion of an ***almost scalar*** matrix, i.e., a matrix for which all but one eigenvalues coincide.

Theorem 5.4. *For every $\varepsilon > 0$, the problem of ε-accurately computing the endpoints $\underline{y}$ and $\overline{y}$ of the range $f([\underline{x}_1, \overline{x}_1], \ldots, [\underline{x}_n, \overline{x}_n])$ for a given quadratic polynomial $f(x_1, \ldots, x_n) = a_0 + \sum a_i \cdot x_i + \sum a_{ij} \cdot x_i \cdot x_j$ with an almost scalar matrix a_{ij} and for given intervals $[\underline{x}_i, \overline{x}_i]$ is NP-hard.*

Third try: bilinear functions

The third generalization is a generalization of *linear* functions. A natural generalization is the notion of a *bilinear function*, i.e., a quadratic function of the type $\sum a_{ij} \cdot x_i \cdot y_j$.

Theorem 5.5. *For every $\varepsilon > 0$, the problem of ε-accurately computing the endpoints $\underline{y}$ and $\overline{y}$ of the range $f([\underline{x}_1, \overline{x}_1], \ldots, [\underline{x}_n, \overline{x}_n], [\underline{y}_1, \overline{y}_1], \ldots, [\underline{y}_n, \overline{y}_n])$ for a given bilinear function $f(x_1, \ldots, x_n, y_1, \ldots, y_n) = \sum a_{ij} \cdot x_i \cdot y_j$ and for given intervals $[\underline{x}_i, \overline{x}_i]$ and $[\underline{y}_i, \overline{y}_i]$ is NP-hard.*

These three NP-hardness results are also true for narrow input intervals

Comment. In Chapter 3, we have shown that for every $\delta > 0$, the basic problem of interval computations is NP-hard even if we allow only input intervals that are both absolutely and relatively δ-narrow. In the above three NP-hardness results, we can make the same restriction, and the restricted problems will still be NP-hard.

Proofs

Proof of Theorem 5.2. For *diagonal* matrices, the quadratic function $f(x_1, \ldots, x_n)$ takes the form $\sum a_{ii} \cdot x_i^2 + \sum a_i \cdot x_i + a_0 = a_0 + \sum (a_{ii} \cdot x_i^2 + a_i \cdot x_i)$. The restrictions on the values of x_i are independent on each other ($x_i \in \mathbf{x}_i$); therefore, the function $f(x_1, \ldots, x_n)$ attains its smallest value $\underline{y}$ (or, its largest value $\overline{y}$) if and only if each of the terms $a_{ii} \cdot x_i^2 + a_i \cdot x_i$ attains its smallest (correspondingly, its largest) value. In other words, $\underline{y} = a_0 + \underline{y}_1 + \ldots + \underline{y}_n$ and $\overline{y} = a_0 + \overline{y}_1 + \ldots + \overline{y}_n$, where by $\underline{y}_i$ and $\overline{y}_i$, we denoted the minimum (correspondingly, the maximum) of the expression $f_i(x_i) = a_{ii} \cdot x_i^2 + a_i \cdot x_i$.

The values $\underline{y}_i$ and $\overline{y}_i$ can be easily computed if we consider two possible cases:

- If $a_{ii} = 0$, then $f_i(x_i) = a_i \cdot x_i$, and hence, $[\underline{y}_i, \overline{y}_i] = a_i \cdot \mathbf{x}_i$.
- If $a_{ii} \neq 0$, then

$$f_i(x_i) = a_{ii} \cdot \left(x_i^2 + \frac{a_i}{a_{ii}} x \right) = a_{ii} \cdot \left(x_i + \frac{1}{2} \cdot \frac{a_i}{a_{ii}} \right)^2 - \frac{1}{4} \cdot \frac{a_i^2}{a_{ii}}.$$

For $x_i \in [\underline{x}_i, \overline{x}_i]$, we can easily compute the exact bounds for this expression:

- the bounds for $z_i = x_i + (1/2) \cdot (a_i/a_{ii})$ are $\underline{z}_i = \underline{x}_i + (1/2) \cdot (a_i/a_{ii})$ and $\overline{z}_i = \overline{x}_i + (1/2) \cdot (a_i/a_{ii})$;
- the bounds for $u_i = z_i^2$ are: $[(\underline{z}_i)^2, (\overline{z}_i)^2]$ when $0 \le \underline{z}_i \le \overline{z}_i$; $[(\overline{z}_i)^2, (\underline{z}_i)^2]$ when $\underline{z}_i \le \overline{z}_i \le 0$; and $[0, \max\{(\underline{z}_i)^2, (\overline{z}_i)^2\}]$ when $\underline{z}_i \le 0 \le \overline{z}_i$;
- the bounds for $f_i(x_i)$ are $[\underline{y}_i, \overline{y}_i] = a_{ii} \cdot [\underline{u}_i, \overline{u}_i] - (1/4) \cdot (a_i^2/a_{ii})$.

In both cases, for each i from 1 to n, we need *finitely many* computation steps to compute $\underline{y}_i$ and $\overline{y}_i$, so, the total computation of $\underline{y}$ and $\overline{y}$ can be completed in a time that is *linear* in n. The theorem is proven.

Proof of Theorem 5.3. Let us first prove that the *exact* computation of the range is NP-hard for 3-band matrices. To prove this result, we will use another known NP-hard problem *PARTITION*: given integers $s_1, \ldots, s_n$, check whether there exist values $x_i \in \{-1, 1\}$ for which $\sum s_i \cdot x_i = 0$. We will reduce this problem to the problem of computing the range.

For each instance of *PARTITION*, we will consider the problem of computing the range of the following quadratic polynomial of $2n+1$ variables $x_1, \ldots, x_n, t_0, t_1, \ldots, t_n$:

$$f(x_1, \ldots, x_n, t_0, t_1, \ldots, t_n) =$$

$$\sum_{i=1}^{n} (1 - x_i^2) + t_0^2 + \sum_{i=1}^{n} (t_i - x_i \cdot s_i - t_{i-1})^2 + t_n^2$$

on the intervals $\mathbf{x}_i = [-1, 1]$ and $\mathbf{t}_i = [-S, S]$, where we denoted $S = \sum |s_i|$. If we order the variables in the order $t_0, x_1, t_1, x_2, t_2, \ldots, x_n, t_n$, then, as one can easily see, the only cross-terms are between the variables that are either immediate neighbors (t_i and x_i or x_i and t_{i-1}) or neighbors to immediate neighbors (t_i and t_{i-1}). Hence, the corresponding matrix is indeed 3-band.

Let us show that the lower endpoint $\underline{y}$ of the desired range is equal to 0 if and only if the given instance of *PARTITION* has a solution. Indeed, if this instance has a solution x_i, then we take these x_i, $t_0 = 0$, and $t_i = s_1 \cdot x_1 + \ldots + s_i \cdot x_i$. Vice versa, let $\underline{y} = 0$. Since $x_i \in [-1, 1]$, each term $1 - x_i^2$ is non-negative. Since every other term is a square, the entire function $f(x_1, \ldots, x_n, t_0, \ldots, t_n)$ is non-negative, and hence, the lower endpoint $\underline{y}$ of its range is non-negative too. If this minimum is 0, this means that $f(x_1, \ldots, x_n, t_0, \ldots, t_n) = 0$ for some values x_i and t_i. The sum of non-negative terms can be equal to 0 if and only if every term in this sum is equal to 0. Thus, for every i, $x_i^2 = 1$ (i.e., $x_i \in \{-1, 1\}$), $t_0 = 0$, $t_i = t_{i-1} + s_i \cdot x_i$, and $t_n = 0$. From these equations, we conclude, by induction, that $t_i = s_1 \cdot x_1 + \ldots + s_i \cdot x_i$, and therefore, the equation $t_n = 0$ means that $s_1 \cdot x_1 + \ldots + s_n \cdot x_n = 0$, i.e., that x_i's are indeed the solution to the given instance to the *PARTITION* problem.

This reduction proves that the problem of computing the range *exactly* is NP-hard. Let us now show that the problem of computing this range *approximately* is also NP-hard. For that, let us first show that for appropriate positive $\delta < 1$, if, for the above range estimation problem, $\underline{y} \leq \delta^2$, then the given instance of *PARTITION* has a solution. Indeed, let $\underline{y} = f(x_1, \ldots, x_n, t_0, \ldots, t_n) \leq \delta^2$. Since $f(x_1, \ldots, x_n, t_0, \ldots, t_n)$ is the sum of non-negative terms, we can conclude that each of these terms is $\leq \delta^2$.

In particular, we have $1 - x_i^2 \leq \delta^2$, hence, $x_i^2 \geq z = 1 - \delta^2$, and since $\sqrt{z} \geq z$ for $z \leq 1$, we have $|x_i| \geq \sqrt{z} \geq z = 1 - \delta^2$. Since we assumed that $x_i \in \mathbf{x}_i = [-1, 1]$, we conclude that $|x_i| \leq 1$. Thus, if we define the *sign* of x_i by $\tilde{x}_i = \text{sign}(x_i) = \pm 1$, we conclude that $|x_i - \tilde{x}_i| \leq \delta^2$. We will show that the values $\tilde{x}_i$ form a solution to the *PARTITION* problem, i.e., that the (integer-valued) sum $s = \sum s_i \cdot \tilde{x}_i$ is equal to 0.

Indeed, for other terms q^2 from $f(x_1, \ldots, x_n, t_0, \ldots, t_n)$, from $q^2 \leq \delta^2$, we conclude that $|q| \leq \delta$. In particular, $|t_0| \leq \delta$, $|t_i - x_i \cdot s_i - t_{i-1}| \leq \delta$, and $|t_n| \leq \delta$. We can represent the sum $s_1 \cdot x_1 + \ldots + s_n \cdot x_n$ as follows:

$$\sum_{i=1}^{n} s_i \cdot x_i = t_0 + \sum_{i=1}^{n} (t_i - x_i \cdot s_i - t_{i-1}) - t_n.$$

Therefore,

$$\left| \sum_{i=1}^{n} s_i \cdot x_i \right| \leq |t_0| + \sum_{i=1}^{n} |t_i - x_i \cdot s_i - t_{i-1}| + |t_n|.$$

Each of the terms in the right-hand side is bounded by δ, so $|\sum s_i \cdot x_i| \leq (n+2) \cdot \delta$. Similarly, from

$$s = \sum s_i \cdot \tilde{x}_i = \sum s_i \cdot x_i + \sum s_i \cdot (\tilde{x}_i - x_i),$$

we conclude that

$$|s| \leq \left| \sum s_i \cdot x_i \right| + \sum |s_i| \cdot |\tilde{x}_i - x_i|.$$

Replacing $|\sum s_i \cdot x_i|$ and $|\tilde{x}_i - x_i|$ by their known upper bounds, we conclude that

$$|s| \le (n+2) \cdot \delta + \sum |s_i| \cdot \delta^2 = (n+2) \cdot \delta + S \cdot \delta^2.$$

Thus, if we take δ for which $(n+2) \cdot \delta + S \cdot \delta^2 < 1$, we will be able to conclude that $|s| < 1$, and, since s in an integer, to conclude that $s = 0$.

It is easy to choose such δ: e.g., we can choose δ for which $(n+2) \cdot \delta \le 1/3$ (i.e., $\delta \le 1/(3 \cdot (n+2))$) and $S \cdot \delta^2 \le 1/3$, i.e., $\delta \le \sqrt{1/(3S)}$. To satisfy both inequalities, we can take, e.g., the smallest of the required upper bounds: $\delta = \min\{1/(3 \cdot (n+2)), \sqrt{1/(3S)}\}$.

For this δ, we have proven that if $\underline{y} \le \delta^2$, then the corresponding instance of *PARTITION* problem has a solution. We already know that in this case, $\underline{y} = 0$. So, if the original instance of the *PARTITION* problem has a solution, then $\underline{y} = 0$; otherwise, $\underline{y} > \delta^2$. Hence, if we can compute $\underline{y}$ with an accuracy $\varepsilon_0 = 0.5 \cdot \delta^2$, we will be able to tell whether the given instance of *PARTITION* problem has a solution or not. Therefore, for this ε_0, computing the bounds $\underline{y}$ and $\overline{y}$ with an accuracy ε_0 is an NP-hard problem. Then, for an *arbitrary* $\varepsilon > 0$, we can consider a function $(\varepsilon/\varepsilon_0) \cdot f(x_1, \ldots, x_n, t_0, \ldots, t_n)$. Computing the lower endpoint for this function with accuracy ε is equivalent to computing the lower endpoint for the original function $f(x_1, \ldots, x_n, t_0, \ldots, t_n)$ with an accuracy ε_0 and is, therefore, also NP-hard. The theorem is proven.

Proof of Theorem 5.4. To prove this theorem, we will use reduction of the same *PARTITION* problem as in the proof of Theorem 5.3. For each instance $s_1, \ldots, s_n$ of *PARTITION*, we consider the following quadratic function of n variables:

$$f(x_1, \ldots, x_n) = \sum_{i=1}^{n} (1 - x_i^2) + (s_1 \cdot x_1 + \ldots + s_n \cdot x_n)^2$$

for $x_i \in \mathbf{x}_i = [-1, 1]$. One can easily see that the corresponding matrix is almost scalar.

Similarly to the proof of Theorem 5.3, this function $f(x_1, \ldots, x_n)$ is the sum of non-negative terms and therefore, $\underline{y} \ge 0$. If $x_1, \ldots, x_n$ form a solution to the *PARTITION* problem, then for these x_i, $f(x_1, \ldots, x_n) = 0$ and hence, $\underline{y} = 0$. Vice versa, if $\underline{y} = 0$, then $f = 0$ for some values $x_1, \ldots, x_n$, for which, therefore, $x_i \in \{-1, 1\}$ and $\sum s_i \cdot x_i = 0$. So, *PARTITION* is indeed reducible to the problem of *exactly* computing the range of $f(x_1, \ldots, x_n)$. Similarly to the proof of the previous theorem, this reduction also works for *approximate* computations. Thus, the problem of computing the range is indeed NP-hard for quadratic functions with almost scalar matrices. The theorem is proven.

Proof of Theorem 5.5. This theorem easily follows from the general results about matrices that we will prove in Chapters 21 and 22. Namely, let us consider an arbitrary bilinear function and intervals $\mathbf{x}_i = \mathbf{y}_i = [-1,1]$. A bilinear function is, by definition, linear in each of the variables y_i. Thus, for fixed $x_1, \ldots, x_n$, the largest possible value of $\sum_{i,j} a_{ij} \cdot x_i \cdot y_j = \sum_i y_i \cdot (\sum_j a_{ij} \cdot x_j)$ over $y_i \in [-1,1]$ is equal to $\sum_i |\sum_j a_{ij} \cdot x_j|$. In algebraic terms, if we take a matrix A with elements a_{ij}, this expression is the l^1-norm $\|Ax\|_1$ of the vector Ax. Thus, the desired maximum of the bilinear form is equal to the largest possible value of this expression for all vectors $x \in [-1,1]^n$. In Chapter 21, we show that this largest possible value coincides with a matrix norm $\|A\|_{\infty,1}$ of the matrix A. Thus, the maximum of a bilinear function for $\mathbf{x}_i = \mathbf{y}_i = [-1,1]$ is equal to the matrix norm. In Chapters 21 and 22, it is shown that computing this norm (even approximately) is NP-hard. The theorem is proven.

Proof of the comment about narrow intervals. To prove Theorem 3.2, we reduced the variables x_i that take values from *arbitrary* intervals $\mathbf{x}_i$ to new variables y_i that take values from *narrow* intervals $\mathbf{y}_i$ (that are either degenerate or equal to $[1-\delta, 1+\delta]$); the corresponding reduction was linear: $x_i = p_i \cdot y_i + q_i$ for some rational numbers p_i and q_i. If we substitute these expressions for x_i into a quadratic function $f(x_1, \ldots, x_n) = a_0 + \sum a_i \cdot x_i + \sum a_{ij} \cdot x_i \cdot x_j$, we get a new quadratic function $F(y_1, \ldots, y_n) = a'_0 + \sum a'_i \cdot y_i + \sum a'_{ij} \cdot y_i \cdot y_j$, with $a'_{ij} = a_{ij} \cdot p_i \cdot p_j$. It is easy to see that if the original matrix a_{ij} is 3-band or bilinear, then the resulting matrix a'_{ij} is also 3-band, or, correspondingly, bilinear. Thus, for narrow input intervals, the basic problem of interval computations for 3-*band* (correspondingly, for *bilinear*) matrices are NP-hard.

To prove a similar result about almost scalar matrices, we will take into consideration that in our proof of Theorem 5.4, all the variables x_i take the values from the same interval $[-1,1]$. Therefore, for all i, we have the same linear transformation: $p_1 = \ldots = p_n = p$. The resulting new matrix a'_{ij} is, therefore, equal to $p^2 \cdot a_{ij}$. Hence, whenever the original matrix was almost scalar, the new matrix is almost scalar as well. Thus, the basic problems of interval computations for *almost scalar* matrices and narrow input intervals is also NP-hard. The comment is proven.

6

BASIC PROBLEM OF INTERVAL COMPUTATIONS FOR POLYNOMIALS WITH BOUNDED COEFFICIENTS

In Chapter 3, we showed that the basic problem of interval computations is NP-hard for polynomials $f(x_1, \ldots, x_n)$. In the proof, we did not restrict the values of the coefficients. In this chapter, we show that the problem remains NP-hard if we only consider polynomials with coefficients 0, 1, 2, *and* 3.

The main result of this chapter was obtained in collaboration with G. Heindl.

Theorem 6.1.

- *For quadratic polynomials, the basic problem of interval computations remains NP-hard if we allow only polynomials with coefficients* 0, 1, 2, *and* 3.
- *For cubic polynomials, the basic problem of interval computations remains NP-hard if we allow only polynomials with coefficients* 0 *and* 1.

Comment. We do not know whether for quadratic polynomials, we can use only 0 and 1 (or at least 0, 1, and 2) and still get NP-hardness.

Proof of Theorem 6.1. For cubic polynomials, the result is easy to prove: for every m, we can take the cubic polynomial

$$f_n(z_1, \ldots, z_m, a_0, a_1, \ldots, a_m, a_{11}, \ldots, a_{1m}, \ldots, a_{m1}, \ldots, a_{mm}) =$$

$$a_0 + a_1 \cdot z_1 + \ldots + a_m \cdot z_m + a_{11} \cdot z_1 \cdot z_1 + a_{12} \cdot z_1 \cdot z_2 + \ldots + a_{mm} \cdot z_m \cdot z_m$$

with $n = m + 1 + m + m^2$ $(= m^2 + 2m + 1)$ variables $x_1, \ldots, x_n$:

$$x_1 = z_1, \ldots, x_m = z_m, x_{m+1} = a_0, x_{m+2} = a_1, \ldots, x_{2m+1} = a_m,$$

$$x_{2m+2} = a_{11}, x_{2m+3} = a_{12}, \ldots, x_n = a_{mm}.$$

All the coefficients of this polynomial are equal to 0 or to 1. When a_i and a_{ij} are numbers, i.e., degenerate intervals, then the basic problem for the function $f(x_1, \ldots, x_n)$ reduces to the basic problems for a generic quadratic polynomial, which is known to be NP-hard. The result about *cubic* polynomials is thus proven.

Let us prove the result about *quadratic* polynomials. Similarly to the proof of Theorem 3.1, we can show that a 3-CNF formula $F = F_1 \& \ldots \& F_k$ with v Boolean variables $z_1, \ldots, z_v$ (where each F_j is a disjunction of two or three literals), is satisfiable if and only if $\underline{y} = 0$, where $\underline{y}$ is the lower endpoint of the polynomial

$$f(x_1, \ldots, x_v, p_1, \ldots, p_k, q_1, \ldots, q_k) = \sum_{i=1}^{v} x_i \cdot (1 - x_i) + \sum_{j=1}^{k} f[F_j]^2,$$

for the intervals $\mathbf{x}_i = \mathbf{p}_j = \mathbf{q}_j = [0, 1]$, where:

- for $F_j = a \vee b$, $f[F_j] = f[a] + f[b] + p_j - 2$;
- for $F_j = a \vee b \vee c$, $f[F_j] = f[a] + f[b] + f[c] + p_j + q_j - 3$,

$f[z_i] = x_i$, and $f[\neg z_i] = 1 - x_i$.

Comment. The only difference between this polynomial and the polynomial described in the proof of Theorem 3.1 is that there, we had $f[F_j] = f[a] + f[b] + f[c] + 2p_j - 3$ for disjunction F_j of the type $a \vee b \vee c$.

If we substitute the expressions for $f[a]$ into $f[F_j]$, we conclude that each term $f[F_j]$ has either the form $p_j + \varepsilon(j,1) \cdot x_{i(j,1)} + \varepsilon(j,2) \cdot x_{i(j,2)} + c_j$ or the form $p_j + q_j + \varepsilon(j,1) \cdot x_{i(j,1)} + \varepsilon(j,2) \cdot x_{i(j,2)} + \varepsilon(j,3) \cdot x_{i(j,3)} + c_j$ for $\varepsilon(j,k) = \pm 1$ and for some constant c_j.

The coefficients of the above-described polynomial can be arbitrarily large: e.g., the constant coefficient is the sum of several terms. To decrease these coefficients to the desired values, we will change the names of v variables x_i to x_{i0}, introduce new variables:

- for every $j = 1, \ldots, k$, a new variable k_j whose interval value is $\mathbf{k}_j = [c_j, c_j]$;
- for every $i = 1, \ldots, v$, and for every $j = 1, \ldots, k$, two new variables x_{ij} and y_{ij}, with $\mathbf{x}_{ij} = [0, 1]$ and $\mathbf{y}_{ij} = [-1, 0]$,

and consider the following new polynomial:

$$G(\{x_{ij}\}, \{y_{ij}\}, \{p_j\}, \{q_j\}, \{k_j\}) =$$

$$\sum_{i=1}^{v} x_{i0} \cdot (1 - x_{i0}) + \sum_{i=1}^{v} \sum_{j=1}^{k} (x_{i,j-1} + y_{ij})^2 + \sum_{i=1}^{v} \sum_{j=1}^{k} (y_{i,j} + x_{ij})^2 + \sum_{j=1}^{k} G_j^2,$$

where, depending on the number of lierals in F_j,

$$G_j = p_j + r_{i(j,1),j} + r_{i(j,2),j} + k_j$$

or

$$G_j = p_j + q_j + r_{i(j,1),j} + r_{i(j,2),j} + r_{i(j,3),j} + k_j,$$

and r_{ij} denotes either x_{ij} or y_{ij} depending on whether the corresponding term ε is equal to 1 or to -1.

Let us first show that if we open all the parentheses, we will get a polynomial with coefficients 0, 1, 2, and 3. Indeed, this is a quadratic polynomial, and it has no constant terms; its only linear terms are x_{i0} (with coefficient 1), and all other terms are either products of two different variables, or squares.

- Each *product* of different variables comes from some square. For every two different variables, there is at most one squared term with these two terms, so, the coefficient at this product is either 0 (if there is no such term at all) or 2 (if there is exactly one such term).
- The *square* of each variable p_j, q_j, and k_j occurs only once; so, these squares come with coefficient 1.
- Each square x_{ij}^2 comes from no more than 3 terms: $(y_{i,j} + x_{ij})^2$, $(x_{i,j} + y_{i,j+1})^2$, and, possibly, G_j^2. Thus, the coefficient at x_{ij}^2 is equal to 0, 1, 2, or 3.
- Similarly, the coefficient at y_{ij}^2 is equal to 0, 1, 2, or 3.

Thus, G is the polynomial with the desired values of the coefficients.

Let us show that the lower endpoint $\underline{y}$ of the range of this polynomial G is equal to 0 if and only the original formula F is satisfiable. Indeed, since $x_{i0} \in [0, 1]$, this polynomial is always non-negative. Therefore, the lower endpoint $\underline{y}$ of its range is always ≥ 0. It can be equal to 0 if and only if each of the non-negative terms that sum up to G are equal to 0, i.e., if:

- $x_{i0} \cdot (1 - x_{i0}) = 0$;
  - hence, $x_{i0} = 0$ or $x_{i0} = 1$;
- $y_{ij} = -x_{i,j-1}$ and $x_{ij} = -y_{ij}$;
  - hence, $x_{ij} = -(-x_{i,j-1}) = x_{i,j-1}$, and therefore, $x_{ij} = x_{i0}$ and $y_{ij} = -x_{i0}$;
- $G_j = 0$;
  - hence, due to the definitions of G_j and $f[F_j]$, we can conclude that $f[F_j] = 0$.

Similarly to the proof of Theorem 3.1, one can then prove that the formula F is satisfiable.

Vice versa, if the formula F is satisfiable, we can take p_j and x_j as in Theorem 3.1, and then take $q_j = p_j$, $k_j = c_j$, $x_{ij} = x_{i0} = x_i$, and $y_{ij} = -x_{i0}$. One can easily check that for these values, $G = 0$ and therefore, $\underline{y} = 0$.

So, we have reduced the NP-hard propositional satisfiabilty problem to the problem of computing the range for quadratic polynomials with coefficients 0, 1, 2, and 3. Thus, this range computing problem is also NP-hard. The theorem is proven.

7

FIXED DATA PROCESSING ALGORITHMS, VARYING DATA: STILL NP-HARD

In Chapter 3, we showed that if we consider arbitrary polynomials and arbitrary input intervals, then the corresponding interval computation problem is NP-hard. In this chapter, we show that this problem remain NP-hard if we fix a sequence of polynomials $f_n(x_1, \ldots, x_n)$ (one for each n) and consider arbitrary (narrow) input intervals.

In terms of data processing, this means that we fix the data processing algorithm and consider arbitrary input data.

Motivations. In Chapter 3, we showed that if we consider *arbitrary* polynomials and *arbitrary* input intervals, then the corresponding interval computation problem is NP-hard. In data processing terms, polynomials correspond to data processing algorithms, and input intervals represent input data. In these terms, the result from Chapter 3 means that if we allow *arbitrary* data processing algorithms and *arbitrary* input data, then the corresponding problem is NP-hard.

A natural question is: what if we *fix* a data processing algorithm and allow arbitrary input data; will the problem still remain NP-hard? Of course, if we fix a sequence of *linear* functions $f_n(x_1, \ldots, x_n)$, then the corresponding interval computation problem is feasible (even linear time). So, the real question is: Is it possible to choose a sequence $f_n(x_1, \ldots, x_n)$ in such a way that for this sequence, the interval computation problem remains NP-hard? Our answer is: Yes, it is possible.

Theorem 7.1. *There exists a sequence of cubic polynomials $f_n(x_1, \ldots, x_n)$ with rational coefficients for which the basic problem of interval computations is NP-hard.*

Comments.

- In other words, the problem of computing the range $f_n(\mathbf{x}_1, \ldots, \mathbf{x}_n)$ for given intervals $\mathbf{x}_1, \ldots, \mathbf{x}_n$ is NP-hard.
- This problem remains NP-hard even if we only allow narrow intervals $\mathbf{x}_i$ and if we only want to compute the results approximately:

Theorem 7.2. *There exists a sequence of cubic polynomials $f_n(x_1, \ldots, x_n)$ with rational coefficients such that for every $\varepsilon > 0$ and for every $\delta > 0$, the ε–approximate basic problem of interval computations is NP-hard for polynomials $f_n(x_1, \ldots, x_n)$ and for intervals $\mathbf{x}_i$ that are absolutely and relatively δ-narrow.*

	Arbitrary $f_n(x_1, \ldots, x_n)$, arbitrary intervals $\mathbf{x}_i$	Fixed $f_n(x_1, \ldots, x_n)$, abritrary intervals $\mathbf{x}_i$
Linear $f_n(x_1, \ldots, x_n)$	Linear time	Linear time
Quadratic $f_n(x_1, \ldots, x_n)$	NP-hard	?
Cubic $f_n(x_1, \ldots, x_n)$	NP-hard	NP-hard

Proof of Theorem 7.1. As a desired sequence of cubic functions $f_n(x_1, \ldots, x_n)$, we will take the functions

$$f_n(z_1, \ldots, z_m, a_0, a_1, \ldots, a_m, a_{11}, \ldots, a_{1m}, \ldots, a_{m1}, \ldots, a_{mm}) =$$

$$a_0 + a_1 \cdot z_1 + \ldots + a_m \cdot z_m + a_{11} \cdot z_1 \cdot z_1 + a_{12} \cdot z_1 \cdot z_2 + \ldots + a_{mm} \cdot z_m \cdot z_m$$

with $n = m + 1 + m + m^2$ $(= m^2 + 2m + 1)$ variables $x_1, \ldots, x_n$:

$$x_1 = z_1, \ldots, x_m = z_m, x_{m+1} = a_0, x_{m+2} = a_1, \ldots, x_{2m+1} = a_m,$$

$$x_{2m+2} = a_{11}, x_{2m+3} = a_{12}, \ldots, x_n = a_{mm}.$$

When a_i and a_{ij} are numbers, i.e., degenerate intervals, then the basic problem for the function $f_n(x_1, \ldots, x_n)$ reduces to the basic problems for a generic quadratic polynomial, which is known to be NP-hard (Theorem 3.1). The theorem is proven.

Proof of Theorem 7.2. This theorem similarly follows from Theorem 3.2, if we take into consideration that every degenerate interval is (automatically) absolutely and relatively δ-narrow.

8

FIXED DATA, VARYING DATA PROCESSING ALGORITHMS: STILL INTRACTABLE

In Chapter 3, we showed that if we consider arbitrary polynomials and arbitrary input intervals, then the corresponding interval computation problem is NP-hard. In the previous chapter, we showed that if we fix polynomials and allow arbitrary intervals, then the problem remains NP-hard. In this chapter, we show that if we instead fix intervals and allow arbitrary polynomials, then the problem also remains intractable.

In terms of data processing, this means that if we fix the data and consider arbitrary data processing algorithms, the basic problem is NP-hard.

Motivations. In Chapter 3, we showed that if we consider *arbitrary* polynomials and *arbitrary* input intervals, then the corresponding interval computation problem is NP-hard. In data processing terms, polynomials correspond to data processing algorithms, and input intervals represent input data. In these terms, the result from Chapter 3 means that if we allow *arbitrary* data processing algorithms and *arbitrary* input data, then the corresponding problem is NP-hard.

In the previous chapter, we have shown that this problem remains NP-hard if we *fix* a sequence of data processing algorithms and allow *arbitrary* input data. In this chapter, we show that the problem remains intractable if, instead, we *fix* the input data (i.e., the intervals $\mathbf{x}_i$) and allow *arbitrary* data processing algorithms.

Theorem 8.1. *There exists a sequence of intervals $\mathbf{x}_1^{(0)}, \ldots, \mathbf{x}_n^{(0)}, \ldots$ with rational endpoints for which the basic problem of interval computations is NP-hard for quadratic polynomials $f(x_1, \ldots, x_n)$ with rational coefficients.*

Comments.

- In other words, for these intervals $\mathbf{x}_i^{(0)}$, the problem of computing the range $f\left(\mathbf{x}_1^{(0)}, \ldots, \mathbf{x}_n^{(0)}\right)$ of a given quadratic polynomial $f(x_1, \ldots, x_n)$ is NP-hard.
- This problem remains NP-hard even if we only allow *narrow* intervals $\mathbf{x}_i^{(0)}$, and if we only want to compute the results *approximately*:

Theorem 8.2. *For every $\delta > 0$, there exists a sequence of intervals $\mathbf{x}_1^{(0)}, \ldots, \mathbf{x}_n^{(0)}, \ldots$ with rational endpoints, each of which is absolutely and relatively δ-narrow and for which, for every $\varepsilon > 0$, the ε-approximate basic problem of interval computations for quadratic polynomials $f(x_1, \ldots, x_n)$ with rational coefficients is NP-hard.*

The results from this chapter and from the previous Chapter 7 can be represented by the following table:

	Arbitrary f_n, arbitrary $\mathbf{x}_i$	Fixed f_n, arbitrary $\mathbf{x}_i$	Arbitrary f_n, fixed $\mathbf{x}_i^{(0)}$
Linear $f_n(x_1, \ldots, x_n)$	Linear time	Linear time	Linear time
Quadratic $f_n(x_1, \ldots, x_n)$	NP-hard	?	NP-hard
Cubic $f_n(x_1, \ldots, x_n)$	NP-hard	NP-hard	NP-hard

Proof of Theorems 8.1 and 8.2. We can take $\mathbf{x}_i^{(0)} = [1 - \delta, 1 + \delta]$ for every i; for this choice, Theorems 8.1 and 8.2 follow from the proof of Theorem 3.2. The theorems are proven.

9

WHAT IF WE ONLY ALLOW SOME ARITHMETIC OPERATIONS IN DATA PROCESSING?

In the previous chapters, we analyzed the basic problem of interval computations for polynomials $f(x_1, \ldots, x_n)$. A polynomial can be defined as a function obtained from (rational) numerical constants and variables by using arithmetic operations $+$, $-$, and $*$. A natural question analyzed in this chapter is: What if we only allow some of these operations?

9.1. Definitions and the Main Results

For general polynomials $f(x_1, \ldots, x_n)$, the basic problem of interval computations is NP-hard. Polynomials can be defined as functions obtained from (rational) numbers and variables $x_1, \ldots, x_n$ by applying three arithmetic operations: addition $+$, subtraction $-$, and multiplication $*$. It is therefore natural to ask: What if we only allow *some* of these operations? Will the problem remain intractable? In this chapter, we will answer this question.

Definition 9.1. *Let $\mathcal{O} \subseteq \{+, -, *\}$ be a non-empty set of arithmetic operations. By an **$\mathcal{O}$-polynomial**, we mean a function that can be obtained from rational constants and from the variables $x_1, \ldots, x_n$ by applying operations from the set $\mathcal{O}$.*

In the set $\{+, -, *\}$ of all possible arithmetic operations, there are seven non-empty subsets $\mathcal{O}$: $\{+\}$, $\{-\}$, $\{*\}$, $\{+, -\}$, $\{+, *\}$, $\{-, *\}$, and $\{+, -, *\}$. The following theorem describes the complexity of interval computations for the corresponding $\mathcal{O}$-polynomials:

Theorem 9.1.

- *If the set $\mathcal{O}$ contains multiplication and at least one more operation, then for $\mathcal{O}$-polynomials, the basic problem of interval computations is NP-hard.*
- *For every other set $\mathcal{O}$, there exists a polynomial-time algorithm that solves the basic problem of interval computations for all $\mathcal{O}$-polynomials.*

If, instead of *arbitrary* rational constants, we only allow *non-negative* constants (and, correspondingly, *positive* intervals, i.e., intervals with non-negative endpoints), the results are slightly different:

Definition 9.2. *Let $\mathcal{O} \subseteq \{+, -, *\}$ be a non-empty set of arithmetic operations. By a positive $\mathcal{O}$-polynomial, we mean a function that can be obtained from non-negative rational numbers and from the variables $x_1, \ldots, x_n$ by applying operations from $\mathcal{O}$.*

Theorem 9.2.

- *For positive $\{-, *\}$-polynomials and positive intervals, the basic problem of interval computations is NP-hard.*
- *For positive $\{+, -, *\}$-polynomials and positive intervals, the basic problem of interval computations is NP-hard.*
- *For every other set $\mathcal{O}$, there exists a polynomial-time algorithm that solves the basic problem of interval computations for all positive $\mathcal{O}$-polynomials and for all positive intervals.*

The corresponding results can be described by the following table:

Set $\mathcal{O}$ of possible operations	Arbitrary $\mathcal{O}$-polynomials and arbitrary intervals	Positive $\mathcal{O}$-polynomials and positive intervals
$\{+\}$	Linear time	Linear time
$\{-\}$	Linear time	Linear time
$\{*\}$	Polynomial time	Polynomial time
$\{+,-\}$	Linear time	Linear time
$\{+,*\}$	NP-hard	Polynomial time
$\{-,*\}$	NP-hard	NP-hard
$\{+,-,*\}$	NP-hard	NP-hard

Proofs

Proof of Theorem 9.1. If we only allow + and −, then the resulting $\mathcal{O}$-polynomials are linear functions, and for linear functions, the basic problem of interval computations can be solved in linear time.

If we only allow multiplication, i.e., if $\mathcal{O} = \{*\}$, then the general $\mathcal{O}$-polynomial takes the form $f(x_1, \ldots, x_n) = r \cdot x_1^{a_1} \cdot \ldots \cdot x_n^{a_n}$ for some non-negative integers a_i. For each i, we can easily describe the set $\mathbf{X}_i$ of all possible values of $X_i = x_i^{a_i}$ when $x_i \in [\underline{x}_i, \overline{x}_i]$:

- If a_i is odd, or if a_i is even and $\underline{x}_i \geq 0$, then the function x^{a_i} is strictly *increasing* on the intervals $\mathbf{x}_i = [\underline{x}_i, \overline{x}_i]$, and therefore, $\mathbf{X}_i = [(\underline{x}_i)^{a_i}, (\overline{x}_i)^{a_i}]$.
- If a_i is even and $\overline{x}_i \leq 0$, then the function x^{a_i} is strictly *decreasing* on the interval $\mathbf{x}_i$, and therefore, $\mathbf{X}_i = [(\overline{x}_i)^{a_i}, (\underline{x}_i)^{a_i}]$.
- If a_i is even and $\underline{x}_i \leq 0 \leq \overline{x}_i$, then $\mathbf{X}_i = [0, \max\{(\underline{x}_i)^{a_i}, (\overline{x}_i)^{a_i}\}]$.

The desired interval $\mathbf{y}$ of possible values of $y = r \cdot X_1 \cdot \ldots \cdot X_n$ can be obtained if we apply naive interval computations: $\mathbf{y} = r \cdot \mathbf{X}_1 \cdot \ldots \cdot \mathbf{X}_n$.

If we allow + and $*$, then we can get an arbitrary polynomial with rational coefficients, and for arbitrary polynomials, the basic problem of interval computations is NP-hard. (Therefore, the problem is also NP-hard if we allow − in addition to + and $*$.)

If $\mathcal{O} = \{-, *\}$, then we also get NP-hardness, because by using $-$ and $*$, we can get $+$ as $a+b = a-(0-b)$ and thus, every polynomial with rational coefficients can also be represented by using only $-$ and $*$. The theorem is proven.

Proof of Theorem 9.2. Since positive $\mathcal{O}$-polynomials form a particular case of $\mathcal{O}$-polynomials, polynomial-time algorithms known for arbitrary $\mathcal{O}$-polynomials are applicable to positive ones as well. Thus, from Theorem 9.1, we get the feasibility for $\mathcal{O} = \{+\}$, $\mathcal{O} = \{-\}$, $\mathcal{O} = \{+, -\}$, and $\mathcal{O} = \{*\}$.

If we only allow $+$ and $*$, then the resulting polynomials are monotonely non-decreasing in each variable; therefore, when $x_i \in [\underline{x}_i, \overline{x}_i]$, the smallest possible value of $f(x_1, \ldots, x_n)$ is attained when $x_i = \underline{x}_i$ for all i, and the largest possible value of x_i is attained when $x_i = \overline{x}_i$ for all i. Thus, the desired range is easy to compute:

$$f([\underline{x}_1, \overline{x}_1], \ldots, [\underline{x}_n, \overline{x}_n]) = [f(\underline{x}_1, \ldots, \underline{x}_n), f(\overline{x}_1, \ldots, \overline{x}_n)].$$

For $\mathcal{O} = \{+, -, *\}$, we can get an arbitrary polynomial with rational coefficients, and for arbitrary polynomials and positive intervals, NP-hardness was proven in Theorem 3.2. If $\mathcal{O} = \{-, *\}$, then we also get NP-hardness, because (as we have mentioned in the proof of Theorem 9.1) by using $-$ and $*$, we can get $+$ as $a + b = a - (0 - b)$ and thus, every combination of $+$, $-$, and $*$ can also be represented by using only $-$ and $*$. The theorem is proven.

Comment. In our proofs, we used *monotonicity* of the function $f(x_1, \ldots, x_n)$. Monotonicity has indeed been successfully used in interval computations: see, e.g., Collatz [70, 71], Lakshmikantham *et al.* [247], Walter [423], Harrison [141], Moore [291], Schröder [383], Rall [335], Mannshardt [270], Dimitrova *et al.* [91], Markov [273].

10

FOR FRACTIONALLY-LINEAR FUNCTIONS, A FEASIBLE ALGORITHM SOLVES THE BASIC PROBLEM OF INTERVAL COMPUTATIONS

In Chapter 3, we have shown that while the basic problem of interval computations is feasible for linear functions $f(x_1, \ldots, x_n)$, for quadratic functions, this problem is already NP-hard. In this chapter, we show that for another natural generalization of linear functions, namely, for fractional-linear functions, the situation is much better: there exists a feasible algorithm that solves the basic problem of interval computations for such functions.

We have already mentioned that the basic problem of computing an interval $f(\mathbf{x}_1, \ldots, \mathbf{x}_n)$ is feasible for linear functions $f(x_1, \ldots, x_n)$. Can we generalize this result? There are two natural generalizations of linear functions:

- *quadratic* functions; for them, as we have shown, the basic problem is NP-hard;
- *fractionally linear* functions, i.e., functions of the type

  $$y = f(x_1, \ldots, x_n) = \frac{a_0 + a_1 x_1 + \ldots + a_n x_n}{b_0 + b_1 x_1 + \ldots + b_n x_n};$$

  these functions occur in many practical applications such as measuring instruments, intelligent control, etc. (see, e.g., Krotkov *et al.* [240, 241], Gerasimov *et al.* [123, 124], Mazin *et al.* [278, 279], Trejo *et al.* [415], Lea *et al.* [249]); let us show that for such functions, the basic problem is feasible.

Definition 10.1. *By a basic interval computation problem for fractionally linear functions, we mean the following problem:*

GIVEN:

- *a positive integer* $n > 0$;
- $n+1$ *rational numbers* $a_0, a_1, \ldots, a_n$;
- $n+1$ *rational numbers* $b_0, b_1, \ldots, b_n$;
- n *intervals* $\mathbf{x}_1, \ldots, \mathbf{x}_n$ *with rational endpoints;*

COMPUTE: the range $\mathbf{y} = f(\mathbf{x}_1, \ldots, \mathbf{x}_n)$ *of the function*

$$y = f(x_1, \ldots, x_n) = \frac{a_0 + a_1 x_1 + \ldots + a_n x_n}{b_0 + b_1 x_1 + \ldots + b_n x_n}.$$

Definition 10.2. *We say that an instance of the basic interval computation problem for a fractionally linear function is non-degenerate if*

$$0 \notin b_0 + b_1 \cdot \mathbf{x}_1 + \ldots + b_n \cdot \mathbf{x}_n.$$

Comment. If an instance is not non-degenerate, then the interval of possible values of the denominator contains 0, and therefore, the set of all possible values of the fractions contains ∞. So, the range is a (finite) interval only if the instance is non-degenerate.

Definition 10.3. *We say that an algorithm computes the optimal enclosure for a fractionally linear problem, if it computes the endpoints* $\underline{y}$ *and* $\overline{y}$ *of the range* $\mathbf{y} = f(\mathbf{x}_1, \ldots, \mathbf{x}_n)$.

Theorem 10.1. (Lea *et al.* [249]) *There exists an algorithm that computes the optimal enclosure for an arbitrary non-degenerate fractionally linear problem in quadratic time (i.e., in computation time* $\le Cn^2$*).*

Comment. So, the *basic problem of interval computations is feasible for fractionally linear functions.* In Lea *et al.* [249], this result is applied to intelligent control problems.

This result can be represented as a table:

Function $f(x_1,\ldots,x_n)$	Computational complexity of the corresponding basic problem of interval computations
Linear f	Linear time
Quadratic f	NP-hard
Fractionally-linear f	Quadratic time

Let us first describe the algorithm. This algorithm consists of 6 steps, which we will number Steps 0 through 5. In this algorithm, we will assume that the expression a/b has a meaning not only for $b \neq 0$, but for $b = 0$ as well: for $a < 0$, $a/0$ means $-\infty$; for $a > 0$, it means $+\infty$.

ALGORITHM.

0. *Eliminating irrelevant variables.* If for some i, we have $a_i = b_i = 0$, we eliminate the variable x_i. (And we denote the resulting new number of variables again by n.)

1. *Making a denominator positive.* If $b_0 + \sum b_i \cdot \underline{x}_i < 0$, change the signs of all the coefficients, i.e., set $a_i^{\text{new}} = -a_i$ and $b_i^{\text{new}} = -b_i$ for all i.

2. *Making the coefficients in the denominator non-negative.* For all $i = 1,\ldots,n$, if $b_i < 0$, replace x_i with the new variable $x_i^{\text{new}} = -x_i$, change the signs of the coefficients a_i and b_i ($a_i^{\text{new}} = -a_i$ and $b_i^{\text{new}} = -b_i$), and change $\mathbf{x}_i = [\underline{x}_i, \overline{x}_i]$ to $\mathbf{x}_i^{\text{new}} = [-\overline{x}_i, -\underline{x}_i]$.

3. *Eliminating degenerate variables.* If $a_i/b_i = a_j/b_j$ for some $i \neq j$, and $|b_i| \geq |b_j|$, replace two variables x_i and x_j with a single variable x_i^{new}, for which b_i and a_i stay the same as before ($b_i^{\text{new}} = b_i$ and $a^{\text{new}} = a_i$), and for which $\mathbf{x}_i^{\text{new}} = \mathbf{x}_i + (a_j/a_i)\mathbf{x}_j$. (And we denote the resulting new number of variables again by n.)

4. *Ordering the variables.* Order the variables x_i in the increasing order of the corresponding ratios a_i/b_i, so that:

$$\frac{a_1}{b_1} < \frac{a_1}{b_2} < \ldots < \frac{a_n}{b_n}.$$

5. *Computing $\underline{y}$ and $\overline{y}$.* Compute $\overline{y} = \max(\overline{y}_0, \overline{y}_1, ..., \overline{y}_n)$, where

$$\overline{y}_k = \frac{a_0 + a_1 \cdot \underline{x}_1 + \ldots + a_k \cdot \underline{x}_k + a_{k+1} \cdot \overline{x}_{k+1} + \ldots + a_n \cdot \overline{x}_n}{b_0 + b_1 \cdot \underline{x}_1 + \ldots + b_k \cdot \underline{x}_k + b_{k+1} \cdot \overline{x}_{k+1} + \ldots + b_n \cdot \overline{x}_n}.$$

Compute $\underline{y} = \min(\underline{y}_0, \underline{y}_1, \ldots, \underline{y}_n)$, where

$$\underline{y}_k = \frac{a_0 + a_1 \cdot \overline{x}_1 + \ldots + a_k \cdot \overline{x}_k + a_{k+1} \cdot \underline{x}_{k+1} + \ldots + a_n \cdot \underline{x}_n}{b_0 + b_1 \cdot \overline{x}_1 + \ldots + b_k \cdot \overline{x}_k + b_{k+1} \cdot \underline{x}_{k+1} + \ldots + b_n \cdot \underline{x}_n}.$$

Example. Let $n = 2$, $\mathbf{x}_1 = \mathbf{x}_2 = [1, 2]$,

$$f(x) = \frac{1 + x_1 + x_2}{1 + x_1 - 4x_2}.$$

In this case, the interval $1 + [1, 2] - 4[1, 2] = [-6, -1]$ does not contain 0, so, the problem is non-degenerate. Let us apply the above algorithm:

0. This function does not have any irrelevant variables, so we move directly to Step 1.

1. Since $b_0 + \sum b_i \cdot \underline{x}_i = 1 + 1 - 4 = -2 < 0$, we change the signs of all the coefficients. As a result, we arrive at the following problem: $\mathbf{x}_1 = \mathbf{x}_2 = [1, 2]$,
$$f(x) = \frac{-1 - x_1 - x_2}{-1 - x_1 + 4x_2}.$$

2. The coefficient b_i is negative for $i = 1$, so for this i, we introduce the new variable, and correspondingly change the coefficients a_1, b_1 and the interval $\mathbf{x}_1$. As a result, we get the following problem: $\mathbf{x}_1 = [-2, -1]$, $\mathbf{x}_2 = [1, 2]$,
$$f(x) = \frac{-1 + x_1 - x_2}{-1 + x_1 + 4x_2}.$$

3. The values $a_1/b_1 = 1$ and $a_2/b_2 = -1/4$ are different, so, we do nothing at this step.

4. Since $a_1/b_1 > a_2/b_2$, we change the order of the variables. As a result, we get the following problem: $\mathbf{x}_1 = [\underline{x}_1, \overline{x}_1] = [1, 2]$, $\mathbf{x}_2 = [\underline{x}_2, \overline{x}_2] = [-2, -1]$,
$$f(x) = \frac{-1 - x_1 + x_2}{-1 + 4x_1 + x_2}.$$

5. We compute $\overline{y}$ as $\overline{y} = \max(\overline{y}_0, \overline{y}_1, \overline{y}_2)$, where:

$$\overline{y}_0 = \frac{-1-2+(-1)}{-1+8+(-1)} = \frac{-4}{6} = -\frac{2}{3},$$

$$\overline{y}_1 = \frac{-1-1+(-1)}{-1+4+(-1)} = \frac{-3}{2} = -\frac{3}{2},$$

$$\overline{y}_2 = \frac{-1-1+(-2)}{-1+4+(-2)} = \frac{-4}{1} = -4.$$

Hence, $\overline{y} = -(2/3)$. We also compute $\underline{y}$ as $\underline{y} = \min(\underline{y}_0, \underline{y}_1, \underline{y}_2)$, where:

$$\underline{y}_0 = \frac{-1-1+(-2)}{-1+4+(-2)} = \frac{-4}{1} = -4,$$

$$\underline{y}_1 = \frac{-1-2+(-2)}{-1+8+(-2)} = \frac{-5}{5} = -1,$$

$$\underline{y}_2 = \frac{-1-2+(-1)}{-1+8+(-1)} = \frac{-4}{6} = -\frac{2}{3}.$$

So, $\underline{y} = -4$, and $\mathbf{y} = [-4, -(2/3)]$.

Proof. Let us denote the numerator of the function $f(x_1, \ldots, x_n)$ by N, and its denominator by D. Let us first prove that Steps 1–4 do not change the problem:

- *Step* 1. If we change the signs of all the coefficients a_i and b_i, then both numerator and denominator will change signs, and the ratio will remain unchanged.

- *Step* 2. If we rename the variable $x_i = -x_i^{\text{new}}$, then the values a_i and b_i, and the interval of possible values of x_i^{new} must be changed accordingly.

- *Step* 3. If $a_i/b_i = a_j/b_j$, then $a_j/a_i = b_j/b_i$. Therefore,

$$b_i \cdot x_i + b_j \cdot x_j = b_i \cdot (x_i + (b_j/b_i) \cdot x_j) = b_i \cdot (x_i + (a_j/a_i) \cdot x_j),$$

  and $a_i \cdot x_i + a_j \cdot x_j = b_i \cdot (x_i + (a_j/a_i) \cdot x_j)$. Therefore, we can replace the terms $a_i \cdot x_i + a_j \cdot x_j$ and $b_i \cdot x_i + b_j \cdot x_j$ that depend on x_i and x_j with $a_i \cdot x_i^{\text{new}}$ and $b_i \cdot x_i^{\text{new}}$, where the new variable x_i^{new} is equal to $x_i^{\text{new}} = x_i + (a_j/a_i)x_j$. If $x_i \in \mathbf{x}_i$ and $x_j \in \mathbf{x}_j$, then the set $\mathbf{x}_i^{\text{new}}$ of possible values of the new variable is equal to $\mathbf{x}_i^{\text{new}} = \mathbf{x}_i + (a_j/a_i) \cdot \mathbf{x}_i$.

■ *Step* 4. Renaming the variables does not change the problem.

In view of Steps 1–4, we can assume that $b_i \geq 0$, and that the ratio a_i/b_i is increasing as i increases.

After Step 1, we can be sure that the value of the denominator D is positive at least for one combination of $x_i \in \mathbf{x}_i$; since the problem is non-degenerate, the denominator cannot attain 0, and hence, it is always positive.

The function $f(x_1, \ldots, x_n)$ is a continuous function defined on a compact set

$$\mathbf{x}_1 \times \ldots \times \mathbf{x}_n.$$

Therefore, its maximum $\overline{y}$ is attained at some point $(x_1^{\text{opt}}, \ldots, x_n^{\text{opt}})$:

$$\overline{y} = f(x_1^{\text{opt}}, \ldots, x_n^{\text{opt}}) = \frac{D^{\text{opt}}}{N^{\text{opt}}}.$$

The function $f(x_1, \ldots, x_n)$ is smooth; therefore, if for some i, the value x_i^{opt} is *inside* the interval $[\underline{x}_i, \overline{x}_i]$ (i.e., $x_i^{\text{opt}} \in (\underline{x}_i, \overline{x}_i)$), then i-th partial derivative must be equal to 0:

$$\frac{\partial f}{\partial x_i}(x_1, \ldots, x_n)_{|x_1 = x_1^{\text{opt}}, \ldots, x_n = x_i^{\text{opt}}} = 0.$$

Applying the formula for the derivative of the fraction, we conclude that

$$a_i \cdot D^{\text{opt}} - b_i \cdot N^{\text{opt}} = 0,$$

hence, $a_i/b_i = N^{\text{opt}}/D^{\text{opt}} = \overline{y}$. In this case, if we replace x_i with $\underline{x}_i$, then, the new value $D' = D(x_1^{\text{opt}}, \ldots, x_{i-1}^{\text{opt}}, \underline{x}_i, x_{i+1}^{\text{opt}}, \ldots, x_n^{\text{opt}})$ of the denominator D will be equal to

$$D' = D(x_1^{\text{opt}}, \ldots, x_{i-1}^{\text{opt}}, x_i^{\text{opt}}, x_{i+1}^{\text{opt}}, \ldots, x_n^{\text{opt}}) + b_i \cdot (\underline{x}_i - x_i^{\text{opt}})$$

$$= D^{\text{opt}} + b_i \cdot (\underline{x}_i - x_i^{\text{opt}}).$$

Similarly, the new value $N' = N(x_1^{\text{opt}}, \ldots, x_{i-1}^{\text{opt}}, \underline{x}_i, x_{i+1}^{\text{opt}}, \ldots, x_n^{\text{opt}})$ of the numerator N will be equal to $N' = N^{\text{opt}} + a_i \cdot (\underline{x}_i - x_i^{\text{opt}})$. Since $N^{\text{opt}} = D^{\text{opt}} \cdot \overline{y}$ and $a_i = b_i \cdot \overline{y}$, we conclude that

$$D' = D^{\text{opt}} \cdot \overline{y} + b_i \cdot \overline{y} \cdot (\underline{x}_i - x_i^{\text{opt}}) = \overline{y} \cdot (D^{\text{opt}} + b_i \cdot (\underline{x}_i - x_i^{\text{opt}})) = \overline{y} \cdot N'.$$

Hence, the value $f(x_1^{\text{opt}}, \ldots, x_i^{\text{opt}}, \underline{x}_i, x_{i+1}^{\text{opt}}, \ldots, x_n^{\text{opt}}) = D'/N'$ is equal to the maximum $\overline{y}$ of the function $f(x_1, \ldots, x_n)$.

Therefore, if for some i, $x_i^{\rm opt} \in (\underline{x}_i, \overline{x}_i)$, we can change this value $x_i^{\rm opt}$ to $\underline{x}_i$ and still get a point at which the function $f(x_1, \ldots, x_n)$ attains its maximum. So, without losing generality, we can assume that for every i, $x_i^{\rm opt}$ is either equal to $\underline{x}_i$, or to $\overline{x}_i$.

If $x_i^{\rm opt} = \underline{x}_i$, then, since the function $f(x_1, \ldots, x_n)$ attains its *maximum* for $x_i^{\rm opt}$, an increase in x_i can either decrease the value of the function $f(x_1, \ldots, x_n)$, or keep this value unchanged. So, at the optimal point $(x_1^{\rm opt}, \ldots, x_n^{\rm opt})$, the function $f(x_1, \ldots, x_n)$ must have a non-negative i-th partial derivative $\partial f/\partial x_i$. This partial derivative is equal to $(a_i \cdot D^{\rm opt} - b_i \cdot N^{\rm opt})/(D^{\rm opt})^2$, so, the fact that this derivative is non-negative, means that $a_i \cdot D^{\rm opt} - b_i \cdot N^{\rm opt} \le 0$, which is equivalent to $a_i \cdot D^{\rm opt} \le b_i \cdot N^{\rm opt}$. Since $D^{\rm opt} > 0$ and $b_i \ge 0$, we can divide both sides of this inequality by $b_i \cdot D^{\rm opt}$, resulting in $a_i/b_i \le N^{\rm opt}/D^{\rm opt} = \overline{y}$.

Similarly, if $x_i^{\rm opt} = \overline{x}_i$, a decrease in x_i can either decrease the value of the function $f(x_1, \ldots, x_n)$, or keep it unchanged. So, we will get $\partial f/\partial x_i \ge 0$, and $a_i/b_i \ge \overline{y}$.

So, for every i, either $x_i^{\rm opt} = \underline{x}_i$ and $a_i/b_i \le \overline{y}$, or $x_i^{\rm opt} = \overline{x}_i$ and $a_i/b_i \ge \overline{y}$. Hence, if $a_i/b_i < \overline{y}$, we have $x_i^{\rm opt} = \underline{x}_i$, and if $a_i/b_i > \overline{y}$, we have $x_i^{\rm opt} = \overline{x}_i$. If $a_i/b_i = \overline{y}$, then, as above, we can switch from $x_i^{\rm opt}$ to $\underline{x}_i$ without changing the value of the function $f(x_1, \ldots, x_n)$.

Since after Step 4, the variables are ordered in the order of the ratio a_i/b_i, this means that for all variables $x_1, \ldots, x_k$ up to some k-th one, we have $x_i^{\rm opt} = \underline{x}_i$, and for the other variables $x_{k+1}, \ldots, x_n$, we have $x_i^{\rm opt} = \overline{x}_i$. In other words, we conclude that $\overline{y} = \overline{y}_k$ for some k. Hence,

$$\overline{y} \le \max(\overline{y}_0, \ldots, \overline{y}_n).$$

On the other hand, each value $\overline{y}_k$ is a possible value of the function $f(x_1, \ldots, x_n)$ and is therefore, not exceeding the maximum $\overline{y}$ of the function $f(x_1, \ldots, x_n)$. So, $\overline{y}_k \le \overline{y}$ for all k; hence,

$$\max(\overline{y}_0, \ldots, \overline{y}_n) \le \overline{y}.$$

From these two inequalities, we conclude that $\overline{y} = \max(\overline{y}_0, \ldots, \overline{y}_n)$.

The proof for $\underline{y}$ is similar.

To complete the proof, we must now show that our algorithm requires quadratic time. Indeed, initial steps 1 and 2 require the number of operations that is linear in n. Sorting (Step 4) can be done in time $n \log_2(n) \ll n^2$ (see, e.g., Cormen *et al.* [76]), and the final step requires us to compute $2(n+1)$ expressions $\underline{y}_k$ and $\overline{y}_k$, each of which requires $4n+1$ arithmetic operations: $2n$ multiplications, $2n$ additions, and 1 division. Totally, we need $\le 2(n+1)(4n+1) = O(n^2)$ arithmetic operations. The theorem is proven.

11

SOLVING INTERVAL LINEAR SYSTEMS IS NP-HARD

In the previous chapter, we showed how to compute the range of a fractionally linear function $f(x_1, \ldots, x_n)$ on given intervals $\mathbf{x}_1, \ldots, \mathbf{x}_n$ in polynomial time. A fractionally linear function $f(x_1, ..., x_n)$ can be described as a solution of a linear equation $(b_0 + \sum b_i x_i) \cdot f = a_0 + \sum a_i x_i$. The problem of finding the range of f is thus equivalent to finding the set of all possible solutions of this equation when x_i take the values in their respective intervals. A natural generalization is, therefore, the solution of a system of linear equations *$\sum a_{ij} \cdot f_j = b_i$, where the coefficients a_{ij} and b_i are linear functions of the variables that are defined with interval uncertainty.*

In this chapter and in the next Chapter 12, we analyze the computational complexity and feasibility of solving such interval linear equations. In most formulations, this problem is NP-hard, but some particular cases of this problem turn out to be feasible.

11.1. Introduction

In the previous chapters, we analyzed the computational complexity and feasibility of the problem of estimating the range $\mathbf{y} = f(\mathbf{x}_1, \ldots, \mathbf{x}_n)$ of an *explicitly* given function $y = f(x_1, \ldots, x_n)$ on given intervals $\mathbf{x}_1, \ldots, \mathbf{x}_n$. In terms of *data processing*, the range estimation problem describes the situation of *indirect measurement* of y, when we have measured the quantities x_i, and from the intervals $\mathbf{x}_i$ of possible values of x_i we find the interval $\mathbf{y}$ of possible values of $y = f(x_1, \ldots, x_n)$. If there are several quantities $y_1 = f_1(x_1, \ldots, x_n), \ldots, y_m = f_m(x_1, \ldots, x_n)$ that we want to measure, then we have to estimate the range of each of these variables.

In many real-life situations, we do not have the *explicit* formula that expresses each of these variables y_j in terms of x_i; instead, we have *implicit* formulas, i.e., a system of *equations* $F_k(x_1, \ldots, x_n, y_1, \ldots, y_m) = 0$ that relate y_j and x_i. What is the computational complexity and feasibility of the range estimation problem for such *implicitly* defined functions? This is the question that we will try to answer in this chapter. Namely, we will show that this problem is NP-hard even in the simplest case of systems of *linear* equations.

A general system of linear equations with the unknowns $y_1, \ldots, y_m$ has the form $\sum_j a_{ij} \cdot y_j = b_i$. The situation in which we only know the *intervals* $\mathbf{a}_{ij}$ and $\mathbf{b}_i$ of the possible values of each of the coefficients is called a *system of interval linear equations* and denoted by $\sum_j \mathbf{a}_{ij} \cdot y_j = \mathbf{b}_i$. The matrix formed by the intervals $\mathbf{a}_{ij}$ is called an *interval matrix* and denoted by $\mathbf{A}$; the vector formed by coefficients $\mathbf{b}_i$ is called an *interval vector* and denoted by $\mathbf{b}$. In terms of $\mathbf{A}$ and $\mathbf{b}$, a system of interval linear equations can be rewritten as $\mathbf{A}y = \mathbf{b}$, where we denoted $y = (y_1, \ldots, y_m)$.

If we only know the intervals of possible values of the coefficients, then a vector $y = (y_1, \ldots, y_m)$ is possible if and only if $\sum a_{ij} \cdot y_j = b_i$ (i.e., $Ay = b$) for some $a_{ij} \in \mathbf{a}_{ij}$ and $b_i \in \mathbf{b}_i$. (In the following text, we will sometimes describe these component-wise inclusions in a shortened form, as $A \in \mathbf{A}$ and $b \in \mathbf{b}$.) The set of all possible vectors is called a *solution set* of the system of interval linear equations. In the interval computations framework, we would like, for each j from 1 to m, to describe the *interval* that contains all possible values of y_j. The narrower this interval, the better. Ideally, therefore, we should compute the interval $\mathbf{y}_j = [\underline{y}_j, \overline{y}_j]$, where $\underline{y}_j$ is the smallest possible value of y_j for $y \in Y$, and $\overline{y}_j$ is the largest possible value of y_j. If we cannot compute this interval exactly, then we would like to compute an *enclosure* for this interval, i.e., an interval $\widetilde{\mathbf{y}}_j = [\underline{\widetilde{y}}_j, \overline{\widetilde{y}}_j] \supseteq \mathbf{y}_j$.

When all the coefficients are precisely known, then we get a problem of solving a system of linear equations which can be solved by known polynomial time algorithms; see, e.g., Schrijver [382], Cormen [76]. There also exist many good algorithms for solving systems of *interval* linear equations; see, e.g., Alefeld *et al.* [10], Rump [369], Neumaier [303]. It turns out, however, that each algorithm that computes the *exact* ranges $\mathbf{y}_j$ sometimes takes an *exponentially long* time. This fact led to the suspicion that the general problem of computing the exact range is not feasible. This problem indeed turned out to be NP-hard.

Moreover, it is even somewhat more difficult than the problem of computing the range for explicit functions; there are two reasons for that:

- First, for explicit functions, computing the *exact* range is difficult, but we can always use naive interval computations, and easily get ***an enclosure*** for the range. For interval linear systems, even computing an ***enclosure*** is a difficult task.

- Second, before we start finding the range, it is desirable to check whether the interval system is consistent at all: it could be that our model is wrong and the system is inconsistent, in which case the enclosure makes no practical sense (it can also happen that no finite enclosure is possible because the solution set is unbounded). It turns out that not only ***computing*** the enclosures is NP-hard, but even ***checking consistency*** is NP-hard.

Historical comment. The first NP-hardness result for the problem of solving interval linear systems was proved by Kreinovich, Lakeyev, and Noskov in 1993 [221, 245, 246]: the problem of computing the bounds ***exactly*** for arbitrary ***rectangular*** (not necessarily square) matrices is NP-hard. Later in 1993, for arbitrary ***rectangular*** matrices, it was shown (by the same authors) that the problem of estimating the range with a given ***accuracy*** $\varepsilon > 0$ is also NP-hard [222]. In that same year (1993), Rohn *et al.* proved [360] that the problem of computing the bounds ***exactly*** is NP-hard even for ***square regular*** matrices (regular means that every matrix $a_{ij} \in \mathbf{a}_{ij}$ is nonsingular). On hearing about these two results, A. Neumaier conjectured that the problem of computing the bounds with a given ***accuracy*** is NP-hard even for ***square regular*** matrices. This conjecture was proven correct in 1995 by Rohn, Lakeyev *et al.* [355, 220]. Finally, it has been recently proven that for every $\varepsilon > 0$ and $\delta > 0$, the problem of computing the bounds with an ***accuracy*** ε for ***square regular*** interval matrices made of intervals of width $\leq \delta$ is also NP-hard: Kahl [167], Rohn [358].

11.2. Definitions and Main NP-Hardness Results

Definitions and NP-Hardness of Checking Consistency

Definition 11.1. *By an interval linear system, we mean a tuple $\langle m, n, \mathbf{A}, \mathbf{b}\rangle$, where m and n are positive integers, $\mathbf{A}$ is a $(m \times n)$-interval matrix, and $\mathbf{b}$ is an m-dimensional interval vector. This system will also be written as $\mathbf{A}y = \mathbf{b}$.*

Definition 11.2.

- *We say that a vector* $(y_1, \ldots, y_m)$ *is a* ***possible solution*** *of an interval linear system* $\langle m, n, \mathbf{A}, \mathbf{b}\rangle$ *if, for some values* $a_{ij} \in \mathbf{a}_{ij}$ *and* $b_i \in \mathbf{b}_i$, *we have*

$$\sum_{j=1}^{m} a_{ij} \cdot y_j = b_i \quad \text{for all } i = 1, \ldots, n$$

- *We say that an interval linear system is* ***consistent*** *if it has a possible solution.*

Theorem 11.1. *Checking consistency of interval linear systems is NP-hard.*

Similarly to Chapter 3, this NP-hardness results is also true for *narrow* intervals $\mathbf{a}_{ij}$ and $\mathbf{b}_i$:

Theorem 11.2. (Kahl [167]) *For every* $\delta > 0$, *the problem of checking consistency of interval linear systems with intervals* $\mathbf{a}_{ij}$ *and* $\mathbf{b}_i$ *that are both absolutely and relatively* δ*-narrow is NP-hard.*

	Arbitrary absolute accuracy of $\mathbf{a}_{ij}$ and $\mathbf{b}_i$	Absolutely δ-narrow $\mathbf{a}_{ij}$ and $\mathbf{b}_i$
Arbitrary relative accuracy of $\mathbf{a}_{ij}$ and $\mathbf{b}_i$	NP-hard	NP-hard
Relatively δ-narrow $\mathbf{a}_{ij}$ and $\mathbf{b}_i$	NP-hard	NP-hard

Definition of the Solution Interval

Definition 11.3. *Let* $\langle m, n, \mathbf{A}, \mathbf{b}\rangle$ *be an interval linear systems, and let* $j \le m$ *be a positive integer. By* ***j-th solution interval*** $\mathbf{y}_j$, *we mean a (possibly infinite) interval* $\mathbf{y}_j = [\underline{y}_j, \overline{y}_j]$, *where:*

- $\underline{y}_j$ *is the smallest possible value of* y_j *for all possible solutions* $y = (y_1, \ldots, y_{j-1}, y_j, y_{j+1}, \ldots, y_m)$ *of the given interval linear system.*
- $\overline{y}_j$ *is the largest possible value of* y_j *for all possible solutions* $y = (y_1, \ldots, y_{j-1}, y_j, y_{j+1}, \ldots, y_m)$ *of the given interval linear system.*

Comment. The words "possibly infinite" are added to this definition because the set of possible solutions can be unbounded: the trivial 1×1 example is a system consisting of a single equation $[-1, 1] \cdot y_1 = [1, 1]$, for which the set of possible values of y_1 is $(-\infty, -1] \cup [1, \infty)$.

Computing Possibly Infinite Solution Intervals is NP-Hard

As we have just mentioned, the set of possible solutions can be unbounded; in this case, the desired answers $\underline{y}_j$ and $\overline{y}_j$ are infinite. It turns out (see below) that computing these (possibly infinite) values *exactly* is NP-hard. Moreover, it is NP-hard to produce *any* enclosure to the solution as long as we require that for a finite interval, the enclosure should be finite:

Definition 11.4. *We say that an algorithm produces a possibly finite enclosure to the solution* $\mathbf{y}_j = [\underline{y}_j, \overline{y}_j]$ *of an interval linear system if it produces an enclosure* $\widetilde{\mathbf{y}}_j \supseteq \mathbf{y}_j$ *that is finite if the solution interval* $\mathbf{y}_j$ *is finite.*

Theorem 11.3. (Rohn [352]) *The problem of computing a possibly finite enclosure of a given interval linear system is NP-hard.*

Corollary. *The problem of exactly computing solution intervals of a given interval linear system is NP-hard.*

Theorem 11.4. (Kahl [167]) *For every* $\delta > 0$*, the problem of computing a possibly finite enclosure of a given interval linear system with intervals* $\mathbf{a}_{ij}$ *and* $\mathbf{b}_i$ *that are both absolutely and relatively* δ*-narrow is NP-hard.*

Corollary. *For every* $\delta > 0$*, the problem of exactly computing solution intervals of a given interval linear system with intervals* $\mathbf{a}_{ij}$ *and* $\mathbf{b}_i$ *that are both absolutely and relatively* δ*-narrow is NP-hard.*

Computing Finite Solution Intervals Exactly is Also NP-Hard

In most practical cases, we are interested in the *finite* solution intervals; in other words, we are interested in solving the following problem:

Definition 11.5. *By a problem of solving interval linear systems,* we mean *the following problem:*

GIVEN:

- *an interval linear system* $\langle m, n, \mathbf{A}, \mathbf{b} \rangle$*; and*
- *a positive integer* $j \le m$ *for which* j*-th solution interval* $\mathbf{y}_j = [\underline{y}_j, \overline{y}_j]$ *is finite;*

COMPUTE: the endpoints of j*-th solution interval.*

Comment. In short, we want our algorithm to work *if* the solution interval is finite. (It is worth mentioning here that, according to the proof of Theorem 11.3, there is no easy algorithm to check *whether* a solution interval is finite.)

Theorem 11.5. *The problem of solving interval linear systems is NP-hard.*

Theorem 11.6. (Kahl [167]) *For every* $\delta > 0$*, the problem of solving interval linear systems with intervals* $\mathbf{a}_{ij}$ *and* $\mathbf{b}_i$ *that are both absolutely and relatively* δ*-narrow is NP-hard.*

Computing Finite Solution Intervals Approximately is NP-Hard

Since computing finite solution intervals *exactly* is NP-hard, the natural question is: is computing the solution intervals *approximately* feasible or NP-hard? In general, it turns out to be NP-hard.

Finiteness is guaranteed, if, e.g., the interval matrix $\mathbf{A}$ is *regular* in the sense that for every $A \in \mathbf{A}$, and for every vector b, there is exactly one vector y for which $Ay = b$. This condition is *difficult* to check (as we will see in the chapters about the properties of interval matrices, it is even *NP-hard* to check), but there are verifiable properties that guarantee regularity. One of such properties is the following property of *strong regularity*:

Definition 11.6. *A square interval matrix* $\mathbf{A} = [\tilde{A}-\Delta, \tilde{A}+\Delta]$ *is called strongly regular if* $\varrho(|(\tilde{A})^{-1}|\Delta) < 1$ *(where* $|M|$ *denotes a matrix with elements* $|m_{ij}|$*, and* $\rho(M)$ *denotes the spectral radius of a matrix* M*).*

We will show that the problem of approximately computing the solution intervals is NP-hard even if we consider only interval linear systems with strongly regular matrices $\mathbf{A}$:

Theorem 11.7. *Suppose for some real number* $\varepsilon > 0$*, there exists a polynomial-time algorithm which for each strongly regular* $n \times n$ *interval matrix* $\mathbf{A}$ *and each* $\mathbf{b}$ *(both with rational bounds) computes rational enclosures* $[\widetilde{\underline{y}}_j, \widetilde{\overline{y}}_j]$ *of the solution intervals* $\mathbf{y}_j$ *for which*

$$\left|\frac{\widetilde{\overline{y}}_j - \overline{y}_j}{\overline{y}_j}\right| \le \varepsilon$$

for each j *with* $\overline{y}_j \neq 0$*. Then* P=NP.

Theorem 11.8. *Suppose for some real number* $\varepsilon > 0$*, there exists a polynomial-time algorithm which for each strongly regular* $n \times n$ *interval matrix* $\mathbf{A}$ *and each* $\mathbf{b}$ *(both with rational bounds) computes rational enclosures* $[\widetilde{\underline{y}}_j, \widetilde{\overline{y}}_j]$ *of the solution intervals* $\mathbf{y}_j$ *for which* $|\widetilde{\overline{y}}_j - \overline{y}_j| \le \varepsilon$ *for all* j*. Then* P=NP.

So, the problem of approximately computing solution intervals is NP-hard if we understand "approximately" both in terms of *absolute* and in terms of *relative* accuracy:

Computing exact solution intervals	Computing absolutely ε-approximate solution intervals	Computing relatively ε-approximate solution intervals
NP-hard	NP-hard	NP-hard

Comments.

- As we explained in Chapter 1, due to book size limitations, we had to omit some easily accessible proofs. In particular, we do not present the proofs of this theorem and of the following theorems. These proofs are described, in detail, in the paper [220] published in *Reliable Computing.*

- If P$\neq$NP, then for absolute accuracy, not only we cannot compute enclosures with one and the same accuracy (i.e., with one and the same bound for absolute overestimation) for all n in reasonable time, but even if we allow accuracy to decrease polynomially with n, we still will not be able to compute these "relaxed-accuracy" enclosures:

Theorem 11.9. *Suppose for some polynomial $\varepsilon(n) > 0$, there exists a polynomial-time algorithm which for each strongly regular $n \times n$ interval matrix $\mathbf{A}$ and each $\mathbf{b}$ (both with rational bounds) computes rational enclosures $[\widetilde{\underline{y}}_j, \widetilde{\overline{y}}_j]$ of the solution intervals $\mathbf{y}_j$ for which $|\widetilde{\overline{y}}_j - \overline{y}_j| \le \varepsilon(n)$ for all j. Then* P=NP.

Comments.

- In these three theorems, we can additionally assume that all intervals from the matrix $\mathbf{A}$ and the vector $\mathbf{b}$ are absolutely δ-narrow for a given $\delta > 0$.
- Since the *general* problem of solving interval linear systems is NP-hard, it is desirable to find *feasible subclasses*. We will look for such subclasses in the next chapter.

11.3. Interval Linear Systems with Symmetric Matrices

Motivation, Definitions, and a Brief History

In some cases, we know that the actual (unknown) matrix A is symmetric ($a_{ij} = a_{ji}$). In this case, every known interval bound on an element a_{ij} also bounds a_{ji}, and vice versa. Thus, the *interval* matrix $\mathbf{A}$ (formed by such interval bounds) is also symmetric ($\mathbf{a}_{ij} = \mathbf{a}_{ji}$), and it is natural to consider the set $Y^{\rm sym}$ of all solutions that correspond to systems with symmetric matrices only:

$$Y^{\rm sym} = \{y | Ay = b \text{ for some } A \in \mathbf{A},\ b \in \mathbf{b},\ A \text{ symmetric}\}.$$

For this problem, we want to compute the *symmetric* solution interval $\mathbf{y}_j^{\rm sym} = [\underline{y}_j^{\rm sym}, \overline{y}_j^{\rm sym}]$, where

$$\underline{y}_j^{\rm sym} = \min_{Y^{\rm sym}} y_j, \quad \text{and} \quad \overline{y}_j^{\rm sym} = \max_{Y^{\rm sym}} y_j$$

Enclosure methods for the symmetric case were given by Jansson [163] and Alefeld and Mayer [9] (see also Alefeld *et al.* [5, 6, 7, 8]).

NP-Hardness Results for Interval Linear Systems with Symmetric Matrices

J. Rohn has shown, in [356], that computing the exact endpoints of the symmetric solution intervals $\mathbf{y}_j^{\text{sym}}$ is an NP-hard problem. In Rohn, Lakeyev *et al.* [355, 220], it is shown that *approximate* computation of these symmetric solution intervals is also NP-hard, even for strongly regular matrices:

Theorem 11.10. *Suppose for some real number $\varepsilon > 0$, there exists a polynomial-time algorithm which for each symmetric strongly regular $n \times n$ interval matrix $\mathbf{A}$ and each $\mathbf{b}$ (both with rational bounds) computes rational enclosures $[\underset{\sim}{\underline{y}}_j^{\text{sym}}, \widetilde{\overline{y}}_j^{\text{sym}}]$ of the symmetric solution intervals $\mathbf{y}_j^{\text{sym}}$ for which*

$$\left| \frac{\widetilde{\overline{y}}_j^{\text{sym}} - \overline{y}_j^{\text{sym}}}{\overline{y}_j^{\text{sym}}} \right| \leq \varepsilon$$

for each j with $\overline{y}_j^{\text{sym}} \neq 0$. Then P=NP.

Theorem 11.11. *Suppose for some polynomial $\varepsilon(n) > 0$, there exists a polynomial-time algorithm which for each symmetric strongly regular $n \times n$ interval matrix $\mathbf{A}$ and each $\mathbf{b}$ (both with rational bounds) computes rational enclosures $[\underset{\sim}{\underline{y}}_j^{\text{sym}}, \widetilde{\overline{y}}_j^{\text{sym}}]$ of the symmetric solution intervals $\mathbf{y}_j^{\text{sym}}$ for which $|\widetilde{\overline{y}}_j^{\text{sym}} - \overline{y}_j^{\text{sym}}| \leq \varepsilon$ for all j. Then* P=NP.

Comment. In these Theorems 11.10 and 11.11, we can additionally assume that all intervals from the matrix $\mathbf{A}$ and the vector $\mathbf{b}$ are absolutely δ-narrow for a given $\delta > 0$.

Proofs

Proof of Theorem 11.1. To prove this result, we will reduce the *PARTITION* problem (known to be NP-hard) to this problem. In the *PARTITION* problem, we are given a sequence of integers $s_1, \ldots, s_m$, and we must check whether there exist values $y_1, \ldots, y_m$ for which $y_j \in \{-1, 1\}$ and $s_1 \cdot y_1 + \ldots + s_m \cdot y_m = 0$. For every sequence $s_1, \ldots, s_m$, we consider the interval linear system that consists of the following equations: $[-1,1]\cdot y_j=[1,1]$ $(1\le j\le m)$, $[1,1]\cdot y_j=[-1,1]$ $(1\le j\le m)$, and $[s_1,s_1]\cdot y_1+\ldots+[s_m,s_m]\cdot y_m=[0,0]$.

If the given instance has a solution $y_1,\ldots,y_m$, then, as one can easily see, these y_j form a possible solution to the interval linear system. Vice versa, let the interval linear system be consistent, i.e., have a possible solution $(y_1,\ldots,y_m)$. Then, for every j, from the first equation, we conclude that $a_j\cdot y_j=1$ for some $a_j\in[-1,1]$; hence, $y_j=1/a_j$ and therefore, either $y_j\le -1$, or $y_j\ge 1$. From the second equation, it follows that $y_j=b_j\in[-1,1]$, i.e., that $y_j\in[-1,1]$. Together with the previously derived inequalities, we conclude that $y_j=-1$ or $y_j=1$. Hence, the third equation implies that $\sum s_j\cdot y_j=0$, i.e., that this sequence $y_1,\ldots,y_m$ is a solution to the given instance. We have proven the reduction, and thus, the theorem is proven.

Proof of Theorem 11.2. We will use the following result from the proof of Theorem 3.2: every quantity a whose possible values form a (non-degenerate) interval $\mathbf{a}$ can be represented, for some real numbers k and l, as $a=k\cdot a'+l$, where possible values of a' form an interval $[1-\delta,1+\delta]$.

Let us use this result to replace, one-by-one, all non-degenerate coefficient intervals $\mathbf{a}_{ij}$ and $\mathbf{b}_i$ by δ-narrow ones. First, we can move all intervals $\mathbf{b}_i$ to the left-hand side by introducing a new variable y_0 and a new equation $y_0=1$, and replacing each equation $\sum \mathbf{a}_{ij}\cdot y_j=\mathbf{b}_i$ by a new equation $\mathbf{a}_{ij}\cdot y_j-\mathbf{b}_i\cdot y_0=0$. This system is clearly equivalent to the original one. Therefore, without losing generality, we can assume that the only non-degenerate intervals are among the coefficients $\mathbf{a}_{ij}$.

For each such interval $\mathbf{a}_{ij}$, we first compute the values k_{ij} and l_{ij} for which $a_{ij}=k_{ij}\cdot a'_{ij}+l_{ij}$ and a'_{ij} takes the values from the interval $[1-\delta,1+\delta]$. Then, $a_{ij}\cdot y_j=k_{ij}\cdot a'_{ij}\cdot y_j+l_{ij}\cdot y_j$. So, we can introduce a new variable y_{ij}, add a new equation $y_{ij}-\mathbf{a}'_{ij}\cdot y_j=0$, and replace the term $\mathbf{a}_{ij}\cdot y_j$ in the original equation by a combination $k_{ij}\cdot y_{ij}+l_{ij}\cdot y_j$. One can easily see that for the new system, possible values of the old variables y_j are exactly the same as for the old one. Thus, we can step-by-step eliminate all non-narrow intervals, and get a system with narrow intervals whose consistency is equivalent to the consistency of the original system. Thus, we have reduced the NP-hard problem of checking consistency of *arbitrary* interval linear systems to checking consistency of *narrow* interval linear systems. Thus, checking consistency of narrow interval systems is also NP-hard. The theorem is proven.

Proof of Theorem 11.3. In our proof, we will consider interval linear systems of the type $\mathbf{A}y = 0$, where $\mathbf{A}$ is a square matrix, and 0 is a vector consisting of all 0's. For every $A \in \mathbf{A}$, if A is a regular matrix, then $y = 0$ is the only solution, and if A is not regular, then the solution set is unbounded (a line, or, more general, a subspace). Thus, if the interval matrix $\mathbf{A}$ only contains regular matrices, the solution set consists of a single point 0; otherwise, the solution set is unbounded.

It is known (see the proof in chapters about interval matrices) that checking whether a given interval matrix $\mathbf{A}$ contains a singular matrix is NP-hard. If we could always compute a possibly finite enclosure, then we would be able to check whether the set of all possible solutions is bounded, and therefore, whether $\mathbf{A}$ contains a regular matrix. Thus, the problem of computing an enclosure is indeed NP-hard. The theorem is proven.

Proof of Theorem 11.4. This theorem is proven by the same reduction as we used in the proof of Theorem 11.2.

Proof of Theorem 11.5. To prove this result, we will reduce the *PARTITION* problem (known to be NP-hard) to this problem. In the *PARTITION* problem, we are given a sequence of integers $s_1, \ldots, s_m$, and we must check whether there exist values $y_1, \ldots, y_m$ for which $y_j \in \{-1, 1\}$ and $s_1 \cdot y_1 + \ldots + s_m \cdot y_m = 0$. For every sequence $s_1, \ldots, s_m$, we consider the interval linear system that consists of the following equations: $[-1, 1] \cdot y_j = [1, 1]$ $(1 \le j \le m+1)$, $[1, 1] \cdot y_j = [-1, 1]$ $(1 \le j \le m+1)$, and $[s_1, s_1] \cdot y_1 + \ldots + [s_m, s_m] \cdot y_m + [s_{m+1}, s_{m+1}] \cdot y_{m+1} = [s_{m+1}, s_{m+1}]$, where we denoted $s_{m+1} = -0.5 \cdot (s_1 + \ldots + s_m)$.

Similarly to the proof of Theorem 11.1, each of the variables y_j can only take the values $y_j \in \{-1, 1\}$.

- The value $y_{m+1} = -1$ is always a possible solution, because the values $y_1 = \ldots = y_m = 1$ and $y_{m+1} = -1$ satisfy all the equations.
- On the other hand, if there is a possible solution with $y_{m+1} = 1$, then, from the last equation, we can conclude that $s_1 \cdot y_1 + \ldots + s_m \cdot y_m = 0$, i.e., that the original instance of *PARTITION* problem has a solution. Vice versa, if this instance has a solution $y_1, \ldots, y_m$, then we can add $y_{m+1} = 1$ and get a possible solution of the interval linear system.

Hence:

- If the given instance of *PARTITION* has a solution, we get $\overline{y}_{m+1} = 1$.
- If the given instance of *PARTITION* has no solutions, we get $\overline{y}_{m+1} = -1$.

Thus, if we were able to compute $\overline{y}_j$, we would thus be able to check whether the given instance of *PARTITION* is solvable. This reduction shows that our problem is NP-hard. The theorem is proven.

Proof of Theorem 11.4. This theorem is proven by the same reduction as we used in the proof of Theorem 11.2.

Proof of the Comments after Theorems 11.9 and 11.11. If we multiply both $\mathbf{A}$ and $\mathbf{b}$ by the same positive constant k, we will not change either strong regularity property or the solution set, but if k is small enough, all resulting intervals will be absolutely δ-narrow. The comment is proven.

12

INTERVAL LINEAR SYSTEMS: SEARCH FOR FEASIBLE CLASSES

In the previous chapter, we showed that in general, solving interval systems is NP-hard. In this chapter, we look for feasible classes of interval linear systems.

12.1. General Idea

Since the *general* problem of solving interval linear systems is NP-hard, it is desirable to find *feasible subclasses*. We will exploit two different approaches to finding such subclasses:

First approach. The first approach is based on the following idea: Solving linear systems is feasible when all interval coefficients are *numerical* (i.e., degenerate intervals); the difficulty is caused by *non-degenerate* intervals. Thus, we will consider subclasses in which *some* of the coefficient *intervals are degenerate*, hoping that this restriction will make computations easier.

Second approach. The second approach is based on a slightly different idea: Solving generic *non-interval linear system* is *feasible* but still somewhat *time-consuming*: e.g., for a square $n \times n$ system, traditional Gaussian elimination method takes $\approx n^3$ computational steps. If we *add interval uncertainty* to this complexity, we end up with *NP-hardness* (i.e., in effect, with *exponential time*). It is therefore reasonable to consider *subclasses* of (*non-interval*) *linear systems* for which *faster algorithms* are known, and hope that when we add interval uncertainty to thus simplified systems, we will get feasible algorithms.

The original complexity of solving systems of linear equations comes from the fact that we need lots of input data: n^2 rational numbers a_{ij}. The more numbers we need to input, and the more bits we need to describe each number, the larger the input length. Thus, to simplify the problem of solving linear systems, we can do one of three things:

- limit the *number* of non-zero elements, i.e., consider *sparse* matrices, in which some of the coefficients a_{ij} are zeros;
- limit the possible *values* of the coefficients a_{ij};
- if none of the above restriction works, we may want to try to limit *both* the *number* of non-zero elements *and* the possible *values* of the coefficients.

Our plans. In this chapter, we will pursue all these approaches.

12.2. First Approach: Interval Systems in Which Some Intervals are Degenerate

Since complexity is caused by interval uncertainty, let us consider the systems in which some intervals are degenerate. The first natural idea is as follows: In the above results, we allowed both the coefficients $\mathbf{a}_{ij}$ of the $(m \times n)$-coefficient matrix and the right-hand sides $\mathbf{b}_i$ to be arbitrary intervals. What if we only allow interval uncertainty in $\mathbf{a}_{ij}$? or only in $\mathbf{b}_i$?

Theorem 12.1. *Checking consistency of interval linear systems* $\mathbf{A}y = b$ *with interval matrices* $\mathbf{A}$ *and real-number vectors* b *is NP-hard.*

Theorem 12.2. *The problem of solving interval linear systems* $\mathbf{A}y = b$ *with an interval matrix* $\mathbf{A}$ *and a real-number vector* b *is NP-hard.*

Comment 1. Both theorems 12.1 and 12.2 hold for interval matrices in which all components are absolutely and relatively *δ-narrow.*

Comment 2. Theorem 12.2 holds even if want to compute the solution intervals *approximately* (with absolute or relative ε-accuracy, for a given ε).

Comment 3. We will see from the proof that Theorem 12.2 holds even if we consider only vectors $e^{(i)} = (0, \ldots, 0, 1, 0, \ldots, 0)$ with one (i-th) element equal to 1 and all other elements equal to 0. For an arbitrary matrix A, the solutions $y^{(i)}$ to the corresponding systems $Ay = e^{(i)}$ form the *inverse matrix* to the matrix A. Thus, this result shows that *computing exact (or even approximate) bounds on elements of an inverse interval matrix is NP-hard* (Coxson [78]).

Comment 4. In these two theorems, we assumed that all the intervals $\mathbf{b}_i$ from the *right*-hand sides of the linear equations are numerical; if we assume instead that all the coefficient intervals $\mathbf{a}_{ij}$ from the *left*-hand sides are numerical, we get a *feasible* problem:

Theorem 12.3. *There exists a polynomial-time algorithm that checks consistency of interval linear systems $Ay = \mathbf{b}$ with numerical matrices A and interval vectors $\mathbf{b}$.*

Theorem 12.4. *There exists a polynomial-time algorithm that solves interval linear systems $Ay = \mathbf{b}$ with numerical matrices A and interval vectors $\mathbf{b}$.*

	Numerical A	Interval $\mathbf{A}$
Numerical b	Polynomial time	NP-hard
Interval $\mathbf{b}$	Polynomial time	NP-hard

Theorems 12.3 and 12.4 are true even if for each equation $\sum \mathbf{a}_{ij} \cdot y_j = \mathbf{b}_i$, we allow *at most one* interval coefficient:

Definition 12.1. *An interval linear system $\sum \mathbf{a}_{ij} \cdot y_j = \mathbf{b}_i$ is called almost numerical if for each equation i, at most one of intervals $\mathbf{a}_{i1}, \ldots, \mathbf{a}_{im}, \mathbf{b}_i$, is non-degenerate.*

Theorem 12.5. *Checking consistency of almost numerical interval linear systems $\sum \mathbf{a}_{ij} \cdot y_j = b_i$ (with numerical right-hand sides) is NP-hard.*

Theorem 12.6. *The problem of solving almost numerical interval linear systems $\sum \mathbf{a}_{ij} \cdot y_j = b_i$ (with numerical right-hand sides) is NP-hard.*

12.3. Second Approach, First Try: Band and Sparse Matrices

Band Matrices

For non-interval linear systems, simple algorithms are possible for *w-band* matrices, i.e., for square matrices for which $a_{ij} = 0$ for $|i-j| \geq w$. These systems are of great practical importance: For example, a natural way of solving a system of ordinary linear differential equations $dx_i/dt = \sum c_{ij} \cdot x_j + d_i$, $1 \leq i \leq p$, is *time discretization* when we consider the moments of time t_0, $t_1 = t_0 + \Delta t$, ..., $x_k = t_0 + k \cdot \Delta t$, and the following system of equations: $x_i(t_{k+1}) = x_i(t_k) + \Delta t \cdot \sum c_{ij} \cdot x_j(t_k) + \Delta t \cdot d_i$. If we order the resulting variables $x_i(t_k)$ in *chronological* order $x_1(t_0), \ldots, x_n(t_0), x_1(t_1), \ldots, x_n(t_1), \ldots$, then we get a linear system with a $2p$-band matrix.

What is the complexity of solving interval linear systems with w-band *interval* matrices, i.e., for square interval matrices, for which $\mathbf{a}_{ij} = 0$ for $|i-j| \geq w$? When $w = 1$, we get *diagonal* matrices for which $\mathbf{a}_{ij} = 0$ for $i \neq j$. For these matrices, each equation is of the form $\mathbf{a}_{jj} \cdot y_j = \mathbf{b}_j$; from this equation, we can easily find the set of all possible values of y_j. For larger w, the problem is NP-hard:

Theorem 12.7. *For every $w \geq 3$, checking consistency of interval linear systems $\mathbf{A}y = \mathbf{b}$ with w-band interval matrices $\mathbf{A}$ is NP-hard.*

Theorem 12.8. *For every $w \geq 3$, the problem of solving interval linear systems $\mathbf{A}y = \mathbf{b}$ with w-band interval matrices $\mathbf{A}$ is NP-hard.*

Comment 1. We do not know whether a feasible algorithm is possible for 2-band matrices:

Diagonal (1-band) matrices	2-band matrices	3-band matrices
Linear time	?	NP-hard

Comment 2. The same results are true for interval linear systems in which all intervals are *absolutely* δ-narrow for all δ. For interval linear systems whose elements are *both absolutely* and *relatively* δ-narrow, we can only prove NP-hardness for 4-band matrices:

Theorem 12.9. *For every $w \geq 4$, checking consistency of interval linear systems $\mathbf{A}y = \mathbf{b}$ with w-band interval matrices $\mathbf{A}$, and with intervals $\mathbf{a}_{ij}$ and $\mathbf{b}_i$ that are both absolutely and relatively δ-narrow, is NP-hard.*

Theorem 12.10. *For every $w \geq 3$, the problem of solving interval linear systems $\mathbf{A}y = \mathbf{b}$ with w-band interval matrices $\mathbf{A}$, and with intervals $\mathbf{a}_{ij}$ and $\mathbf{b}_i$ that are both absolutely and relatively δ-narrow, is NP-hard.*

	Diagonal (1-band) matrices	2-band matrices	3-band matrices	4-band matrices
Arbitrary intervals $\mathbf{a}_{ij}$, $\mathbf{b}_i$	Linear time	?	NP-hard	NP-hard
Absolutely δ-narrow intervals $\mathbf{a}_{ij}$, $\mathbf{b}_i$	Linear time	?	NP-hard	NP-hard
Absolutely and relatively δ-narrow intervals $\mathbf{a}_{ij}$, $\mathbf{b}_i$	Linear time	?	?	NP-hard

Sparse Matrices

w-band matrices are a specific class of *sparse* matrices, in which most elements are equal to 0. For many classes of sparse matrices, simpler algorithms are indeed possible (see, e.g., Schendel [377]). In addition to w-band matrices, another important class of sparse matrices is the class of *d-sparse* matrices, in which in each row i, at most d elements a_{ij} are different from 0. This class is also very important in practical applications:

- A natural way to solve a linear *partial* differential equation, e.g., a simple equation $\partial f(t,x)/\partial t = \partial^2 f(t,x) \partial x^2 + g(t,x)$, is to consider discrete values t_k and x_k, and the corresponding system

  $$\frac{f(t_{k+1}, x_l) - f(t_k, x_l)}{\Delta t} = \frac{f(t_k, x_{l-1}) - 2f(t_k, x_l) + f(t_k, x_{l+1})}{\Delta x^2} + g(t_k, x_l).$$

  This is a system of linear equations with unknowns $f(t_k, x_l)$. A specific feature of such systems, as opposed to *generic* linear systems, is that each equation contains only a *few* variables (four in the above example) with non-zero coefficients, so, it is 4-sparse.

- Another case where sparse matrices appear is *robotic vision* (see, e.g., Elguea *et al.* [99]). To act reasonably, a robot must make a 3-D interpretation of the 2-D visual picture, in order to decide which visible faces correspond to which objects. The resulting interpretations must satisfy certain restrictions of the type: "a face must go through a certain point", or "two faces must not have an intersection", etc. The total number of variables in the description of a 3-D scheme may be huge, but each restriction is about a face and a point, or two faces, etc, and therefore, the corresponding equation contains only a few variables. In other words, the corresponding system is also sparse.

What happens if we consider *d-sparse interval* linear systems, i.e., systems in which for every i, at most d coefficients $\mathbf{a}_{ij}$ are different from $[0, 0]$?

When $d = 1$, then each equation has the form $\mathbf{a}_{ij} \cdot y_j = \mathbf{b}_i$. From each equation of this type, we can easily describe the set of possible values of y_j; if for some j, two or more different equations contain y_j, we take the *intersection* of the corresponding sets. This can be done by a *linear-time* algorithm. For $d \geq 2$, the problem becomes NP-hard:

Theorem 12.11. *For every integer $d \geq 2$, checking consistency of interval linear systems with d-sparse matrices is NP-hard.*

Theorem 12.12. *For every integer $d \geq 2$, the problem of solving interval linear systems with d-sparse matrices is NP-hard.*

1-sparse matrices	2-sparse matrices	d-sparse matrices, $d \geq 3$
Linear time	NP-hard	NP-hard

Comment. The problem remains NP-hard even if we combine the requirement of 2-sparseness with the requirement that almost all intervals are degenerate:

Theorem 12.13. *For every integer $d \geq 2$, checking consistency of almost numerical interval linear systems with d-sparse matrices is NP-hard.*

Theorem 12.14. *For every integer $d \geq 2$, the problem of solving almost numerical interval linear systems with d-sparse matrices is NP-hard.*

12.4. Second Approach, Second Try: Restricting the Size of the Coefficients a_{ij}

Theorem 12.15. (Heindl *et al.* [143]) *Checking consistency of regular interval systems, in which every endpoint of every interval is either* 0 *or* 1, *is NP-hard.*

Theorem 12.16. (Heindl *et al.* [143]) *The problem of solving regular interval systems, in which every endpoint of every interval is either* 0 *or* 1, *is NP-hard.*

In other words, the problem is NP-hard for regular linear interval systems in which each interval coefficient is equal to $[0,0]$, to $[1,1]$, or to $[0,1]$.

Comment. It is known that NP-hard numerical problems can be, crudely speaking, of two types (see, e.g., Garey *et al.* [121], Section 4.2):

- Problems that are, in general, NP-hard, but for which a polynomial-time algorithm is possible if we restrict ourselves to instances in which the lengths of all numerical coefficients are bounded by a constant C. Such problems are called *pseudo-polynomial.*
- Problems that remain NP-hard even if we restrict ourselves to instances in which the lengths of all numerical coefficients are bounded by some constant C. Such problems are called *NP-hard in the strong sense.*

In these terms, our result shows that *the problem of finding exact (or ε-approximate) component-wise bounds for the solution set of a linear interval system is NP-hard in the strong sense.*

12.5. Second Approach, Third Try: Sparse Matrices with Restricted Coefficients

Since neither of the first two tries (i.e., considering sparse matrices and restricting the size of the coefficients) led to feasible algorithms, let us try imposing *both* restrictions. In this case, we do get a feasible algorithm:

Theorem 12.17. *Let w be a positive integer, and let S be a finite set of rational numbers. Then, there exists a polynomial-time algorithm that checks whether a given interval linear systems in which $\mathbf{a}_{ij}$ is a w-band matrix, and all endpoints of all interval coefficients $\mathbf{a}_{ij}$ and $\mathbf{b}_i$ belong to the set S, is consistent.*

Theorem 12.18. *Let w be a positive integer, and let S be a finite set of rational numbers. Then, there exists a polynomial-time algorithm that solves all interval linear systems in which $\mathbf{a}_{ij}$ is a w-band matrix, and all endpoints of all interval coefficients $\mathbf{a}_{ij}$ and $\mathbf{b}_i$ belong to the set S.*

12.6. Additional Results: In Brief

Other NP-hardness results. Several other NP-hardness results for interval linear systems are presented in Lakeyev *et al.* [244]. In particular, it is proven that this problem is NP-hard for interval linear systems with finitely many solutions, etc.

Other feasibility results. In some practical problems, we know the *signs* $s_j = \text{sign}(y_j)$ of all the variables y_j. Under this assumption, it is possible to design *polynomial-time* algorithms for solving interval linear systems; see, e.g., Rohn [343] and Rohn [349] (Theorems 2.2. and 2.3).

In some other cases, we do not know the signs, but we know that only a *few sign combinations* $s = (s_1, \ldots, s_m)$ are possible. If there are indeed no more than polynomially many possible sign combinations, then we can make polynomially many calls of the above algorithm and get a polynomial-time algorithm for this case as well (Jansson [164]).

12.7. Different Notions of a Solution: Computational Complexity and Feasibility

In the previous sections, we considered interval linear systems $\mathbf{A}y = \mathbf{b}$ that originate from *data processing* and *indirect measurements.* In these problems, we are interested in the set Y of all possible values of y for which $Ay = b$ for some $A \in \mathbf{A}$ and $b \in \mathbf{b}$. This solution set is also called a *united* solution set, because it unites all possible vectors y for which $Ay = b$ for possible A and b.

In some other practical problems, e.g., in *control*, we get similarly-looking linear interval systems for which, however, we are interested in *different* solution sets.

Tolerance Solution

An important example of such a problem comes from the economic *planning* problem. Global economy is often described by a Leontieff-type *input-output model*, in which the consumption level b_i of each item is a linear function of the productions $y_1, \ldots, y_m$ of different products: $b_i = a_{i1} \cdot y_1 + \ldots + a_{im} \cdot y_m$. Our objective is to *plan* the production, i.e., to find the values of y_j that guarantee the desired consumption.

- If we know *exactly* the desired consumption level b_i of each product, and if we know exactly the coefficients a_{ij}, then the *planning problem* (i.e., the problem of determining the production levels y_j) becomes a simple (easily solvable) linear equation.
- In practice, we often do not know the exact values of b_i and a_{ij}. Instead, we only know the *intervals* $\mathbf{b}_1, \ldots, \mathbf{b}_n$ that describe the desired consumption of each item, and the *intervals* $\mathbf{a}_{ij}$ of possible values of each coefficient a_{ij}. In this situation, we would like to set up the production levels $y_1, \ldots, y_m$ in such a way that for all possible values of the coefficients $a_{ij} \in \mathbf{a}_{ij}$, we get the desired consumption levels of all the products.

In other words, we are interested in finding the set Y of all vectors y for which $Ay \in \mathbf{b}$ for *all* $A \in \mathbf{A}$. This new solution set is called a *tolerance solution* set (it is also sometimes called a *tolerable solution* set). For this set, we can formulate similar problems of checking consistency and of finding the smallest and largest values of y_j.

The corresponding economic problem was first considered and analyzed in Rohn [340, 341, 345]; the notion of a tolerance solution set was also analyzed by Shaidurov and Shary [389], Neumaier [303], Shary [390, 393], Lakeyev and Noskov [245, 246], and others. In contrast to the previous notion of a solution set, for this solution set, the above problems are solvable in polynomial time:

Theorem 12.19. *There exists a polynomial-time algorithm that checks, for every interval linear system, whether its tolerance solution set is non-empty.*

Theorem 12.20. *There exists a polynomial-time algorithm that, given an interval linear system $\langle m, n, \mathbf{A}, \mathbf{b}\rangle$, and a positive integer $j \le m$, computes the smallest $\underline{y}_j$ and the largest $\overline{y}_j$ values of y_j for all vectors y from the tolerance solution set.*

Controlled Solution

Other control problems lead to another notion of solution, called *controlled solution set.* In many control problems, it is relatively *easy to maintain* steady control flow $y_1, \ldots, y_m$, but it is *difficult to modify* it.

- For example, in *space exploration*, it is relatively easy to ignite a rocket engine and guarantee a stable flow, but it is very difficult to stop it or adjust it if something goes wrong.
- In *radar astronomy*, it is relatively easy to generate a mighty radio signal sent to a planet, but it is difficult to change the parameters of this signal.

Since it is difficult to *change* the values y_j, we fix these values, and *tune* the resulting control by applying an appropriate transformation of the original control $y_1, \ldots, y_m \rightarrow b_1, \ldots, b_n$. In the reasonable linear approximation, we get *linear* equations $b_i = \sum a_{ij} \cdot y_j$. The coefficients a_{ij} are adjustable, but adjustable within certain limits; in other words, we know the *intervals* $\mathbf{a}_{ij}$ within which we can change these coefficients. For every i, we also know the interval $\mathbf{b}_i$ of possible values that we may want to achieve. Our goal is to set up the initial control values y_j in such a way that we will be able to achieve every vector $b \in \mathbf{b}$ by choosing appropriate coefficients $a_{ij} \in \mathbf{a}_{ij}$.

In mathematical terms, we again have an interval linear system $\mathbf{A}y = \mathbf{b}$, but this time, we are interested in finding all the vectors y for which, for every $b \in \mathbf{b}$, there exists an $A \in \mathbf{A}$ such that $Ay = b$. The set of all such vectors is called a *controlled solution set.* This set was studied by Khlebalin and Shokin [177], Lakeyev and Noskov [245, 246], and Shary [393].

Theorem 12.21. (Lakeyev *et al.* [246]) *Checking whether an interval linear system has a non-empty controlled solution set is NP-hard.*

Theorem 12.22. (Lakeyev *et al.* [246]) *The problem of computing, for every interval linear system $\langle m, n, \mathbf{A}, \mathbf{b}\rangle$ with a non-empty controlled solution set, an element from this set, is NP-hard.*

Name of solution set	United	Tolerance	Controlled
Description of solution set	$\exists A \exists b (Ay = b)$	$\forall A \exists b (Ay = b)$	$\forall b \exists A (Ay = b)$
Feasibility	NP-hard	Polynomial time	NP-hard

Proofs

Proof of Theorem 12.1. This proof follows the same idea that we used in Theorem 11.2: that we can move all intervals $\mathbf{b}_i$ to the left-hand side if we introduce a new variable y_0, add a new equation $y_0 = 1$, and replace each equation $\sum \mathbf{a}_{ij} \cdot y_j = \mathbf{b}_i$ with a new equation $\mathbf{a}_{ij} \cdot y_j - \mathbf{b}_i \cdot y_0 = 0$ with no intervals in the right-hand side. The theorem is proven.

Proof of Theorem 12.2 is done by a similar reduction.

Proof of Theorem 12.4. If all the coefficients a_{ij} are numbers, then the problems of finding the smallest value of y_j for which this system has a solution becomes a linear programming problem:

$$y_j \to \max$$

under the conditions

$$\underline{b}_i \leq \sum a_{ij} \cdot y_j \leq \overline{b}_i, \ 1 \leq i \leq n.$$

Hence, we can apply known polynomial-time algorithms of solving linear programming problems (see, e.g., Karmarkar [169]). The theorem is proven.

Proof of Theorem 12.3 is done by a similar reduction.

Proof of Theorem 12.5. To prove this theorem, we will show that the problem of checking consistency of an arbitrary interval linear system

$$\sum_{j=1}^{m} \mathbf{a}_{ij} \cdot y_j = b_i, \ 1 \leq i \leq n,$$

with numerical right-hand sides (for which NP-hardness was proved in Theorem 12.3) can be reduced to systems of this type.

Indeed, for each equation i, we introduce m new variables z_{ij}, $1 \le j \le m$, and replace the original equation with the following $m+1$ equations:

$$\sum_{j=1}^{m} z_{ij} = b_i;$$

$$\mathbf{a}_{ij} \cdot y_j - z_{ij} = 0, \ 1 \le j \le m.$$

In this new system, each equation has at most one interval coefficient, and since z_{ij} is equal to $\mathbf{a}_{ij} \cdot y_j$, these two systems are clearly equivalent to each other. The theorem is proven.

Proof of Theorem 12.6 is done by a similar reduction.

Proof of Theorem 12.7. For this proof, we will use a reduction to *PARTITION* problem that is similar to the proof of Theorem 5.3. Namely, for each instance $s_1, \ldots, s_m$ of *PARTITION*, we will introduce $3m+1$ variables $t_1, \ldots, t_m$, $y_1, \ldots, y_m$, and $z_1, \ldots, z_m, z_{m+1}$, that will be presented in the following order: $t_0, y_1, z_1, t_1, y_2, z_2, \ldots, t_{m-1}, y_m, z_m, t_m$, and the following $3m+1$ equations:

- $s_1 \cdot y_1 - t_1 = 0;\ \ y_1 = [-1,1];\ \ [-1,1] \cdot y_1 = [1,1];$
- $t_1 + s_2 \cdot y_2 - t_2 = 0;\ \ y_2 = [-1,1];\ \ [-1,1] \cdot y_2 = [1,1];$
- $\ldots$
- $t_{k-1} + s_k \cdot y_k - t_k = 0;\ \ y_k = [-1,1];\ \ [-1,1] \cdot y_k = [1,1];$
- $\ldots$
- $t_{m-1} + s_m \cdot y_m - t_m = 0;\ \ y_m = [-1,1];\ \ [-1,1] \cdot y_m = [1,1];$
- $t_m = 0.$

The corresponding interval matrix is 3-band. (Note that the variables t_0 and z_i are "fictitious" variables that are not present in the equations and whose only purpose is to make the matrix square.)

One can easily see (similar to the proof of Theorems 5.3 and 11.1) that for every solution of this system, $y_j \in \{-1,1\}$, $t_k = s_1 \cdot y_1 + \ldots + s_k \cdot y_k$, and $t_m = s_1 \cdot y_1 + \ldots + s_m \cdot y_m = 0$. Vice versa, if y_j is a solution to the given instance of *PARTITION* problem, we can take these y_j, $t_k = s_1 \cdot y_1 + \ldots + s_k \cdot y_k$, and arbitrary z_j, and get a solution to our linear interval system. Thus, our system has a solution if and only if the given instance of *PARTITION* has a solution. This reduction shows that solving 3-band linear systems is indeed NP-hard.

For every $w \geq 3$, every 3-band matrix is also w-band; hence, solving w-band linear systems is also NP-hard. The theorem is proven.

Proof of Theorem 12.8 is done by a similar reduction.

Proof of Theorem 12.9. For this proof, we will also use a reduction of *PARTITION* problem. Namely, for each instance $s_1, \ldots, s_m$ of *PARTITION*, we will introduce $5m$ variables $t_1, \ldots, t_m$, $y_1, \ldots, y_m$, $y'_1, \ldots, y'_m$, $z_1, \ldots, z_m$, and $u_1, \ldots, u_m$, that will be presented in the following order: $y_1, y'_1, z_1, u_1, t_1, y_2, y'_2, z_2, u_2, t_2, \ldots, y_m, y'_m, z_m, u_m, t_m$, and the following m groups of 5 equations in each $(1 \leq k \leq m)$:

- $[-(1/\delta), -(1/\delta)] \cdot y_k + [1-\delta, 1+\delta] \cdot y'_k = [1,1]$;
- $[1,1] \cdot t_{k-1} + [s_k, s_k] \cdot y_k - [1,1] \cdot t_k = [0,0]$;
- $[-(1/\delta), -(1/\delta)] \cdot y_k + [1,1] \cdot y'_k = [0,0]$;
- $[\delta, \delta] \cdot y_k + [1,1] \cdot z_k = [1-\delta, 1+\delta]$;
- $[1,1] \cdot z_k = [1,1]$.

For $k = 1$ and $k = m$, the second equation takes a simplified form:

- for $k = 1$, we take $[s_1, s_1] \cdot y_1 - [1,1] \cdot t_1 = [0,0]$;
- for $k = m$, we take $[1,1] \cdot t_{m-1} + [s_m, s_m] \cdot y_m = [0,0]$.

(This simplification is formally equivalent to taking $t_0 = t_m = 0$.) The corresponding interval matrix is 4-band.

If the values $y_1, \ldots, y_m$ form a solution to the given instance of *PARTITION* problem, then we can take these values y_k, $z_k = 1$, $y'_k = y_k/\delta$, $t_k = s_1 \cdot x_1 + \ldots + s_k \cdot y_k$, and arbitrary u_k, and get a solution of the above interval system.

Vice versa, let us assume that the above interval linear system is consistent. Then:

- From the fifth equation, we conclude that $z_k = 1$; therefore, from the fourth equation, we get $\delta \cdot y_k \in [-\delta, \delta]$ and hence, $y_k \in [-1, 1]$.
- Similarly, from the third equation, we conclude that $y'_k = y_k/\delta$, and therefore, the first equation implies that $a \cdot y_k = 1$ for some $a \in [-1, 1]$; hence, either $y_k \leq -1$ or $y_k \geq 1$.
- Together with the previous conclusion, we get $y_k \in \{-1, 1\}$.
- The second equation implies that $t_k = t_{k-1} + s_k \cdot y_k$, where $t_0 = 0$ (and $t_m = 0$). Therefore, $t_k = s_1 \cdot y_1 + \ldots + s_k \cdot y_k$, and the condition $t_m = 0$ implies that $s_1 \cdot y_1 + \ldots + s_m \cdot y_m = 0$ for $y_j \in \{-1, 1\}$. Thus, the given instance of *PARTITION* has a solution.

So, our system has a solution if and only if the given instance of *PARTITION* has a solution. This reduction shows that solving 3-band linear systems is indeed NP-hard. The theorem is proven.

Proof of Theorem 12.10 is done by a similar reduction, with $s_1 \cdot y_1 + \ldots + s_m \cdot y_m + s_{m+1} \cdot y_{m+1} = s_{m+1}$ instead of $s_1 \cdot y_1 + \ldots + s_m \cdot y_m = 0$.

Proof of Theorem 12.11.

1°. For every $d \geq 2$, every 2-sparse system is d-sparse. Thus, it is sufficient to prove that the problem is NP-hard for 2-sparse matrices.

2°. To prove NP-hardness of 2-sparse problems, we will reduce propositional satisfiability problem for 3-CNF formulas, (that is known to be NP-hard) to this problem.

3°. For this reduction, we will use the idea similar to what naive interval computation does to estimate the value of the expressions: we will "unfold" the computation of F, and take the results of all intermediate steps as new (auxiliary) variables.

Computing $F = F_1 \& \ldots \& F_k$ means that we first compute $F_1 \& F_2$, then we compute $F_1 \& F_2 \& F_3$, etc., until we reach $F_1 \& \ldots \& F_k$. So, we will need auxiliary variables to stand for these intermediate expressions $F_1 \& \ldots \& F_j$ for $j = 2, \ldots, k$.

For each of the expressions F_k, if $F_k = a \vee b$, then the only intermediate expression we need is F_k itself, so, for such expressions, we will introduce one additional variable that will stand for the truth value of F_k.

If $F_k = a \vee b \vee c$, then we need *two* expression: first, we compute $a \vee b$, and only then, F_k.

Also, each negative literal $\neg z_i$ is also an intermediate result, so, we introduce a new Boolean variable $\tilde{z}_i$ with the meaning of $\neg z_i$, and add the equation $\tilde{z}_i \equiv \neg z_i$.

As a result, we get the following new variables and equations:

- For each negative literal $\neg z_i$, we add a new variable $\tilde{z}_i$ and a new formula $\tilde{z}_i \equiv \neg z_i$.
- For each expression F_j of the type $F_j = a \vee b$, we will add a new variable V_j that will represent the truth value of this expression. For each such variable, we add the formula $V_j \equiv a \vee b$.
- For each expression F_j of the type $a \vee b \vee c$, we will introduce two new variables:
  - V_j will represent the truth value of this expression;
  - W_j will represent the truth value of $a \vee b$.

  In terms of these new variables, the formula $F_j = a \vee b \vee c$ will be represented by the following equivalent set of two formulas: $W_j \equiv a \vee b$ and $V_j \equiv W_j \vee c$.
- We will also introduce "global" propositional variables $f_2, \ldots, f_k$ (i.e., variables that correspond not to one of the expressions F_k, but to the formula in general): a variable f_j will mean the truth value of the partial conjunction $F_1 \& F_2 \& \ldots \& F_j$ (in particular, f_k will mean the truth value of the formula F itself). In terms of these new variables, the formula $F = F_1 \& \ldots \& F_k$ will be represented by the following sequence of formulas: $f_2 \equiv V_1 \& V_2$ and $f_j \equiv f_{j-1} \& V_j$ $(3 \leq j \leq k)$.

As a result, the original formula F is reformulated as a set of formulas of the type $a \equiv \neg b$, $a \equiv b \vee c$, and $a \equiv b \& c$ for some Boolean variables a, b, and c. The objective is to make $f_k =$"true", where f_k is one of these Boolean variables.

Example. Let us illustrate the above construction on the example of a formula $F = (z_1 \vee z_2 \vee z_3)\&(z_1 \vee \neg z_2)$. For this formula, $k = 2$, $F_1 = z_1 \vee z_2 \vee z_3$, and $F_2 = z_1 \vee \neg z_2$; therefore, in the reformulated formula, we will have the Boolean variables V_1, W_1, V_2, f_2, and $\tilde{z}_2$. The formula itself will be equivalent to the conjunction of the following formulas:

$$\tilde{z}_2 \equiv \neg z_2; \quad W_1 \equiv z_1 \vee z_2; \quad V_1 \equiv W_1 \vee z_3; \quad V_2 \equiv z_1 \vee \tilde{z}_2; \quad f_2 \equiv V_1 \& V_2.$$

The goal is to make $f_2 =$"true".

4°–8°. Let us now describe the reformulated propositional formula in terms of real-number variables and interval linear equations.

4°. First, for each Boolean variable a, we introduce a corresponding real-number variable x_a. We want to guarantee that this new variable x_a can only take two values: 0 and 1 (they will correspond to "false" and "true"). The fact that a variable x_a takes only values 0 and 1 will be represented in a way similar to [221]: Namely, we introduce a new variable y_a, and add the following interval linear equations:

- $[-2, 2] \cdot y_a = [1, 1]$.

  It is easy to show that this equation is equivalent to $|y_a| \geq 0.5$, i.e., to $y_a \in (-\infty, -0.5] \cup [0.5, \infty)$, or, to $y_a \notin (-0.5, 0.5)$.

- $y_a = [-0.5, 0.5]$.

  This equation guarantees that $y_a \in [-0.5, 0.5]$; together with $y_a \notin (-0.5, 0.5)$, this equation thus guarantees that $y_a \in \{-0.5, 0.5\}$.

- $x_a - y_a = [0.5, 0.5]$.

  This equation says that $x_a = y_a + 0.5$. Since we already know that $y_a = -0.5$ or $y_a = 0.5$, we can conclude that x_a is equal either to 0, or to 1.

5°. The formula $b \equiv \neg a$ can be easily reformulated in terms of the real-number variables x_a and x_b, as $x_a + x_b = [1, 1]$ (so, $x_b = 1 - x_a$, which for $x_a \in \{0, 1\} = \{$"true", "false"$\}$ means exactly that $b = \neg a$).

6°. For each formula $a \equiv b \vee c$ or $a \equiv b\&c$, we will introduce a new variable x_{bc} with the intended value $2x_b + x_c$. In other words, this new variable will contain the information about the truth values of two Boolean variables b and c; the variable x_{bc} takes only the values 0, 1, 2, and 3, and in its binary representation, the first binary digit is equal to x_b (i.e., to the truth value of b), and the second binary digit is equal to x_c (i.e., to the truth value of c).

6.1°. Let us first describe the fact that the variable x_{bc} can only take the values 0, 1, 2, and 3. To describe this fact, we will introduce three auxiliary variables y_{1bc}, y_{2bc}, and y_{3bc}, and add the following linear interval equations:

- $[-2,2]\cdot y_{1bc} = [1,1]$; this equation leads to $y_{1bc} \notin (-0.5, 0.5)$.
- $x_{bc} - y_{1bc} = [0.5, 0.5]$; together with the previous equation, this one leads to $x_{bc} \notin (0,1)$.
- $[-2,2]\cdot y_{2bc} = [1,1]$; this equation leads to $y_{2bc} \notin (-0.5, 0.5)$.
- $x_{bc} - y_{2bc} = [1.5, 1.5]$; together with the previous equation, this one leads to $x_{bc} \notin (1,2)$.
- $[-2,2]\cdot y_{3bc} = [1,1]$; this equation leads to $y_{3bc} \notin (-0.5, 0.5)$.
- $x_{bc} - y_{3bc} = [2.5, 2.5]$; together with the previous equation, this one leads to $x_{bc} \notin (2,3)$.
- $x_{bc} = [0,3]$; this equation means that $x_{bc} \in [0,3]$. Together with $x_{bc} \notin (0,1)$, $x_{bc} \notin (1,2)$, and $x_{bc} \notin (2,3)$, this means that $x_{bc} \in \{0,1,2,3\}$.

6.2°. Let us now represent the fact that the first digit in the binary representation of x_{bc} must coincide with x_b. In other words, we want to represent the following two facts:

- if $x_b = 0$, then $x_{bc} = 00_2 = 0$ or $x_{bc} = 01_2 = 1$; and
- if $x_b = 1$, then $x_{bc} = 10_2 = 2$ or $x_{bc} = 11_2 = 3$.

Enumerating all four possible combinations of truth values of b and c, one can see that the above two facts are captured by the equation $x_{bc} - 2x_b = [0,1]$.

6.3°. We must now describe the fact that the second digit in the binary expansion of x_{bc} coincides with x_c. In other words, we must represent the following two facts:

- if $x_c = 0$, then $x_{bc} = 00_2 = 0$ or $x_{bc} = 10_2 = 2$; and
- if $x_c = 1$, then $x_{bc} = 01_2 = 1$ or $x_{bc} = 11_2 = 3$.

These two facts will be represented by introducing two new variables $\tilde{x}_{1bc}$ and $\tilde{x}_{2bc}$, and the following five interval equations:

$$x_{bc} - x_c = [0,2];\ \tilde{x}_{1bc} + x_{bc} = [1,1];\ \tilde{x}_{2bc} + x_{bc} = [2,2];$$

$$[-1,1] \cdot \tilde{x}_{1bc} + [0,1] \cdot x_c = [1,1];\ [-1,1] \cdot \tilde{x}_{2bc} + [0,1] \cdot \tilde{x}_c = [1,1].$$

Let us show that this is indeed an equivalent representation. In other words, we will show that if $x_c \in \{0,1\}$ and $x_{bc} \in \{0,1,2,3\}$, then this system of interval linear equations is solvable if and only if the second binary digit in the binary expansion of x_{bc} coincides with x_c. Indeed:

- If x_{bc} is the desired value, then the difference $x_{bc} - x_c$ is equal to 0 or to 2 and therefore, belongs to the interval $[0,2]$. The second and third equations are automatically satisfied by taking $\tilde{x}_{1bc} = 1 - x_{bc}$ and $\tilde{x}_{2bc} = 2 - x_{bc}$. To show that the fourth and the fifth equations can also be satisfied, we must consider two cases:

  - If $x_c = 1$, then we can satisfy the fourth equation by taking

    $$0 \cdot \tilde{x}_{1bc} + 1 \cdot x_c = 1.$$

    In this case, $x_{bc} = 1$ or 3, so, $\tilde{x}_{2bc} = 2 - x_{bc} = \pm 1$, and the fifth equation can be satisfied by taking $(\pm 1) \cdot \tilde{x}_{2bc} + 0 \cdot \tilde{x}_c = 1$.

  - If $x_c = 0$, then $\tilde{x}_c = 0$, and either $x_{bc} = 0$, or $x_{bc} = 2$. Therefore, the value of $\tilde{x}_{1bc} = 1 - x_{bc}$ is either 1, or -1. Hence, we can satisfy the fourth equation by taking $(\pm 1) \cdot \tilde{x}_{1bc} + 0 \cdot x_c = 1$, and the fifth equation by taking $0 \cdot \tilde{x}_{2bc} + 1 \cdot \tilde{x}_c = 1$.

- Vice versa, let us assume that these five interval equations are satisfied. We already know that $x_c \in \{0,1\}$. So, let us consider these two possibilities one by one:

  - If $x_c = 0$, then from the first equation, we conclude that $x_{bc} \in [0,2]$. Since x_{bc} can only take the values 0, 1, 2, and 3, we can thus conclude that it takes the value 0, 1, or 2. To complete the proof of correctness of our representation, we only need to show that x_{bc} cannot be equal to 1. Indeed, if $x_{bc} = 1$, then, due to the second equation, $\tilde{x}_{1bc} =$

$1-1=0$. Since x_c is also 0, the left hand side of the fourth equation is 0 and therefore, cannot be equal to the right-hand side that is 1. So, $x_{bc} \in \{0,2\}$.

- If $x_c = 1$, then from the first equation, we conclude that $x_{bc} \in [1,3]$. Since x_{bc} can only take the values 0, 1, 2, and 3, we can thus conclude that it takes the value 1, 2, or 3. To complete the proof of correctness of our representation, we only need to show that x_{bc} cannot be equal to 2. Indeed, if $x_{bc} = 2$, then, due to the third equation, $\tilde{x}_{2bc} = 2-2=0$. Since x_c is also 0, the left hand side of the fifth equation is 0 and therefore, cannot be equal to the right-hand side that is 1. So, $x_{bc} \in \{1,3\}$.

7°. Let us now show how to represent the relation $a \equiv b \vee c$, assuming that we have already guaranteed that the values of the variables x_a, x_b, and x_c belong to the set $\{0,1\}$, and that the variable x_{bc} takes the desired value.

In this case, we must describe the following conditions:

- If $x_{bc} = 0$, then $x_a = 0$.
- If $x_{bc} \in \{1,2,3\}$, then $x_a = 1$.

We will show that these two conditions are equivalent to the following two linear interval equations: $x_{bc} - x_a = [0,2]$ and $3x_a - x_{bc} = [0,2]$. Let us prove this equivalence.

- If $a \equiv b \vee c$, then, as one can easily check, in all four cases (corresponding to four possible values of the pair (b,c)), we have $x_{bc} - x_a \in [0,2]$ and $3x_a - x_{bc} \in [0,2]$. Therefore, both interval equations are satisfied.
- Vice versa, let us assume that both interval equations are satisfied, and let us show that then $a \equiv b \vee c$. Indeed:
  - If $x_{bc} = 0$, then from the first equation, we conclude that $x_a = x_{bc} - [0,2] \le x_{bc} \le 0$. Since we know that $x_a = 0$ or $x_a = 1$, the only possibility for $x_a \le 0$ is when $x_a = 0$.
  - Let $x_{bc} \ge 1$. From the second equation, we conclude that $x_a = (1/3) \cdot (x_{bc} + [0,2]) \ge (1/3) \cdot x_{bc}$. Since $x_{bc} \ge 1$, we conclude that $x_a \ge (1/3) \cdot x_{bc} \ge (1/3) \cdot 1$. So, $x_a \ne 0$ and therefore, $x_a = 1$.

8°. Let us now show how to represent the relation $a \equiv b\&c$, assuming that we have already guaranteed that the values of the variables x_a, x_b, and x_c belong to the set $\{0,1\}$, and that the variable x_{bc} takes the desired value.

In this case, we must describe the following conditions:

- If $x_{bc} \in \{0,1,2\}$, then $x_a = 0$.
- If $x_{bc} = 3$, then $x_a = 1$.

We will show that these two conditions are equivalent to the following two linear interval equations: $x_{bc} - x_a = [0,2]$ and $x_{bc} - 3x_a = [0,2]$. Let us prove this equivalence.

- If $a \equiv b\&c$, then, as one can easily check, in all four cases (corresponding to four possible values of the pair (b,c)), we have $x_{bc} - x_a \in [0,2]$ and $x_{bc} - 3x_a \in [0,2]$. Therefore, both interval equations are satisfied.
- Vice versa, let us assume that both interval equations are satisfied, and let us show that then $a \equiv b\&c$. Indeed:
  - If $x_{bc} = 3$, then from the first equation, we conclude that $x_a = x_{bc} - [0,2] \geq x_{bc} - 2 = 1$. Since we know that $x_a = 0$ or $x_a = 1$, the only possibility for $x_a \geq 1$ is when $x_a = 1$.
  - Let $x_{bc} \leq 2$. From the second equation, we conclude that $x_a = (1/3) \cdot (x_{bc} - [0,2]) \leq (1/3) \cdot x_{bc}$. Since $x_{bc} \leq 2$, we conclude that $x_a \leq (1/3) \cdot x_{bc} \leq (1/3) \cdot 2 = 2/3$. So, $x_a \neq 1$ and therefore, $x_a = 0$.

9°. For the resulting system of linear interval equations, if the formula is not satisfiable, then f_k is always false and therefore, $x_{f_k} = 0$; if the formula is satisfiable, then x_{f_k} can also take the value 1.

So, if the formula is not satisfiable, the upper bound for the interval $\mathbf{x}_{f_k}$ is 0; if it is satisfiable, this bound is 1. So, if we could compute the exact interval bounds for each component of the solution of 2-sparse interval linear equations, we would we would thus be able to decide whether a given propositional formula is satisfiable or not, and this satisfiability problem is known to be NP-hard. Therefore, solution of 2-sparse interval linear systems is also NP-hard. The reduction is complete, so the theorem is proven.

Example. Let us illustrate the above construction on the example of a propositional formula $F = (z_1 \vee z_2 \vee z_3)\&(z_1 \vee \neg z_2)$, that was reduced to the set of formulas:

$$\tilde{z}_2 \equiv \neg z_2; \quad W_1 \equiv z_1 \vee z_2; \quad V_1 \equiv W_1 \vee z_3; \quad V_2 \equiv z_1 \vee \tilde{z}_2; \quad f_2 \equiv V_1 \& V_2.$$

This formula contains eight Boolean variables: z_1, z_2, z_3, $\tilde{z}_2$, W_1, V_1, V_2, and f_2. So, first, we must introduce the real-number variables x_1, x_2, x_3, $\tilde{x}_3$, x_{W_1}, x_{V_1}, x_{V_2}, and X_{f_2}, and add equations that guarantee that these variables only take the values 0 and 1; for that guarantee, we will need 7 new variables $y_1, \ldots$:

$$[-2,2] \cdot y_1 = [1,1]; \quad y_1 = [-0.5, 0.5]; \quad x_1 - y_1 = 0.5;$$

$$[-2,2] \cdot y_2 = [1,1]; \quad y_2 = [-0.5, 0.5]; \quad x_2 - y_2 = 0.5;$$

$$[-2,2] \cdot y_3 = [1,1]; \quad y_3 = [-0.5, 0.5]; \quad x_3 - y_3 = 0.5;$$

$$[-2,2] \cdot y_{W_1} = [1,1]; \quad y_{W_1} = [-0.5, 0.5]; \quad x_{W_1} - y_{W_1} = 0.5;$$

$$[-2,2] \cdot y_{V_1} = [1,1]; \quad y_{V_1} = [-0.5, 0.5]; \quad x_{V_1} - y_{V_1} = 0.5;$$

$$[-2,2] \cdot y_{V_2} = [1,1]; \quad y_{V_2} = [-0.5, 0.5]; \quad x_{V_2} - y_{V_2} = 0.5;$$

$$[-2,2] \cdot y_{f_2} = [1,1]; \quad y_{f_2} = [-0.5, 0.5]; \quad x_{f_2} - y_{f_2} = 0.5.$$

The equation that corresponds to $\neg z_2$ is:

$$\tilde{x}_2 + x_2 = 1.$$

To describe the $\vee$ and & relations, we will need the new variables, for which, the new equations are:

$$[-2,2] \cdot y_{112} = 1; \quad x_{12} - y_{112} = 0.5; \quad [-2,2] \cdot y_{212} = 1;$$

$$x_{12} - y_{212} = 1.5; \quad [-2,2] \cdot y_{312} = 1; \quad x_{12} - y_{312} = 2.5;$$

$$x_{12} - 2x_1 = [0,1]; \quad x_{12} - x_2 = [0,2]; \quad \tilde{x}_{112} + x_{12} = 1;$$

$$\tilde{x}_{212} + x_{12} = 2; \quad [-1,1] \cdot \tilde{x}_{112} + [0,1] x_2 = 1; \quad [-1,1] \cdot \tilde{x}_{212} + [0,1] \cdot \tilde{x}_2 = 1;$$

$$[-2,2] \cdot y_{1,W_1,3} = 1; \quad x_{W_1,3} - y_{1,W_1,3} = 0.5; \quad [-2,2] \cdot y_{2,W_1,3} = 1;$$

$$x_{W_1,3} - y_{2,W_1,3} = 1.5; \quad [-2,2] \cdot y_{3,W_1,3} = 1; \quad x_{W_1,3} - y_{3,W_1,3} = 2.5;$$

$$x_{W_1,3} - 2x_{W_1} = [0,1]; \quad x_{W_1,3} - x_3 = [0,2]; \quad \tilde{x}_{1,W_1,3} + x_{W_1,3} = 1;$$

$$\tilde{x}_{2,W_1,3} + x_{W_1,3} = 2; \quad [-1,1] \cdot \tilde{x}_{1,W_1,3} + [0,1] x_3 = 1; \quad [-1,1] \cdot \tilde{x}_{2,W_1,3} + [0,1] \cdot \tilde{x}_3 = 1;$$

$$\tilde{x}_3 + x_3 = 1;$$

$$[-2,2]\cdot y_{11\bar{2}}=1;\ x_{1\bar{2}}-y_{11\bar{2}}=0.5;\ [-2,2]\cdot y_{21\bar{2}}=1;$$
$$x_{1\bar{2}}-y_{21\bar{2}}=1.5;\ [-2,2]\cdot y_{31\bar{2}}=1;\ x_{1\bar{2}}-y_{31\bar{2}}=2.5;$$
$$x_{1\bar{2}}-2x_1=[0,1];\ x_{1\bar{2}}-\tilde{x}_2=[0,2];\ \tilde{x}_{11\bar{2}}+x_{1\bar{2}}=1;$$
$$\tilde{x}_{21\bar{2}}+x_{1\bar{2}}=2;\ [-1,1]\cdot\tilde{x}_{11\bar{2}}+[0,1]\tilde{x}_2=1;\ [-1,1]\cdot\tilde{x}_{21\bar{2}}+[0,1]x_2=1;$$

$$[-2,2]\cdot y_{1,V_1,V_2}=1;\ x_{V_1,V_2}-y_{1,V_1,V_2}=0.5;\ [-2,2]\cdot y_{2,V_1,V_2}=1;$$
$$x_{12}-y_{2,V_1,V_2}=1.5;\ [-2,2]\cdot y_{3,V_1,V_2}=1;\ x_{V_1,V_2}-y_{3,V_1,V_2}=2.5;$$
$$x_{V_1,V_2}-2x_{V_1}=[0,1];\ x_{V_1,V_2}-x_{V_2}=[0,2];\ \tilde{x}_{1,V_1,V_2}+x_{V_1,V_2}=1;$$
$$\tilde{x}_{2,V_1,V_2}+x_{V_1,V_2}=2;\ [-1,1]\cdot\tilde{x}_{1,V_1,V_2}+[0,1]x_{V_2}=1;\ [-1,1]\cdot\tilde{x}_{2,V_1,V_2}+[0,1]\cdot\tilde{x}_{V_2}=1$$
$$\tilde{x}_{V_2}+x_{V_2}=1.$$

The actual $\vee$ and & relations are now represented as follows:

$$x_{12}-x_{W_1}=[0,2];\quad 3x_{W_1}-x_{12}=[0,2];$$
$$x_{W_1,3}-x_{V_1}=[0,2];\quad 3x_{V_1}-x_{W_1,3}=[0,2];$$
$$x_{1\bar{2}}-x_{V_2}=[0,2];\quad 3x_{V_2}-x_{1\bar{2}}=[0,2];$$
$$x_{V_1,V_2}-x_{f_2}=[0,2];\quad x_{V_1,V_2}-3x_{f_2}=[0,2].$$

End of example.

Proof of Theorem 12.12 is done by a similar reduction.

Proof of Theorem 12.13. Let us show that solving an arbitrary 2-sparse system can be reduced to solving an almost numerical 2-sparse system.

Indeed, each equation of the general 2-sparse system has the form

$$\mathbf{a}\cdot x+\mathbf{b}\cdot y=\mathbf{c},$$

where x and y are unknowns. To reduce this problem to an almost numerical system, we introduce two new variables x' and y', and replace the original equation with the following three equations:

$$\mathbf{a}\cdot x-x'=0;\quad \mathbf{b}\cdot y-y'=0;\quad x'+y'=\mathbf{c}.$$

This system of three equations is clearly equivalent to the original equation, and all three equations from this system have at most one non-numerical interval coefficient. Therefore, combining these systems together, we get an almost equivalent 2-sparse system that is equivalent to the original one.

So, if we could solve 2-sparse almost numerical interval linear systems in polynomial time, we would be able to solve all 2-sparse interval linear systems in polynomial time. But this second problem is already proven to be NP-hard (in Theorem 12.11). Therefore, our problem is also NP-hard. The theorem is proven.

Proof of Theorem 12.14 is done by a similar reduction.

Proof of Theorem 12.15. To prove our result, we will start with the known result (mentioned above; see, e.g., Theorems 11.7–9) that the problem of solving regular linear interval systems with rational interval coefficients is NP-hard. We will then describe a general transformation of such systems into systems in which each interval coefficient is $[0,0]$, $[1,1]$, or $[0,1]$. This transformation will be done in several steps.

On some steps, we will introduce new variables in addition to the variables $y_1,\ldots,y_m$ used in the original system. We will make sure that for each of the original variables y_j, the set of possible values of y_j will remain the same. Thus, we will be sure that the resulting final system has exactly the same bounds for $y_1,\ldots,y_m$ as the original system. Thus, if there is a polynomial-time algorithm that can find these bounds for an arbitrary linear interval system with coefficients $[0,0]$, $[1,1]$, or $[0,1]$, then by applying this algorithm to the transformation result, we would be able to compute the bounds for the original system, and this computation is an NP-hard problem. Thus, the problem of computing solution bounds for linear interval systems with coefficients $[0,0]$, $[1,1]$, or $[0,1]$ is also NP-hard.

We will also make sure that each transformation step preserves the number of non-numerical interval coefficients (i.e., coefficients that are not of the type $[a,a]$), and that when we choose some values inside these intervals, all the variables of the resulting system are uniquely determined. In other words, we will make sure that the systems obtained on each transformation step are *regular*. Thus, the problem of computing solution bounds for *regular* linear interval systems with coefficients $[0,0]$, $[1,1]$, or $[0,1]$ will be proven to be also NP-hard.

1) The first transformation simplifies the right-hand side of the linear equation. Namely, we introduce a new variable y_0, replace each of n equations $\sum \mathbf{a}_{ij} \cdot y_j = \mathbf{b}_i$ by an equation $\mathbf{a}_{i0} \cdot y_0 + \sum \mathbf{a}_{ij} \cdot y_j = 0$ with $\mathbf{a}_{i0} = -\mathbf{b}_i$, and add a new equation $y_0 = 1$.

If the vector $(y_0, \ldots, y_m)$ belongs to the solution set of the new system, then $y_0 = 1$, and thus, the values $y_1, \ldots, y_m$ satisfy the original equations. Vice versa, if the values $y_1, \ldots, y_m$ satisfy the original equations, then for $y_0 = 1$, the transformed equations also hold. Thus, for each of the variables $y_1, \ldots, y_m$, the bounds are the same for the original and for the transformed equations.

If we fix the values $a_{ij} \in \mathbf{a}_{ij}$, then, since the original system was regular, the values $y_1, \ldots, y_m$ would be uniquely determined. The only missing value y_0 can be now uniquely determined from the equation $y_0 = 1$. Thus, the new system is regular.

Hence, this transformation preserves both the bounds on y_j and the uniqueness (regularity) property.

2) For every i, all the bounds of all the coefficients $\mathbf{a}_{ij}$ in i-th equation are rational numbers (i.e., fractions). If we multiply all coefficients of this equation by the least common denominator of the corresponding fractions, we get a new equation in which all bounds of all coefficients are integers.

In the second transformation, we apply this procedure to all equations. As a result, we get an equivalent (thus regular) linear interval system, in which all bounds of all interval coefficients are integers.

In particular, the only equation with a the non-zero right-hand side (namely, the equation $y_0 = 1$) stays the same.

3) On the third step, for each $j = 0, \ldots, m$, we introduce two new variables n_j and p_j (here, n stands for *n*egative, and p for *p*ositive).

For each j, we add two new equations $y_j + n_j = 0$ and $n_j + p_j = 0$. In each original equation, we replace each term $[\underline{a}_{ij}, \overline{a}_{ij}] \cdot y_j$ by one of the following six expressions:

- If the interval $[\underline{a}_{ij}, \overline{a}_{ij}]$ is non-degenerate (i.e., if $\underline{a}_{ij} < \overline{a}_{ij}$), then we use one of the following three expressions:
  - If $\underline{a}_{ij} < 0$, then we replace the term $[\underline{a}_{ij}, \overline{a}_{ij}] \cdot y_j$ by the sum $|\underline{a}_{ij}| \cdot n_j + [0, \overline{a}_{ij} - \underline{a}_{ij}] \cdot y_j$.
  - If $\underline{a}_{ij} = 0$, then we leave the term $[\underline{a}_{ij}, \overline{a}_{ij}] \cdot y_j$ $(= [0, \overline{a}_{ij}] \cdot y_j)$ unchanged.
  - If $\underline{a}_{ij} > 0$, then we replace this term by the sum $\underline{a}_{ij} \cdot p_j + [0, \overline{a}_{ij} - \underline{a}_{ij}] \cdot y_j$.
- If the interval $[\underline{a}_{ij}, \overline{a}_{ij}]$ is degenerate (i.e., if $\underline{a}_{ij} = \overline{a}_{ij}$), then we use one of the following three expressions:
  - If $\underline{a}_{ij} < 0$, then we replace the term $[\underline{a}_{ij}, \overline{a}_{ij}] \cdot y_j$ by $|\underline{a}_{ij}| \cdot n_j$.
  - If $\underline{a}_{ij} = 0$, then we delete the term $[\underline{a}_{ij}, \overline{a}_{ij}] \cdot y_j$ (because it is equal to 0).
  - If $\underline{a}_{ij} > 0$, then we replace the term $[\underline{a}_{ij}, \overline{a}_{ij}] \cdot y_j$ by $\underline{a}_{ij} \cdot p_j$.

As a result, we get a linear interval system in which each interval coefficient is either a positive integer, or an interval of the type $[0, z]$ for some positive integer z.

Let us show that this transformation does not change the bounds of the solution set for $y_1, \ldots, y_m$, and preserves regularity.

Indeed, if $y_0, y_1, \ldots, y_m$ is a solution of the system that we had before this step, then, by adding $n_j = -y_j$ and $p_j = y_j$, we get a solution of the transformed system. Vice versa, if we have a solution of the transformed system, then from the new equations $y_j + n_j = 0$ and $n_j + p_j = 0$, we conclude that $n_j = -y_j$, and $p_j = -n_j = y_j$. If $\underline{a}_{ij} < 0$ and $\underline{a}_{ij} < \overline{a}_{ij}$, then, for every $z_{ij} \in [0, \overline{a}_{ij} - \underline{a}_{ij}]$, we have

$$|\underline{a}_{ij}| \cdot n_j + z_{ij} \cdot y_j = \underline{a}_{ij} \cdot y_j + z_{ij} \cdot y_j = a_{ij} \cdot y_j,$$

where $a_{ij} = \underline{a}_{ij} + z_{ij} \in \underline{a}_{ij} + [0, \overline{a}_{ij} - \underline{a}_{ij}] = [\underline{a}_{ij}, \overline{a}_{ij}]$. Thus, $y_0 = 1, y_1, \ldots, y_m$ form a solution of the system that we had before this step. (A similar proof holds in all five other cases: $\underline{a}_{ij} = 0 < \overline{a}_{ij}$, $0 < \underline{a}_{ij} < \overline{a}_{ij}$, $\underline{a}_{ij} = \overline{a}_{ij} < 0$, $\underline{a}_{ij} = \overline{a}_{ij} = 0$, and $\underline{a}_{ij} = \overline{a}_{ij} > 0$).

Since the system that we had before this step was regular, for each choice of coefficients in their intervals, the variables $y_0 (= 1), y_1, \ldots, y_m$ are uniquely determined; the new variables n_j and p_j are uniquely determined from y_j. Thus, the transformed system is also regular.

4) Let us now describe the fourth (and final) transformation step. To simplify the description of this step, let us first rename the variables y_j, n_j, and p_j into $y_j^{[0]}$, $n_j^{[0]}$, and $p_j^{[0]}$.

Let N denote the largest integer bound in the system obtained after the third step, let d denote the number of binary digits in the binary representation of N (i.e., $d = \lfloor \log_2(N) \rfloor$), and let P denote the set of all pairs (i, j) for which i-th equation of the system that we had before this step contains the term $[0, z_{ij}] \cdot y_j$. The transformed system will have the following variables:

- For each $j = 0, \ldots, m$, and for each $k = 0, \ldots, d$, variables $y_j^{[k]}$, $n_j^{[k]}$, and $p_j^{[k]}$.
- For each pair $(i, j) \in P$, and for each $k = 0, \ldots, d$, variables $y_{ij}^{[k]}$, $n_{ij}^{[k]}$, and $p_{ij}^{[k]}$.

This system will consist of the following equations:

- Equations $y_j^{[k]} + n_j^{[k]} = 0$ and $p_j^{[k]} + n_j^{[k]} = 0$ (for all $j \le m$).
- Equations $y_{ij}^{[k]} + n_{ij}^{[k]} = 0$ and $p_{ij}^{[k]} + n_{ij}^{[k]} = 0$ (for all $(i, j) \in I$).
- Equations $y_j^{[k]} + p_j^{[k]} + n_j^{[k+1]} = 0$ (for all $j \le m$ and $k < d$).
- Equations $y_{ij}^{[k]} + p_{ij}^{[k]} + n_{ij}^{[k+1]} = 0$ (for all $(i, j) \in I$ and $k < d$).
- Equations $n_{ij}^{[0]} + [0, 1] \cdot y_j^{[0]} = 0$ (for all $(i, j) \in I$).
- Equations that are obtained from the equations of the system that we had before this step by the following replacement:
  - Each term of the type $z \cdot n_j$ is replaced by the sum of the terms $n_j^{[k]}$ for all places k on which the binary expansion of z has 1 (i.e., for which $\varepsilon_k = 1$ in the binary expansion $z = \sum \varepsilon_k \cdot 2^k$).
  - Each term of the type $z \cdot p_j$ is replaced by the sum of the terms $p_j^{[k]}$ for all places k on which the binary expansion of z has 1.
  - Each term of the type $[0, z_{ij}] \cdot y_j$ is replaced by the sum of the terms $y_{ij}^{[k]}$ for all places k on which the binary expansion of z_{ij} has 1.

As a result, we get a linear interval system in which each interval coefficient is either 0, or 1, or an interval $[0, 1]$.

Let us show that this transformation does not change the bounds of the solution set for $y_1, \ldots, y_m$, and preserves regularity.

Indeed, if the values y_j, n_j, and p_j form a solution of the system that we had before this step, a solution that corresponds to the coefficients $c_{ij} \in [0, z_{ij}]$, then, as one can check, the values $y_j^{[k]} = p_j^{[k]} = 2^k \cdot y_j$, $n_j^{[k]} = -2^k \cdot y_j$, $c'_{ij} = c_{ij}/z_{ij}$, $y_{ij}^{[k]} = p_{ij}^{[k]} = 2^k \cdot c'_{ij} \cdot y_j$, and $n_{ij}^{[k]} = -2^k \cdot c'_{ij} \cdot y_j$ form a solution of the transformed system, for $c'_{ij} \in [0, 1]$.

Indeed, e.g., for this choice of variables, the sum of the terms $p_j^{[k]}$ for all places k on which the binary expansion of z has 1, is equal to the sum of the terms $2^k y_j$, i.e., to the product of y_j and the sum of the terms 2^k that correspond to all places k on which the binary expansion of z has 1. This sum is exactly the binary expansion of z, and hence, the sum is equal to $z \cdot y_j$.

Vice versa, if we have a solution of the transformed system, for $c_{ij} \in [0, 1]$, then:

- From the equations $y_j^{[k]} + n_j^{[k]} = 0$ and $p_j^{[k]} + n_j^{[k]} = 0$, we conclude that $n_j^{[k]} = -y_j^{[k]} = 0$ and $p_j^{[k]} = -n_j^{[k]} = y_j^{[k]}$.
- From the equations $y_j^{[k]} + p_j^{[k]} + n_j^{[k+1]} = 0$, we can now conclude that $n_j^{[k+1]} = -2y_j^{[k]}$, hence, $y_j^{[k+1]} = -n_j^{[k+1]} = 2y_j^{[k]}$. By induction over k, we can conclude that $y_j^{[k]} = 2^k \cdot y_j^{[0]}$, and hence, that $n_j^{[k]} = -2^k \cdot y_j^{[0]}$ and $p_j^{[k]} = 2^k \cdot y_j^{[0]}$.
- Similarly, we can conclude that $y_{ij}^{[k]} = 2^k \cdot y_{ij}^{[0]}$, $n_{ij}^{[k]} = -2^k \cdot y_{ij}^{[0]}$, and $p_{ij}^{[k]} = 2^k \cdot y_{ij}^{[0]}$.
- From the equation $n_{ij}^{[0]} + [0, 1]y_j^{[0]} = 0$, and from our assumption that the coefficient is equal to $c_{ij} \in [0, 1]$, we conclude that $n_{ij}^{[0]} = -c_{ij} \cdot y_j^{[0]}$, and therefore, that $p_{ij}^{[0]} = -n_{ij}^{[0]} = c_{ij} \cdot y_j^{[0]}$.

Thus, the sums of the terms of the type $n_j^{[k]}$ and $y_{ij}^{[k]}$ reduce to $z \cdot n_j$ and $z \cdot y_{ij} = (z \cdot c_{ij}) \cdot y_j$, where $z \cdot c_{ij} \in z \cdot [0,1] = [0, z]$. Thus, the values $y_0 = 1, y_1, \ldots, y_m$ form a solution of the system that we had before this step.

Since the system that we had before this step was regular, for each choice of coefficients in their intervals, the variables $y_0 (= 1), y_1, \ldots, y_m$ are uniquely determined; the new variables $y_j^{[k]}$, $n_j^{[k]}$, and $p_j^{[k]}$ are uniquely determined by the values y_j. For fixed c_{ij}, the values of the variables $y_{ij}^{[k]}$, $n_{ij}^{[k]}$, and $p_{ij}^{[k]}$ are also uniquely determined by the values y_j. Thus, the transformed system is also regular. The theorem is proven.

Proof of Theorem 12.16 is done by a similar reduction.

Proof of Theorem 12.18. This algorithm is based on the ideas originally proposed in Suvorov [406] and Dantsin [83, 84].

To describe this algorithm, we will describe an *auxiliary algorithm* that would check, for every $j = 1, \ldots, n$, and for every possible value $y_j \in S$ of the variables y_j, whether this particular value can be extended to a possible S-solution y of a given linear system (i.e., to a solution all components of which belong to the set S). If we have such an auxiliary algorithm, then, to find the desired max $\overline{y}_j$, we will simply enumerate all possible values of y_j, check which of these values are possible, and take the largest of these possible values. Similarly, we will be able to compute the smallest of the possible values of y_j, i.e., $\underline{y}_j$.

The resulting algorithm for computing $\overline{y}_j$ and $\underline{y}_j$ consists of applying the auxiliary algorithm s times, where s is the number of elements in the set S, and in finding the largest and the smallest of the resulting values y_j. This finding can be easily accomplished, if on each step, we keep track of the largest-so-far and smallest-so-far values, and for every new possible y_j, compare this new value y_j with these largest-so-far and smallest-so-far values. Therefore, if the running time of the auxiliary algorithm is bounded by a polynomial $P(n)$ of the size n of the problem, then the running time of the resulting algorithm for computing $\underline{y}_j$ and $\overline{y}_j$ is also polynomial time: it takes $\le s \cdot P(n) + 2n$ steps, where $2n$ steps are needed for finding the largest and the smallest elements.

So, to prove our theorem, it is sufficient to design a polynomial-time *auxiliary algorithm*. This algorithm will work as follows. The fact that the system is w-band means that for every i, the i-th equation only contains variables y_j with $j = i - (w-1), i - (w-2), \ldots, i-1, i, i+1, \ldots, i + (w-1)$. In particular:

- the first equation contains only variables $y_1, \ldots, y_w$;
- the second equation contains only variables $y_1, y_2, \ldots, y_{w+1}$;
- $\ldots$;
- $(w-1)$-st equation contains only variables $y_1, y_2, \ldots, y_{2w-1}$.

On the first stage of the proposed auxiliary algorithm, we will try, for each of s^{2w-1} possible S-tuples $(y_1, \ldots, y_{2w-1})$, whether this tuple is a solution to the system of first $w-1$ linear equations (if the desired variable y_j is among one of these $y_1, \ldots, y_{2w-1}$, then we only consider those tuples for which y_j takes the given value). To check each tuple, we will try all possible S-values $a_{ij} \in \mathbf{a}_{ij}$ and $b_i \in \mathbf{b}_i$ for all $i \le w-1$ and $j \le 2w-1$. Let us describe how many coefficients there are:

The first equation has w coefficients a_{ij}, the second one has $w+1$ coefficients a_{ij}, etc. Totally, there are $w + (w+1) + \ldots + (2w-1)$ coefficients a_{ij}, i.e.,

$$w \cdot (w-1) + [1 + \ldots + w] = w \cdot (w-1) + \frac{w \cdot (w+1)}{2} = \frac{w \cdot (3w-1)}{2}$$

coefficients. In addition to that, there are w coefficients b_i, $i \le w$. In total, there are $w \cdot (5w-1)/2$ coefficients. Each of these coefficients is an element of the set S and therefore, it can take no more than s different values. Hence, the entire set of coefficients can take $\le s^{w \cdot (5w-1)/2}$ different values. For each set of coefficients a_{ij} and b_i, and for each set $(y_1, \ldots, y_{2w-1})$, checking whether all of $k+1$ equations are satisfied (i.e., whether $\sum a_{ij} \cdot y_j = b_i$) takes $\le w^2$ straightforward multiplications, additions, and comparisons. To check this for all possible sets of coefficients and for all possible S-tuples, we must, therefore, apply $\le w^2 \cdot s^{2w-1} \cdot s^{w \cdot (5w-1)/2}$ computational steps. This may be a huge number, but s and w are both constants independent of the size n of the entire system. Thus, the number of computations spent on this first step is bounded by a constant.

Let us now describe the further stages of the proposed algorithm. Totally, there will be $n - (w-1)$ of them, and these stages will be marked by numbers $w, w+1, \ldots, n$. On each stage i, we will find the set of all tuples $(y_{i-(w-1)}, \ldots, y_i, \ldots, y_{i+(w-1)})$ that can be extended to an S-solution of the system formed by first i equations. To find such a set, let us start with a set of possible values $(y_{i-w}, \ldots, y_{i-1}, \ldots, y_{i+w-2})$ obtained on the previous stage. By induction assumption, this set contains exactly S-tuples that satisfy the first

$i-1$ equations. So, to move to a new set, we must take the i-th equation into consideration.

The i-th equation contains $2w$ coefficients ($2w-1$ coefficients a_{ij} and one coefficient b_i). Each of these coefficients takes $\le s$ values, and therefore, there are no more than s^{2w} possible sets of these coefficients. So, to describe the new set, we can, for each tuple from the old set, do the following:

- For each of s possible values of $y_{i+(w-1)}$, and for each of s^{2w} possible values of the coefficients a_{ij}, b_i, of i-th equation, we check whether the i-th equation $\sum a_{ij} \cdot y_j = b_i$ is satisfied.
- If it is satisfied, then we add the resulting tuple $(y_{i-(w-1)}, \ldots, y_i, \ldots, y_{i+(w-1)})$ to the new set.

For each of $\le s^{2w}$ possible S-tuples, on i-th stage, we need $\le s \cdot s^{2w}$ checks, and each check takes $\le 2 \cdot (2w-1)$ operations ($2w-1$ multiplications $a_{ij} \cdot y_j$ and $2(w-2)$ additions). Hence, the total number of operations on each stage is bounded by $s^{2w-1} \cdot s^{2w} \cdot (4w-2)$. This upper bound depends only on s and w, and it is independent on n. Hence, from the viewpoint of dependency on n, this is simply a constant.

Hence, the total number of computational steps on all n stages is bounded by constant$\cdot n$. Hence, we have described the desired linear-time algorithm. The theorem is proven.

Proof of Theorem 12.17 is done by a similar reduction.

Proof of Theorem 12.19. We want to find the values y_j for which $\underline{b}_i \le \sum a_{ij} \cdot y_j \le \overline{b}_i$ for all $a_{ij} \in [\underline{a}_{ij}, \overline{a}_{ij}]$. For every j, the largest possible value $\overline{t}_{ij}$ of each linear term $a_{ij} \cdot y_j$ is attained on one of the endpoints of the interval $\mathbf{a}_{ij}$, and a similar property is true for the smallest possible values $\underline{t}_{ij}$ of these terms. Therefore, the above inequality is equivalent to the following system of linear inequalities, with new variables $\underline{t}_{ij}$ and $\overline{t}_{ij}$: $\underline{t}_{ij} \le \underline{a}_{ij} \cdot y_j \le \overline{t}_{ij}$, $\underline{t}_{ij} \le \overline{a}_{ij} \cdot y_j \le \overline{t}_{ij}$, $\underline{b}_i \le \sum \underline{t}_{ij}$, and $\sum \overline{t}_{ij} \le \overline{b}_i$. Thus, we can use known polynomial-time algorithms for solving linear programming problems (i.e., for solving systems of linear inequalities; see, e.g., Karmarkar [169]) to solve our problem. The theorem is proven.

Proof of Theorem 12.20 is similar: e.g., to compute $\underline{y}_j$, we solve the linear problem problem of minimizing y_j under the linear inequalities described in the proof of Theorem 12.19.

Proof of Theorem 12.21. To prove NP-hardness of our problem, we will reduce *PARTITION* to it. In *PARTITION*, we are given a sequence of integers $s_1, \ldots, s_m$, and we want to check whether there exist integers $y_j \in \{-1, 1\}$ for which $\sum s_j \cdot y_j = 0$. For each instance of *PARTITION*, let us construct an interval linear system with $3m$ variables $y_1, \ldots, y_m, z_1, \ldots, z_m, t_1, \ldots, t_m$, and the following $5m + 1$ equations:

- $[-1, 1] \cdot y_j = [1, 1]; j = 1, \ldots, m;$
- $[0, 1] \cdot z_j = [1, 1]; j = 1, \ldots, m;$
- $[0, 1] \cdot t_j = [1, 1]; j = 1, \ldots, m;$
- $y_j + z_j = 2; j = 1, \ldots, m;$
- $y_j - t_j = -2; j = 1, \ldots, m;$
- $s_1 \cdot y_1 + \ldots + s_m \cdot y_m = 0.$

If $y_1, \ldots, y_m$ is a solution to the given instance, then we can take $z_j = 2 - y_j$, $t_j = y_j + 2$, and get an element of the controlled set.

Vice versa, let us assume that the controlled solution set is non-empty, and that $y_1, \ldots$ is an element of this set. Then:

- From the first equation, we can then conclude that there exists an element $a \in [-1, 1]$ for which $a \cdot y_j = 1$. Therefore, $y_j = 1/a$, and hence, $|y_j| \geq 1$, i.e., $y_j \geq 1$ or $y_j \leq -1$.
- From the second equation, we can similarly conclude that $z_j = 1/a$ for some a between 0 and 1, i.e., that $z_j \geq 1$. The fourth equation implies that $y_j = 2 - z_j$ and therefore, that $y_j \leq 2 - 1 = 1$.
- Similarly, from the third and the fifth equations, we conclude that $y_j \geq -1$.
- Since we already know that y_j is either ≥ 1 or ≤ -1, we thus conclude that $y_j \in \{-1, 1\}$.
- In this case, the last equation shows that the values y_j form a solution to the given instance of *PARTITION*.

So, for the above interval linear system, the controlled solution set is non-empty if and only if the original instance of *PARTITION* has a solution. The reduction is proven, and so, checking non-emptiness is indeed NP-hard. The theorem is proven.

Proof of Theorem 12.22. We can reduce *PARTITION* to this problem as follows: for each particular case of *PARTITION*, construct the interval linear system as above. Our problem is to produce something that for the case of non-empty controlled solution set would be its element. For the above system, it is easy to check whether this "something" is an element of the controlled solution set: it is sufficient to check whether $\sum s_k \cdot y_k = 0$ and $y_j \in \{-1, 1\}$. By checking this, we would be able to check whether a given instance of *PARTITION* has a solution. Reduction proves that our problem is indeed NP-hard.

13

PHYSICAL COROLLARY: PREDICTION IS NOT ALWAYS POSSIBLE, EVEN FOR LINEAR SYSTEMS WITH KNOWN DYNAMICS

In the previous chapter, we proved that solving interval linear systems is computationally intractable. In this chapter, we describe a practical consequence of this result: that even for linear systems with known dynamics, prediction can be computationally intractable, i.e., in practical terms, not always possible.

This corollary may be of interest to physicists, meteorologists, etc.

13.1. Informal Introduction to the Problem and Motivations for the Following Definitions

One of the main objectives of physics is to predict the future behavior of different objects and systems. As a basis for this prediction, we can use the results of current and past measurements. In this chapter, we will show that such a prediction is not always possible, even if the dynamics of the object is described by known linear equations.

Before we start the mathematical formulations, let us clarify what we mean by the phrase "prediction is not always possible". If the measurements are *absolutely precise*, then, since we assumed that we know the exact dynamics of the object, we can predict the future values of all its parameters exactly. In real life, however, measurements are *never 100% precise*. As a result, each measurement leads not to the *exact* value of the measured quantity, but to the *interval* of possible values. Since we do not know the exact values of the past characteristics, we cannot predict the exact values of the future ones; even for known dynamics, there is an *interval* of possible values of each future characteristic. The question is: can we compute (the bounds of) this interval? and, if we can, can we compute these bounds *fast enough*?

We emphasized the words *fast enough*, because for prediction problems, computations only make sense if they are finished *before* the event that we are actually trying to predict.

> For *example*, if we have an algorithm that predicts the next day's weather, but that takes a year to finish its computations, then these computations are of no use: by the time when the computations are over, we will have already recorded the weather that we were trying to predict.

For simplicity, we will consider systems with *discrete time*, i.e., we only consider moments of time $1, 2, \ldots$ Let $\vec{s}(t) = (s_1(t), \ldots, s_n(t))$ be the state of the system at a moment t. We will show that prediction problem is NP-hard even for *linear* systems, i.e., for systems with linear dynamics, in which the state $\vec{s}(t+1)$ in the next moment of time is a linear function of the state $\vec{s}(t)$ in the previous moment of time: $\vec{s}(t+1) = \vec{d} + D\vec{s}(t)$ for some vector $\vec{d}$ and for some matrix D. Moreover, we will show that prediction is hard even when we know the dynamics, i.e., when we know both the vector $\vec{d}$ and the matrix D *exactly*.

Predictions are based on the results of measurements. The very fact that $\vec{s}$ is a *state* means that every measured characteristic m of the system is uniquely determined by this state, i.e., in mathematical terms, that it is a *function* $f(s_1, \ldots, s_n)$ of the values s_i of the physical quantities that form the state: $m(t) = f(s_1(t), \ldots, s_n(t))$. We will show that the prediction problem is difficult even for the simplest case when every quantity $m_k(t)$ measured at each moment of time t linearly depends on the state:

$$m_k(t) = b_k + m_{k1}(t) \cdot s_1(t) + \ldots + m_{kn}(t) \cdot s_n(t).$$

We can assume, for simplicity, that the measuring instruments are calibrated in such a way that there is no bias, i.e., that $b_k = 0$. Other parameters of this dependency (i.e., values $m_{ki}(t)$) are, usually, not exactly known; as a result, we only know *intervals* $\mathbf{m}_{ki}(t)$ of possible values of $m_{ki}(t)$. Similarly, since measurements are never absolutely precise, in general, we only know an *interval* $\mathbf{m}_k(t)$ of possible values of the measured quantity $m_k(t)$.

We are now ready to formalize what it means, based on the results of the measurements performed at the moments of time $1, \ldots, T$, to predict the results of measurements at the future moment of time $T + 1$.

13.2. Definitions and the Main Result

Definition 13.1. *By a prediction problem, we mean the tuple*

$$\langle n, \vec{d}, D, T, n_1, \ldots, n_T, \mathbf{M}(1), \ldots, \mathbf{M}(T), \mathbf{M}(T+1), \mathbf{m}(1), \ldots, \mathbf{m}(T)\rangle,$$

where:

- n *is a positive integer called dimension of the system (or simply dimension);*
- $\vec{d}$ *is an* $n-$*dimensional vector, and* D *is an* $n \times n$ *matrix (both, with rational components); the pair* $\langle \vec{d}, D\rangle$ *is called dynamics;*
- T *is a positive integer called the time period;*
- n_t, $1 \le t \le T$, *are positive integers; the integer* n_t *is called the number of quantities measured at time* t*;*
- $\mathbf{M}(t)$, $1 \le t \le T+1$, *are interval matrices of sizes* $n_t \times n$*; here, we denoted* $n_{T+1} = 1$ *(so that the matrix* $\mathbf{M}(T+1)$ *is of size* $1 \times n$*, i.e., actually, an* n*-dimensional interval vector); these* $T+1$ *matrices are called measuring procedures; and*
- $\mathbf{m}(t)$, $1 \le t \le T$, *are interval vectors of size* n_t *called measurement results.*

Definition 13.2. *Let* $\delta > 0$ *be a positive rational number. We say that a prediction problem is based on* δ*-accurate measurements if all the intervals components of the matrices* $\mathbf{M}(t)$ *and the vectors* $\mathbf{m}(t)$ *are absolutely* δ*-accurate.*

Comemnt. Recall that an interval $\mathbf{a} = [\underline{a}, \overline{a}]$ is called *absolutely* δ*-accurate* if its *radius* $(1/2) \cdot (\overline{a} - \underline{a})$ does not exceed δ.

Definition 13.3. *We say that a real number p is a possible prediction for a given prediction problem*

$$\langle n, \vec{d}, D, T, n_1, \ldots, n_T, \mathbf{M}(1), \ldots, \mathbf{M}(T), \mathbf{M}(T+1), \mathbf{m}(1), \ldots, \mathbf{m}(T)\rangle,$$

if there exist vectors $\vec{s}(1), \ldots, \vec{s}(T), \vec{s}(T+1)$, vectors $\vec{m}(t) \in \mathbf{m}(t)$, and matrices $M(t) \in \mathbf{M}(t)$ for which:

- $\vec{s}(t+1) = \vec{d} + D\vec{s}(t)$ *for* $t = 1, \ldots, T$;
- $M(t)\vec{s}(t) = \vec{m}(t)$ *for* $t = 1, \ldots, T$; *and*
- $M(T+1)\vec{s}(T+1) = p$.

The smallest and the largest of the possible predictions will be denoted by $\underline{p}$ and $\overline{p}$.

Definition 13.4. *Let $\varepsilon > 0$. By a problem of computing an ε-accurate prediction, we mean the following problem:*

> *GIVEN a prediction problem.*
>
> *COMPUTE rational numbers $\widetilde{\underline{p}}$ and $\widetilde{\overline{p}}$ that are ε-close to, correspondingly, the smallest $\underline{p}$ and the largest $\overline{p}$ possible predictions (i.e., for which $|\widetilde{\underline{p}} - \underline{p}| \leq \varepsilon$ and $|\widetilde{\overline{p}} - \overline{p}| \leq \varepsilon$).*

Comment. Each possible prediction is a solution of the following interval linear system:

$$\vec{s}(t+1) - D\vec{s}(t) = \vec{d}, \ 1 \leq t \leq T;$$
$$\mathbf{M}(t)\vec{s}(t) = \mathbf{m}(t), \ 1 \leq t \leq T;$$
$$\mathbf{M}(T+1)\vec{s}(T+1) - p = 0.$$

We know that *in general*, the problem of *solving interval linear systems is NP-hard*. We will show that it *remains NP-hard if we only consider linear systems that correspond to prediction problems*:

Theorem 13.1. *For every positive integer T, and for arbitrary positive rational numbers $\varepsilon > 0$ and $\delta > 0$, the problem of computing an ε-accurate prediction based on δ-accurate measurements is NP-hard.*

13.3. Auxiliary Result

In our definitions, we assumed the possibility that *different* measuring procedures are used for different moments of time t. What if we only consider *time-invariant* (*stationary*) situations, in which at all moments of time, we use exactly the same set of measuring instruments? It turns out that in this case, the problem is still NP-hard:

Definition 13.5. *By a prediction problem based on stationary measurements, we mean the tuple*

$$\langle n, \vec{d}, D, T, n_1, \mathbf{M}, \mathbf{M}(T+1), \mathbf{m}(1), \ldots, \mathbf{m}(T)\rangle,$$

where:

- n *is a positive integer called dimension of the system (or simply dimension);*
- $\vec{d}$ *is an* $n-$*dimensional vector, and* D *is an* $n \times n$ *matrix (both, with rational components); the pair* $\langle \vec{d}, D\rangle$ *is called dynamics;*
- T *is a positive integer called the time period;*
- n_1 *is a positive integer; it is called the number of measured quantities;*
- $\mathbf{M}$ *is an interval matrix of size* $n_1 \times n$ *called a measuring procedure;* $\mathbf{M}(T+1)$ *is an* n*-dimensional row of this matrix;*
- $\mathbf{m}(t)$, $1 \le t \le T$, *are interval vectors called measurement results.*

Definition 13.6. *We say that a real number* p *is a possible prediction for a given prediction problem*

$$\langle n, \vec{d}, D, T, n_1, \mathbf{M}, \mathbf{M}(T+1), \mathbf{m}(1), \ldots, \mathbf{m}(T)\rangle$$

if there exist vectors $\vec{s}(1), \ldots, \vec{s}(T), \vec{s}(T+1)$, *vectors* $\vec{m}(t) \in \mathbf{m}(t)$, *and a matrix* $M \in \mathbf{M}$ *for which:*

- $\vec{s}(t+1) = \vec{d} + D\vec{s}(t)$ *for* $t = 1, \ldots, T$;
- $M\vec{s}(t) = \vec{m}(t)$ *for* $t = 1, \ldots, T$; *and*
- $M(T+1)\vec{s}(T+1) = p$, *where* $M(T+1)$ *is the corresponding row of the matrix* M.

Definition 13.7. *Let $\varepsilon > 0$. By a problem of computing an ε-accurate prediction based on stationary measurements,* we mean the following problem:

> *GIVEN* a prediction problem with stationary measurements.
>
> *COMPUTE rational numbers $\widetilde{\underline{p}}$ and $\widetilde{\overline{p}}$ that are ε-close to, correspondingly, the smallest possible prediction $\underline{p}$ and the largest possible prediction $\overline{p}$, (i.e., for which $|\widetilde{\underline{p}} - \underline{p}| \le \varepsilon$ and $|\widetilde{\overline{p}} - \overline{p}| \le \varepsilon$).*

Comment. This problem is ***not exactly*** the problem of solving a system of ***interval linear equations***, because we require that the matrix $M \in \mathbf{M}$ be the same for all T equations (i.e., that the coeficients are not independent). However, it is ***still NP-hard***:

Theorem 13.2. *For every positive integer T, and for arbitrary positive rational numbers $\varepsilon > 0$ and $\delta > 0$, the problem of computing an ε-accurate prediction based on δ-accurate stationary measurements is NP-hard.*

Proofs

Proof of Theorem 13.1. To prove this result, we will reduce the *PARTITION* problem (known to be NP-hard) to our prediction problem. In the *PARTITION* problem, we are given a sequence of integers $v_1, \ldots, v_q$, and we must check whether there exist values $x_1, \ldots, x_q$ for which $x_i \in \{-1, 1\}$ and

$$v_1 \cdot x_1 + \ldots + v_q \cdot x_q = 0.$$

Let us take $\delta_0 = \min\{\delta, \delta/(2\varepsilon)\}$, and let us form the following prediction problem: $n = q + 1$, $\vec{d} = \vec{0} = (0, \ldots, 0)$ is the vector consisting of all 0's, D is the unit matrix ($D = I$), $n_1 = \ldots = n_T = 2n + 1 (= 2q + 3)$. Let us first define, for $t = 1, \ldots, T$, the interval matrices $\mathbf{M}(t)$ (with elements $\mathbf{m}_{ij}(t)$) and interval vectors $\mathbf{m}(t)$ (with elements $\mathbf{m}_i(t)$):

- $\mathbf{m}_{ii}(t) = [-\delta_0, \delta_0]$; $\mathbf{m}_{ij}(t) = [0, 0]$ for all $j \neq i$;
- $\mathbf{m}_{i+n,i}(t) = [\delta_0, \delta_0]$; $\mathbf{m}_{i+n,j}(t) = [0, 0]$ for all $j \neq i$;
- $\mathbf{m}_i(t) = [2\varepsilon \cdot \delta_0, 2\varepsilon \cdot \delta_0]$;
- $\mathbf{m}_{i+n}(t) = [-2\varepsilon \cdot \delta_0, 2\varepsilon \cdot \delta_0]$;

We also take $\mathbf{m}_{2n+1,j}(t) = [v_j, v_j]$ for all j, and $\mathbf{m}_{2n+1}(t) = [2\varepsilon \cdot v_n, 2\varepsilon \cdot v_n]$, where we denoted $v_n = -0.5 \cdot (v_1 + \ldots + v_q)$.

Finally, we define the matrix $\mathbf{M}(T+1)$ as follows: $\mathbf{m}_{1,i}(T+1) = [0, 0]$ for all $i \leq q$, and $\mathbf{m}_{1,n}(T+1) = [1, 1]$. It is easy to check that all the intervals are absolutely δ-accurate (this is why we have chosen δ_0 as we did above), i.e., this problem is indeed based on δ-accurate measurements.

Let us describe all possible predictions. Since $\vec{d} = 0$ and $D = I$, we have $\vec{s}(t+1) = \vec{s}(t)$ and thus, the state does not change with time. So, we will omit t and denote the state vector simply by $\vec{s} = (s_1, \ldots, s_n)$.

The vector equation $M(t)\vec{s}(t) = \vec{m}(t)$ consists of $2n+1$ numerical equations. From the first n equations $M(t)\vec{s}(t) = \vec{m}(t)$, we conclude that $m_{ii}(t) \cdot s_i = 2\varepsilon \cdot \delta_0$ for some $m_{ii}(t) \in [-\delta_0, \delta_0]$; therefore, $2\varepsilon \cdot \delta_0 = |m_{ii}| \cdot |s_i| \leq |s_i| \cdot \delta_0$, and $|s_i| \geq 2\varepsilon$, i.e., $s_i \leq -2\varepsilon$ or $s_i \geq 2\varepsilon$.

From the equations numbers $n+1, \ldots, n+i, \ldots, 2n$, we conclude that for every i, $\delta_0 \cdot s_i = m_i$ for some $m_i \in [-2\varepsilon \cdot \delta_0, 2\varepsilon \cdot \delta_0]$, and therefore, that $s_i \in [-2\varepsilon, 2\varepsilon]$. Since we already know that $s_i \leq -2\varepsilon$ or $s_i \geq 2\varepsilon$, we conclude that for every i, either $s_i = -2\varepsilon$ or $s_i = 2\varepsilon$. In other words, we can say that $s_i = 2\varepsilon \cdot x_i$, where $x_i = -1$ or $x_i = 1$.

Finally, the $(2n+1)$-st equation $M(t)\vec{s}(t) = \vec{m}(t)$ means that $v_1 \cdot s_1 + \ldots + v_q \cdot s_q + v_n \cdot s_n = 2\varepsilon \cdot v_n$. If we substitute the expression $s_i = 2\varepsilon \cdot x_i$ into this equation, and divide both sides by 2ε, we conclude that $v_1 \cdot x_1 + \ldots + v_q \cdot x_q + v_n \cdot x_n = v_n$. This equation has a possible solution $x_1 = \ldots = x_q = -1$, $x_n = -1$, and is, therefore, consistent. The variable x_n has two possible values: -1 and 1. The value -1 is always possible, but the value $x_n = 1$ is possible if and only if $v_1 \cdot x_1 + \ldots + v_q \cdot x_q = 0$ for some $x_i \in \{-1, 1\}$, i.e., if and only if this instance of the *PARTITION* problem is solvable.

In our prediction problem, the predicted quantity p is equal to s_n, i.e., to $2\varepsilon \cdot x_n$. So, depending on whether the given instance of *PARTITION* has a solution or not, the set of possible predictions consists either of a single value -2ε, or of two values $\{-2\varepsilon, 2\varepsilon\}$. Thus, depending on whether this instance is solvable or not, we will get either $\overline{p} = 2\varepsilon$ or $\overline{p} = -2\varepsilon$. If we compute this bound $\overline{p}$ with accuracy ε, we will be able to detect this difference. Since *PARTITION* is known to be NP-hard, our prediction problem is thus also NP-hard. The theorem is proven.

Proof of Theorem 13.2. To prove Theorem 13.2, we will describe a reduction similar to the one described in Theorem 13.1. Namely, we will modify the reduction from the Theorem 13.1 as follows:

- First, we will *add T more quantities* to the description of the *state*, i.e., we will consider T additional variables $s_{q+2}, \ldots, s_{q+1+T}$, to the total of $n = q + 1 + T$.
- For the first $q + 1$ components of the state $\vec{s}$, dynamic equations are the same as in the proof of Theorem 13.1: $s_i(t + 1) = s_i(t)$, $1 \le i \le q + 1$. For the *new* components, the dynamics will be slightly more complicated: $s_{q+i}(t + 1) = s_{q+i-1}(t)$.

  In *algebraic* terms:
  - the vector $\vec{d}$ is still all 0's ($\vec{d} = \vec{0}$), but
  - the matrix D, in addition to the $(q+1) \times (q+1)$-identity-matrix part, has T more non-zero terms $d_{q+1+i,q+i} = 1$, $1 \le i \le T$.
- At each moment of time, we will perform *the same* $2q + 3$ measurements that we did in the proof of Theorem 13.1 (their results, thus, do not depend on the new variables s_{q+1+i}), *plus one more measurement*, with $\mathbf{m}_{2q+4,i} = [0, 0]$ for $i \ne q+1+T$ and $\mathbf{m}_{2q+4,q+1+T} = [1, 1]$. For this new measurement, we will take $\mathbf{m}_{2q+4} = [0, 0]$.

  In *algebraic* terms, to get the new matrix $\mathbf{M}$ from the matrix described in Theorem 13.1, we:
  - add T new columns to the right;
  - add a new row below, and
  - fill new rows and new columns with 0's except for the intersection of the last new row and the new column where we place a (degenerate) interval $[1, 1]$.
- Finally, to describe a *predicted quantity p*, instead of taking the previous vector $\mathbf{M}(T + 1)$, we take a *new* vector (that is actually equal to the last row of the new matrix $\mathbf{M}$): $\mathbf{m}_{1,i}(T + 1) = [0, 0]$ for $i \ne q + 1 + T$ and $\mathbf{m}_{1,q+2}(T + 1) = [1, 1]$.

Similarly to the proof of Theorem 13.1, we can conclude that for every possible prediction, we have $s_i(t+1) = s_i(t) = 2\varepsilon \cdot x_i$ for all $i = 1, \ldots, q+1$, where each of the variables x_i is equal either to -1 or to 1, and $x_{q+1} = 1$ is possible if and only if the given instance of the *PARTITION* problem has a solution.

The predicted quantity is $p = s_{q+1+T}(T+1)$. From the dynamic equations for the new variables, we conclude that:

$$s_{q+1+T}(T+1) = s_{q+T}(T) = s_{q+T-1}(T-1) = \ldots = s_{q+i}(i) =$$

$$\ldots = s_{q+2}(2) = s_{q+1}(1).$$

Thus, in our prediction problem, the predicted quantity p is equal to s_{q+1}, i.e., to $2\varepsilon \cdot x_{q+1}$. So, depending on whether the given instance of *PARTITION* has a solution or not, the set of possible predictions consists either of a single value -2ε, or of two values $\{-2\varepsilon, 2\varepsilon\}$. Thus, depending on whether this instance is solvable or not, we will get either $\overline{p} = 2\varepsilon$ or $\overline{p} = -2\varepsilon$. If we compute this bound $\overline{p}$ with accuracy ε, we will be able to detect this difference. Since *PARTITION* is known to be NP-hard, our prediction problem is thus also NP-hard. The theorem is proven.

14

ENGINEERING COROLLARY: SIGNAL PROCESSING IS NP-HARD

In this chapter, we present another example of a practical problem which is computationally intractable in the presence of interval uncertainty: signal processing.

This chapter was written in collaboration with O. Kosheleva.

14.1. Signal Processing: A Brief Introduction

What is signal processing. Signal processing studies dynamic *signals*, i.e., quantities that change with time. Usually, we do not directly observe the *actual* values $x_1, \ldots, x_n$ of this quantity at different moments of time $1, \ldots, n$. Instead, we measure the values of the *transmitted* signal $m_1, \ldots, m_n$ that depend on x_j: $m_i = m_i(x_1, \ldots, x_n)$. One of the basic problems of signal processing is to *reconstruct* the signal x_j from the measurements $m_1, \ldots, m_n$.

An important *particular case* is when the transmission occurs *within a measuring system*, i.e., when x_i are the measured quantities, and m_i are the actual measurement results. If we measure, e.g., a temperature inside the furnace, or parameters of the Martian soil, then signal transmission is a difficult task, and transmission inaccuracies are inevitable (i.e., $m_i \neq x_i$).

If the signal is *strong*, then it usually passes through the transmission lines practically unchanged (i.e., $m_i \approx x_i$), so there is no need for reconstruction. Reconstruction is only necessary when the signal is *weak*, i.e., when the values x_j are *small*. In this case, we can expand the dependence m_i on $x_1, \ldots, x_n$ in Taylor series, and keep only the *linear* terms, i.e., the constant term b_i (*bias*) and the terms proportional to x_j:

$$m_i = b_i + a_{i1} \cdot x_1 + \ldots + a_{in} \cdot x_n. \tag{14.1}$$

In these terms, *reconstructing the signal* means finding the values x_j from the equations (14.1).

If we know the *exact* values of the parameters b_i and a_{ij} and the measurements that lead to the values m_i are precise, then reconstructing the signal simply means solving a system of linear equations.

Uncertainty. In most real-life situations we do not know the *exact* values of the coefficients b_i and a_{ij} that characterize the transmission line; instead, we only know the *intervals* $\mathbf{b}_i$ and $\mathbf{a}_{ij}$ of possible values of these variables.

> In some cases, we also know *probabilities* of different values inside these intervals, but in this chapter, we will only consider the case when the intervals are the only information we have.

Similarly, as a result of the measurement, we do not get the *exact* values m_i; due to inevitable measurement errors, we can, at best, get an *interval* $\mathbf{m}_i$ of possible values of the measured quantity. With this interval uncertainty, we get *interval linear equations* that relate the (unknown) signal x_j with the measurement results:

$$\mathbf{b}_i + \sum_{j=1}^{n} \mathbf{a}_{ij} \cdot x_j = \mathbf{m}_i, \quad 1 \le i \le n \tag{14.2}$$

meaning that $b_i + \sum a_{ij} \cdot x_j = m_i$ for some $b_i \in \mathbf{b}_i$, $a_{ij} \in \mathbf{a}_{ij}$, and $m_i \in \mathbf{m}_i$. In this case, *reconstructing the signal* means finding all possible values of x_i, i.e., solving the corresponding system of *interval linear equations*.

Duplicate measurements. In the *ideal* case, when we know the *exact* values of b_i, a_{ij}, and m_i, we get the *exact* values of x_j from a *single* sequence of measured values. In the *realistic* case, we get only *approximate* values of x_i. If we are not satisfied with the accuracy of the reconstructed signal, then the natural way to improve this accuracy is to perform *additional* measurements, i.e., to have several *channels*, and to reconstruct the signal from the results coming from all these channels.

For *example*, if we are interested in the temperature inside the furnace, and the existing thermo-sensors do not give the desired accuracy, we may want to complement them with other sensors that measure, e.g., the brightness of the furnace.

If we use several channels, then, instead of the system (14.2), we have a system

$$\mathbf{b}_i^{(c)} + \sum \mathbf{a}_{ij}^{(c)} \cdot x_j = \mathbf{m}_i^{(c)}, \tag{14.3}$$

where $c = 1, 2, \ldots, C$ is the number of the channel, $\mathbf{b}_i^{(c)}$ and $\mathbf{a}_{ij}^{(c)}$ are parameters that describe c-th channel, and $\mathbf{m}_i^{(c)}$ are the measurement results coming from c-th channel.

Measuring instruments are usually stationary. *In principle*, arbitrary interval vectors and matrices are possible, so, from the fact that solving interval linear systems is NP-hard, we can conclude that signal processing is also NP-hard. However, these *arbitrary* parameters b_i and a_{ij} are rare: *in practice*, most of the communication channels and measuring instruments are *time-invariant* (*stationary*). Time-invariance means that all properties of the channel and/or instrument do not depend on time; in particular:

- the interval $\mathbf{b}_i^{(c)}$ that describes possible *bias* at time i should not depend on time at all: $\mathbf{b}_i^{(c)} = \mathbf{b}^{(c)}$;
- the interval $\mathbf{a}_{2,1}^{(c)}$ that describes how the value m_2 of the transmitted signal (that corresponds to moment 2), depends on the input signal x_1 at moment 1, should coincide with the interval $\mathbf{a}_{3,2}^{(c)}$ that describes how the value m_3 of the transmitted signal (that corresponds to moment 3), depends on the input signal x_2 at moment 2, etc. In general, the interval coefficient $\mathbf{a}_{ij}^{(c)}$ should not depend on the *absolute* times i and j, but only on the time $i-j$ that passed between i and j: $\mathbf{a}_{ij}^{(c)} = \mathbf{a}_{i-j}^{(c)}$ (i.e., each interval matrix $\mathbf{a}_{ij}^{(c)}$ is a *Toeplitz* matrix).

In this *stationary* case, we get the following system of equations:

$$\mathbf{b}^{(c)} + \sum \mathbf{a}_{i-j}^{(c)} \cdot x_j = \mathbf{m}_i^{(c)}. \tag{14.4}$$

We will show that signal reconstruction is NP-hard even if we only consider such stationary systems.

Comment. In principle, the measured values $m_1, \ldots, m_n$ may depend not only on the values of the signal $x_1, \ldots, x_n$ during the observation period, but also on the other values of the signal, such as x_0 and x_{n+1}. We will show NP-hardness for the simplest case when the signal x_i can only be different from 0 during the observations, i.e., when $x_0 = x_{-1} = \ldots = x_{n+1} = x_{n+2} = \ldots = 0$.

14.2. Definitions and the Main Result

Definition 14.1. *Let C be a positive integer. By a C-channel signal processing problem, we mean the following problem:*

GIVEN:

- *an integer n,*
- *C rational intervals $\mathbf{b}^{(c)}$, $1 \le c \le C$;*
- *C rational $(2n-1)$-dimensional interval vectors $\mathbf{a}^{(c)} = (\mathbf{a}^{(c)}_{-(n-1)}, \ldots, \mathbf{a}^{(c)}_{n-1})$, $1 \le c \le C$;*
- *C rational n-dimensional interval vectors $\mathbf{m}^{(c)} = (\mathbf{m}^{(c)}_1, \ldots, \mathbf{m}^{(c)}_n)$, $1 \le c \le C$; and*
- *an integer i, $1 \le i \le n$.*

COMPUTE the smallest $\underline{x}_i$ and the largest $\overline{x}_i$ possible values of x_i for which, for some $x = (x_1, \ldots, x_{i-1}, x_i, x_{i+1}, \ldots, x_n)$ and for each c and i,

$$b_i^{(c)} + \sum_{j=1}^{n} a_{ij}^{(c)} \cdot x_j = m_i^{(c)}. \tag{14.5}$$

for some $b_i^{(c)} \in \mathbf{b}_c$, $a_{ij}^{(c)} \in \mathbf{a}^{(c)}_{i-j}$, and $m_i^{(c)} \in \mathbf{m}_i^{(c)}$.

Theorem 14.1. *For $C \ge 2$, C-channel signal processing problem is NP-hard.*

Comments.

- We will see from the proof that not only the problem of computing the *exact* bounds for x_i is NP-hard, but the problem of *approximately* computing these bounds, with a given accuracy, is also NP-hard.
- We do not know whether signal processing is NP-hard for the case of a single channel.

Proof

It is sufficient to prove the theorem for $C = 2$. To prove this result, we will reduce the *PARTITION* problem (known to be NP-hard) to our problem. In the *PARTITION* problem, we are given a sequence of integers $s_1, \ldots, s_t$, and we must check whether there exist values $y_1, \ldots, y_t$ for which $y_i \in \{-1, 1\}$ and $s_1 \cdot y_1 + \ldots + s_t \cdot y_t = 0$. Let us take $n = 2t + 2$, $\mathbf{b}^{(1)} = \mathbf{b}^{(2)} = [0, 0]$, and the following intervals $\mathbf{a}_k^{(c)}$ and $\mathbf{m}_i^{(c)}$ $(1 \le c \le 2)$:

- $\mathbf{a}_0^{(1)} = [1, 1]$ and $\mathbf{a}_k^{(1)} = [0, 0]$ for $k \neq 0$,
- $\mathbf{m}_i^{(1)} = [-1, 1]$ for $1 \le i \le t + 1$, and $\mathbf{m}_i^{(1)} = [0, 0]$ for $i > t + 1$.
- $\mathbf{a}_0^{(2)} = [-1, 1]$, $\mathbf{a}_k^{(2)} = [0, 0]$ for $1 \le k \le t + 1$, $\mathbf{a}_{t+1+k}^{(2)} = [s_{t+1-k}, s_{t+1-k}]$, $1 \le k \le t+1$, where we denoted $s_{t+1} = -0.5 \cdot (s_1 + \ldots + s_t)$; and $\mathbf{a}_k^{(2)} = [0, 0]$ for $k < 0$.
- $\mathbf{m}_i^{(2)} = [-1, 1]$ for $0 \le i \le t + 1$, $\mathbf{m}_i^{(2)} = [-S, S]$ for $t + 1 < i < 2t + 2$, where we denoted $S = |s_1| + \ldots + |s_{t+1}|$, and $\mathbf{m}_{2t+2}^{(2)} = [s_{t+1}, s_{t+1}]$,

For the *first channel*, the condition (14.5) implies that for every $i \le t + 1$, $x_i = m_i^{(1)}$ for some $m_i^{(1)} \in [-1, 1]$, and therefore, that $x_i \in [-1, 1]$. For $i > t + 1$, we similarly conclude that $x_i = m_i^{(1)} \in [0, 0]$, i.e., that $x_{t+2} = x_{t+3} = \ldots = x_{2t+2} = 0$.

From the equations that correspond to the *second channel*, we conclude that for every $i \le t + 1$, we have $a_i^{(2)} \cdot x_i = 1$ for some $a_i^{(2)} \in [-1, 1]$, and therefore, that $x_i \le -1$ or $x_i \ge 1$. Since we already know that $x_i \in [-1, 1]$, we conclude that for every $i \le t + 1$, either $x_i = -1$ or $x_i = 1$.

Let us now consider the equations that correspond to $i > t+1$, i.e., to $i = t+1+k$ for $k > 0$. When $t + 1 < i = t + 1 + k < 2t + 2$ (i.e., when $k < t + 1$), the equation (14.5) takes the form

$$\sum_{j=1}^{2t+2} a_{t+1+k,j}^{(2)} \cdot x_j \in [-S, S].$$

Since we already know that $x_{t+2} = x_{t+3} = \ldots = 0$, we can only retain terms corresponding to $j \leq t+1$:

$$\sum_{j=1}^{t+1} a^{(2)}_{t+1+k,j} \cdot x_j \in [-S, S].$$

In the equation (14.5), $a^{(2)}_{t+1+k,j} \in \mathbf{a}^{(2)}_{t+1+k-j}$. Since $k > 0$ and $j \leq t+1$, we have $t+1+k-j = (t+1-j)+k > 0$. For a positive index q, the interval $\mathbf{a}^{(2)}_q$ is degenerate, i.e., consists of a single value; therefore, each value $a^{(2)}_{t+1+k,j}$ coincides with the corresponding value. This value is non-zero only when $k-j \geq 0$, i.e., when $j \leq k$. Thus, the above formula turns into

$$\sum_{j=1}^{k} s_{t+1-(k-j)} \cdot x_j \in [-S, S].$$

Since $x_j = \pm 1$, we have $|\sum s_{t+1-(k-j)} \cdot x_j| \leq \sum |s_k| = S$, so each of these equations is automatically satisfied.

For $i = 2t+2$, the interval $\mathbf{m}^{(2)}_i$ is degenerate, and thus, the corresponding equation (14.5) takes the form $s_1 \cdot x_1 + \ldots + s_t \cdot x_t + s_{t+1} \cdot x_{t+1} = s_{t+1}$.

This equation has a possible solution $x_1 = \ldots = x_t = -1$, $x_{t+1} = -1$, and is, therefore, consistent. The variable x_{t+1} has two possible values: -1 and 1. The value -1 is always possible, but the value $x_{t+1} = 1$ is possible if and only if $s_1 \cdot x_1 + \ldots + s_t \cdot x_t = 0$ for some $x_i \in \{-1, 1\}$, i.e., if and only if this instance of the *PARTITION* problem is solvable. Thus, depending on whether this instance is solvable or not, we will get either $\overline{x}_{t+1} = 1$ or $\overline{x}_{t+1} = -1$. Since *PARTITION* is known to be NP-hard, our problem is thus also NP-hard.

Thus, if we are able to solve the corresponding interval linear systems with accuracy $\varepsilon < 1$, we can check whether the actual value of $\overline{x}_{t+1}$ is -1 or 1, and thus, solve the given instance of *PARTITION*. For $\varepsilon > 1$, we can consider a similar system with the new measurement intervals $\mathbf{m}^{(c)\text{new}}_i = 2\varepsilon \cdot \mathbf{m}^{(c)}_i$. Solving this system with accuracy ε is equivalent to solving the original system with accuracy 0.5, and thus, it is equivalent to solving the given instance of *PARTITION*. The theorem is proven.

15

BRIGHT SIDES OF NP-HARDNESS OF INTERVAL COMPUTATIONS I: NP-HARD MEANS THAT GOOD INTERVAL HEURISTICS CAN SOLVE OTHER HARD PROBLEMS

In the previous chapters, we have proven that many computational problems of data processing and interval computations are NP-hard.

*The immediate conclusion of these results is **negative**: one cannot expect an algorithm that solves all the problems of data processing and interval computations in reasonable time.*

*However, as we will mention in this chapter, the NP-hardness results also have their **bright sides**: namely, it enables us to apply efficient heuristic methods, originally developed for interval computations, to other complicated problems, and thus, get new heuristics (see also Appendix F for more speculative applications).*

15.1. Possibility

By definition, the fact that a problem $\mathcal{P}$ is NP-hard means, as we have mentioned, that if we can solve the problem $\mathcal{P}$ in polynomial time, then we will be able to solve many other hard problems (those in the class NP) in polynomial time. In other words, if a problem $\mathcal{P}$ is NP-hard, then every instance of every other problem from the class NP can be *reduced* to one or several instances of this problem $\mathcal{P}$. Based on this reduction, the majority of computer scientists believe that there is no algorithm that solves *all* instances of an NP-hard problem $\mathcal{P}$. But this does not prevent us from having good heuristics that solve *many* important instances of $\mathcal{P}$. The reduction mentioned above means that we can then solve many cases of other hard problems.

15.2. Such Heuristics Have Actually Been Proposed

For interval computations, many good heuristics are indeed known. In Traylor *et al.* [414], it is shown that these heuristics lead to good heuristic algorithms for solving another NP-hard problem: ***propositional satisfiability*** problem, as described in Chaoter 2. This problem is considered one of the basic NP-complete problems (see, e.g., Garey *et al.* [121]). The resulting heuristic turned out to be closely connected with similar heuristics that try to simulate how our brain works, be it on the level of *neural networks*, or on the level of *chemical reactions*; see, e.g., Kreinovich *et al.* [198, 217, 226, 211] and Fuentes *et al.* [114, 113].

16

IF INPUT INTERVALS ARE NARROW ENOUGH, THEN INTERVAL COMPUTATIONS ARE ALMOST ALWAYS EASY

In the previous chapters, we have shown that in general, interval computations are NP-hard. This means, crudely speaking, that every algorithm that solves the interval computation problems requires, in some instances, unrealistic exponential time. Thus, the worst-case *computational complexity of the problem is large. A natural question is: is this problem easy "on average" (i.e., are complex instances rare), or is this problem difficult "on average" too?*

In this chapter, we show that "on average", the basic problem of interval computations is easy. To be (somewhat) more precise, we show that **if input intervals are narrow enough, then interval computations are almost always easy.**

16.1. Formulation of the Problem: Are Hard Cases Rare?

Hard cases are inevitable. Ideally, in the basic problem considered above, we would like to compute the *exact* range interval of a function $y = f(x_1, ..., x_n)$. Traditional methods of interval computations do not always give the exact range, but they usually give an *enclosure* of the desired range, i.e., an interval that *contains* the range $\mathbf{y}$. The result that computing the exact range is NP-hard means, crudely speaking, that there is no feasible (polynomial time) algorithm that can *always* compute the exact range. In other words, *every enclosure-computing feasible algorithm sometimes either overestimates, or does not work at all.* In particular, every feasible algorithm that is always applicable sometimes overestimates.

Hard cases can be rare or frequent. Every always-applicable feasible algorithm sometimes overestimates. The natural question is: *how often is this "sometimes"?* There are two possible scenarios here:

- a *pessimistic* one: many particular cases are hard; the most pessimistic case is when every feasible algorithm overestimates in almost all cases;
- an *optimistic* one: a few cases are hard, but the vast majority are feasible.

It turns out that we are in the *optimistic* situation: for small input intervals, *almost all interval computation problems are feasible.*

How to formalize "almost all": standard formalization is not applicable. To formulate this result in precise terms, we need to formalize what "almost all" means. In mathematics, "almost all" usually means "all points, except for points from a set of Lebesgue measure 0" (or, "except for points from a set of a small Lebesgue measure"). In the existing computers, however, only rational numbers are represented. The set of all rational numbers is countable and has, therefore, Lebesgue measure 0; so the standard mathematical notion of "almost all" is not applicable.

A new formalization of "almost all". To formalize the notion of "almost all", we must somehow "extend" our algorithms, that are currently applicable only to *rational* numbers, to arbitrary *real* numbers.

In real life, when we say that "an algorithm is applied to *real* numbers $x_1, \ldots, x_n$", we usually mean that this algorithm is applied to *rational* numbers $r_1, \ldots, r_n$ that are η-close to $x_1, \ldots, x_n$, where η is the computer precision. So, if we fix $\eta > 0$, we can say that an algorithm works fine for n real numbers $x_1, \ldots, x_n$ if it works fine for all real-valued vectors $r = (r_1, \ldots, r_n)$ that are η-close to the vector $x = (x_1, \ldots, x_n)$.

Now, we have *real-valued* inputs on which the algorithm works fine, and real-valued inputs on which it does not. For real-valued inputs, we can apply Lebesgue measure. As a result, we arrive at the following definition:

16.2. Definitions and the Main Result

Definition 16.1. *Let $\varepsilon > 0$ be a real number, $D \subseteq R^n$ be a compact domain with a positive Lebesgue measure $\mu(D) > 0$, and $P(x)$ be a property that is true for some points $x \in D$. We say that P is true for (D, ε)-almost all x if $\mu(\{x \in D | \neg P(x)\}) \leq \varepsilon \cdot \mu(D)$.*

Comment. In other words, a property is true for (D, ε)-almost all x if the portion of the domain D for which this property is false does not exceed ε.

Definition 16.2. *Let $\eta > 0$ be a real number.*

- *We say that an interval $\mathbf{x} = [\tilde{x} - \Delta, \tilde{x} + \Delta]$ is η-close to an interval $\mathbf{r} = [r - d, r + d]$ if $|\tilde{x} - r| \leq \eta$ and $|\Delta - d| \leq \eta$.*

- *We say that an interval vector $\mathbf{r} = (\mathbf{r}_1, \ldots, \mathbf{r}_n)$ is η-close to an interval vector $\mathbf{x} = (\mathbf{x}_1, \ldots, \mathbf{x}_n)$ if for all i from 1 to n, the interval $\mathbf{r}_i$ is η-close to the interval $\mathbf{x}_i$.*

- *Similarly, we say that an interval matrix*

$$\mathbf{R} = (\mathbf{r}_{11}, \ldots, \mathbf{r}_{1n}, \ldots, \mathbf{r}_{m1}, \ldots, \mathbf{r}_{mn})$$

is η-close to an interval matrix

$$\mathbf{X} = (\mathbf{x}_{11}, \ldots, \mathbf{x}_{1n}, \ldots, \mathbf{x}_{m1}, \ldots, \mathbf{x}_{mn})$$

if for all i from 1 to m and for all j from 1 to n, the interval $\mathbf{r}_{ij}$ is η-close to the interval $\mathbf{x}_{ij}$.

In this section, we consider *rational* functions, i.e., functions that can be represented as a fraction of two polynomials:

$$f(x_1, \ldots, x_n) = \frac{P(x_1, \ldots, x_n)}{Q(x_1, \ldots, x_n)}.$$

(Polynomials are a particular case of rational functions, corresponding to $Q(x_1, \ldots, x_n) = 1$.) If both polynomials $P(x_1, \ldots, x_n)$ and $Q(x_1, \ldots, x_n)$ have rational coefficients, then we can represent a rational function $f(x_1, \ldots, x_n) = P(x_1, \ldots, x_n)/Q(x_1, \ldots, x_n)$ in the computer by storing the rational coefficients of both polynomials $P(x_1, \ldots, x_n)$ and $Q(x_1, \ldots, x_n)$.

Definition 16.3. *Let $\mathcal{U}$ be an algorithm for solving the basic problem of interval computations.*

- *Let $\eta > 0$ be a real number, let $f(x_1, \ldots, x_n)$ be a rational function with rational coefficients, and let $\mathbf{x}$ be an n-dimensional interval vector with components $\mathbf{x}_i = [\tilde{x}_i - \Delta_i, \tilde{x}_i + \Delta_i]$, $1 \le i \le n$. We say that $\mathcal{U}$ is η-exact on the function $f(x_1, \ldots, x_n)$ and on the interval vector $\mathbf{x}$ if for every n-dimensional interval vector $\mathbf{r} = (\mathbf{r}_1, \ldots, \mathbf{r}_n)$ whose components are intervals with rational endpoints, and that is η-close to $\mathbf{x}$, the algorithm $\mathcal{U}$ returns the exact endpoints of the interval $f(\mathbf{r}_1, \ldots, \mathbf{r}_n)$.*

- *We say that the algorithm $\mathcal{U}$ is almost always exact for narrow input intervals if for every compact domain D with $\mu(D) > 0$, for every rational function $f(x_1, \ldots, x_n)$ with rational coefficients that is finite on an open set $N \supset D$, and for every $\varepsilon > 0$, there exist $\delta > 0$ and $\eta > 0$ such that for (D, ε)-almost all vectors $\tilde{x} = (\tilde{x}_1, \ldots, \tilde{x}_n)$, if all n input intervals $\mathbf{x}_i = [\tilde{x}_i - \Delta_i, \tilde{x}_i + \Delta_i]$ of an input vector $\mathbf{x} = (\mathbf{x}_1, \ldots, \mathbf{x}_n)$ are absolutely δ-narrow ($\Delta_i \le \delta$), the algorithm $\mathcal{U}$ is η-exact on the function $f(x_1, \ldots, x_n)$ and on the interval vector $\mathbf{x}$.*

Theorem 16.1. (Lakeyev *et al.* [243, 224]) *There exists a polynomial-time algorithm $\mathcal{U}$ that, given n intervals $\mathbf{x}_i = [\tilde{x}_i - \Delta_i, \tilde{x}_i + \Delta_i]$, and a rational function $f(x_1, \ldots, x_n)$ with rational coefficients, returns a (possibly infinite) enclosure for the range $f(\mathbf{x}_1, \ldots, \mathbf{x}_n)$, and that is almost always exact for narrow input intervals.*

Comments.

- As we can see from this formulation, one and the same algorithm works for *all* rational functions.

- As we will see from the proof, this theorem is true for a quite reasonable algorithm $\mathcal{U}$ that has, in effect, been used many times before. In other words, the algorithm is not new. What is new is the observation that this algorithm gives an exact estimate in *almost all* cases.

A similar results shows that solving systems of interval linear equations is also almost always easy:

Definition 16.4. *Let $\mathcal{U}$ be an algorithm that, given an $(n \times n)$-interval matrix $\mathbf{A}$ with components $\mathbf{a}_{ij} = [\tilde{a}_{ij} - \Delta_{ij}, \tilde{a}_{ij} + \Delta_{ij}]$, an n-dimensional interval vector $\mathbf{b}$ with components $\mathbf{b}_i = [\tilde{b}_i - \Delta_i, \tilde{b}_i + \Delta_i]$, and an integer $i \le n$, returns an enclosure for the interval $\mathbf{x}_i$ of possible values of x_i (i.e., values for which $\sum a_{ij} \cdot x_j = b_i$ for some $a_{ij} \in \mathbf{a}_{ij}$ and $b_i \in \mathbf{b}_i$).*

- *Let $\eta > 0$ be a real number, $\mathbf{A}$ be an $(n \times n)$-interval matrix, and $\mathbf{b}$ be an n-dimensional interval vector. We say that $\mathcal{U}$ is η-exact on $\mathbf{A}$ and $\mathbf{b}$, if for every $(n \times n)$-interval matrix $\mathbf{A}^r$ and for every n-dimensional interval vector $\mathbf{b}^r$ that are η-close to, correspondingly, $\mathbf{A}$ and $\mathbf{b}$, and whose component intervals have rational endpoints, the algorithm $\mathcal{U}$ returns the exact endpoints of the solution interval $\mathbf{x}_i$ corresponding to $\mathbf{A}^r$ and $\mathbf{b}^r$.*
- *We say that the algorithm $\mathcal{U}$ is almost always exact for narrow input intervals if for every compact domain D of positive $(n \times n) + n$-dimensional Lebesgue measure (for which all matrices A in an open neighborhood of D are invertible), and for every $\varepsilon > 0$, there exist $\delta > 0$ and $\eta > 0$ such that for (D, ε)-almost all pairs $(\tilde{A}, \tilde{b})$(where $\tilde{A}$ is an $(n \times n)$-matrix and b is a vector), if all $(n \times n)$ component intervals $\mathbf{a}_{ij} = [\tilde{a}_{ij} - \Delta_{ij}, \tilde{a}_{ij} + \Delta_{ij}]$ of the interval matrix $\mathbf{A}$ and all n component intervals $\mathbf{b}_i = [\tilde{b}_i - \Delta_i, \tilde{b}_i + \Delta_i]$ of the interval vector $\mathbf{b}$ are absolutely δ-narrow, then the algorithm $\mathcal{U}$ is η-exact on $\mathbf{A}$ and $\mathbf{b}$.*

Theorem 16.2. (Lakeyev *et al.* [243, 224]) *There exists a polynomial-time algorithm $\mathcal{U}$ that, given an $(n \times n)$-interval matrix $\mathbf{A}$, an n-dimensional interval vector $\mathbf{b}$, and an integer $i \le n$, returns a (possibly infinite) enclosure for the interval $\mathbf{x}_i$ of possible values of x_i, and that is almost always exact for narrow input intervals.*

These results can be represented as a table:

	Worst-case complexity	"Average-case" complexity
Computing the range of a rational function	NP-hard	Polynomial time
Solving a system of interval linear equations	NP-hard	Polynomial time

Proofs

Proof of Theorem 16.1. We will describe an algorithm that consists of the following two stages:

- first, we simplify the function $f(x_1, \ldots, x_n)$;
- then, we compute the desired range $\mathbf{y}$.

Let us start with the *first stage*. In the beginning of this stage, we check whether the given function $f(x_1, \ldots, x_n)$ actually depends on all of its variables. To do that, we will:

- apply analytical differentiation formulas to $f(x_1, \ldots .x_n)$ (see, e.g., Rall [334]), and
- check whether the resulting derivatives are identically 0 or not.

In this chapter, we will use the following notation for partial derivatives that we borrow from physics:

$$f_{,i}(x_1, \ldots, x_n) = \frac{\partial f}{\partial x_i}(x_1, \ldots, x_n).$$

The derivative of a rational function is also rational. Therefore, this rational function can be represented as a ratio of two polynomials: $f_{,i}(x_1, \ldots, x_n) = P(x_1, \ldots, x_n)/Q(x_1, \ldots, x_n)$. This ratio is identically equal to 0 if and only if the polynomial $P(x_1, \ldots, x_n)$ is identically equal to 0, and the polynomial is identically 0 if and only if all its coefficients are zeros. Since all the coefficients of $f(x_1, \ldots, x_n)$ are rational numbers, the coefficients of $f_{,i}(x_1, \ldots, x_n)$ (and thus, of $P(x_1, \ldots, x_n)$) are also rational numbers, and for rational numbers, equality to 0 is easy to check.

If for some i, it turns out that the partial derivative $f_{,i}(x_1, \ldots, x_n)$ is identically 0, then the original function $f(x_1, \ldots .x_n)$ does not depend on this variable x_i at all. So, we can substitute an arbitrary value of x_i into the original formula $f(x_1, \ldots, x_n)$, and get a new expression with one fewer variable.

We can repeat this substitution procedure for all the variables x_i for which $f_{,i}(x_1, \ldots, x_n)$ is identically 0. As a result, we get a new expression for $f(x_1, \ldots, x_n)$ that only contains the variables on which this function

$f(x_1, \ldots, x_n)$ essentially depends (i.e., for which the derivative $f_{,i}(x_1, \ldots, x_n)$ is not identically 0). Stage 1 is over.

Comment. To explain why this procedure is so complicated, let us give an example of a rational function $f(x_1, \ldots, x_n)$ that does not depend on x_3 but whose expression explicitly contains x_3:

$$f(x_1, x_2, x_3) = \frac{x_1^2 + x_1 \cdot x_2 + x_1 \cdot x_3 + x_2 \cdot x_3}{x_1^2 + x_1 \cdot x_3}.$$

If we factorize the numerator and the denominator, we will see that this function $f(x_1, x_2, x_3)$, indeed, does not depend on x_3:

$$f(x_1, x_2, x_3) = \frac{(x_1 + x_2) \cdot (x_1 + x_3)}{x_1 \cdot (x_1 + x_3)} = \frac{x_1 + x_2}{x_1}.$$

However, we do not want to factorize, because factorization can be a very time-consuming procedure. Instead, we *differentiate* with respect to all three variables, and indeed get $f_{,3}(x_1, x_2, x_3) = 0$. As a result, we substitute an arbitrary constant instead of x_3, and get an equivalent expression for $f(x_1, x_2, x_3)$ that contains only two variables. For example, if we substitute $x_3 = 0$, we get the following expression:

$$f(x_1, x_2) = \frac{x_1^2 + x_1 \cdot x_2}{x_1^2}.$$

After Stage 1, we have a rational expression $f(x_1, \ldots, x_m)$ for which for all $i \le m$, the derivative $f_{,i}(x_1, \ldots, x_m)$ is not identically 0. For this expression, we do the following:

- First, for each i from 1 to m, we apply an interval centered form (see, e.g., Ratschek *et al.* [336]) to compute the interval enclosure $\mathbf{e}_i = [\underline{e}_i, \overline{e}_i]$ for $f_{,i}(\mathbf{x}_1, \ldots, \mathbf{x}_m)$.

  If on one of the steps of this procedure, we need division by an interval that contains 0, then we simply return the entire real line as the enclosure for the range $f(\mathbf{x}_1, \ldots, \mathbf{x}_n)$.

- If for some i, the interval $\mathbf{e}_i$ contains 0, then we apply the same interval center form (or any other interval computation technique) to compute an enclosure $\mathbf{e}$ for the desired range $f(\mathbf{x}_1, \ldots, \mathbf{x}_m)$.

- If for all i, the interval $\mathbf{e}_i$ does not contain 0, then we compute $s_i = \text{sign}(\overline{e}_i)$ (where $\text{sign}(a) = 1$ for $a > 0$ and $\text{sign}(a) = -1$ if $a < 0$), and return the interval $[\underline{y}, \overline{y}]$ as the exact value of the range, where

$$\underline{y} = f(\tilde{x}_1 - s_1 \cdot \Delta_1, \ldots, \tilde{x}_m - s_m \cdot \Delta_m),$$
$$\overline{y} = f(\tilde{x}_1 + s_1 \cdot \Delta_1, \ldots, \tilde{x}_m + s_m \cdot \Delta_m).$$

Why is this a correct estimate?

- If $0 \in \mathbf{e}_i$, then the fact that the above-described algorithm returns an enclosure follows from the properties of interval computation methods.
- If $0 \notin \mathbf{e}_i$ for all i, this means that for every i, the function $f(x_1, \ldots, x_n)$ is *monotonic* with respect to each of its variables:
  - If $\mathbf{e}_i > 0$, then $f_{,i}(x_1, \ldots, x_n) > 0$, and $f(x_1, \ldots, x_n)$ is *increasing* with respect to x_i.
  - If $\mathbf{e}_i < 0$, then $f_{,i}(x_1, \ldots, x_n) < 0$, and $f(x_1, \ldots, x_n)$ is *decreasing* with respect to x_i.

Comment. As we have already mentioned, the use of monotonicity is not a novel idea. As examples of successful applications of this idea, we can cite Collatz [70, 71], Lakshmikantham *et al.* [247], Walter [423], Harrison [141], Moore [291], Schröder [383], Rall [335], Mannshardt [270], Dimitrova *et al.* [91], Markov [273].

To complete our proof, we must prove that the described algorithm returns the exact range in almost all cases. Crudely speaking, we will show that the above algorithms returns the exact range for all the points $\tilde{x} = (\tilde{x}_1, \ldots, \tilde{x}_n)$ in which all m partial derivatives $f_{,i}(x_1, \ldots, x_m)$ (with respect to m variables on which the function $f(x_1, \ldots, x_n)$ really depends) are different from 0. (We will show that almost all points $\tilde{x}$ have this property.) If all these partial derivatives are different from 0 at a point $\tilde{x}$, then, due to continuity of these derivatives, each of them keeps the same sign in a sufficiently small neighborhood of the point $\tilde{x}$. The fact that the i-th derivative keeps the same sign means that in this small neighborhood, the function $f(x_1, \ldots, x_m)$ is monotonic in x_i. Since the function is monotonic in each of the variables, we can compute its range over the box $\mathbf{x}_1 \times \ldots \times \mathbf{x}_m$ (provided that the input intervals are narrow enough, so that the resulting box fits into this small neighborhood), by simply computing the values of this function $f(x_1, \ldots, x_m)$ at two appropriately chosen endpoints – which is exactly what the above algorithm does.

This is the main idea of the proof that the above algorithm is almost always exact. This idea would constitute a perfect proof if we could decide, for each box, whether this box is "small enough" in the above sense, i.e., whether the corresponding range of each partial derivative is indeed an interval that does not contain 0. Then, we should resort to enclosure estimation only if one of these ranges does contain 0. In reality, we cannot exactly compute the ranges for the partial derivatives; instead, in the algorithm, we rely on the *enclosures* for these ranges. As a result, for some input intervals $\mathbf{x}_1, \ldots, \mathbf{x}_m$ on which the actual ranges of the partial derivatives do not contain 0 (so that the function $f(x_1, \ldots, x_m)$ is actually monotonic), the enclosure for one of these ranges may contain 0 and therefore, the above algorithm will return an enclosure instead of the exact range $f(\mathbf{x}_1, \ldots, \mathbf{x}_m)$.

So, to complete the proof of the theorem, we must show that in spite of this complication, the above algorithm is indeed almost always exact. This final part of the proof is somewhat of a technical nature: it is not very complicated for a professional mathematician, but it requires some familiarity with the basis results about Lebesgue measure, manifolds, compactness, continuity, and uniform continuity, familiarity that other parts of the book do not require. Readers who so not feel comfortable about these notions can skip this part; hopefully, they have already got the idea of the proof from the previous paragraphs.

For those who are still with us, here are the main steps of this final part of the proof. After the first stage, we have a rational function $f(x_1, \ldots, x_m)$ that depends on each of its m variables. Let us assume that this rational function is finite on some open neighborhood N of a compact set D. Since N is an open neighborhood, there exists a real number $\delta_0 > 0$ such that every point that is $\le \delta_0$-close to the set D (in the sup metric $d(x, y) = \max|x_i - y_i|$) also belongs to N. Let us denote the set of all the points that are δ_0-close to D (i.e., a δ_0-neighborhood of D) by D_0. This set D_0 is closed and bounded (since D is bounded), so, D_0 is a compact.

A rational function $f(x_1, \ldots, x_m)$ is continuous in every point x in which it is finite. Since $f(x_1, \ldots, x_m)$ is finite on all points of N, it is also finite on all points from $D_0 \subseteq N$. Therefore, $f(x_1, \ldots, x_m)$ is continuous on every point from the set D_0.

Since $f(x_1, \ldots, x_m)$ is a rational function, for every i, the partial derivative $f_{,i}(x_1, \ldots, x_m)$ is also a rational function, i.e., a ratio of two polynomials: $f_{,i}(x_1, \ldots, x_m) = P_i(x_1, \ldots, x_m)/Q_i(x_1, \ldots, x_m)$. Therefore, the set S_i of all points $x = (x_1, \ldots, x_m) \in D$ for which $f_{,i}(x_1, \ldots, x_m) = 0$ is a set of all solutions of the polynomial equation $P_i(x_1, \ldots, x_m) = 0$. From the topological viewpoint, the structure of the set of all the solutions of a nondegenerate polynomial equation (i.e., a polynomial equation in which a polynomial $P_i(x_1, \ldots, x_m)$ is not identically 0) is well known: it is either a manifold of dimension $\le m-1$, or a union of finitely many manifolds of dimension $\le m-1$. In both cases, the Lebesgue measure $\mu(S_i)$ of this set is equal to 0: $\mu(S_i) = \mu(\{x \mid f_{,i}(x) = 0\}) = 0$.

In our algorithm, we do not use the exact range of the partial derivatives $f_i(\mathbf{x}_1, \ldots, \mathbf{x}_m)$, we only use an *enclosure* of this range. Therefore, to analyze the behavior of our algorithm, we would like to have some measure estimate that is based not on the exact value of $f_{,i}(x_1, \ldots, x_m)$, but on the interval of possible values of this partial derivative. Such an estimate is reasonably easy to get: Indeed, the set S_i can be represented as an intersection of the sequence of a monotonically decreasing sequence of sets:

$$S_i = \{x \mid f_{,i}(x) = 0\} = \bigcap_{k=1}^{\infty} \{x \mid |f_{,i}(x)| \le 2^{-k}\}.$$

Therefore, due to the known properties of measure,

$$\mu\left(\{x \mid |f_{,i}(x)| \le 2^{-k}\}\right) \to \mu(S_i) = 0$$

as $k \to \infty$. So, there exists a k_i for which

$$\mu\left(\{x \mid |f_{,i}(x)| \le 2^{-k_i}\}\right) \le \frac{\varepsilon \cdot \mu(D)}{m}.$$

If we denote by k the largest of these k_i, i.e., $k = \max(k_1, \ldots, k_m)$, then we will conclude that

$$\mu\left(\{x \mid |f_{,i}(x)| \le 2^{-k}\}\right) \le \mu\left(\{x \mid |f_{,i}(x)| \le 2^{-k_i}\}\right) \le \frac{\varepsilon \cdot \mu(D)}{m}.$$

Let us denote the corresponding set $\{x \mid |f_{,i}(x)| \le 2^{-k}\}$ by A_i. Then, for a union A of these sets

$$A = \bigcup_{i=1}^{m} A_i = \{x \mid |f_{,i}(x)| \le 2^{-k} \text{ for some } i\},$$

we have

$$\mu(A) \le \sum_{i=1}^{m} \mu(A_i) \le \sum_{i=1}^{m} \frac{\varepsilon \cdot \mu(D)}{m} = \varepsilon \cdot \mu(D).$$

We want to find $\delta > 0$ for which the above-described algorithm computes the precise range for all absolutely δ-narrow intervals whose centers $\tilde{x} = (\tilde{x}_1, \ldots, \tilde{x}_n)$ satisfy the condition $\tilde{x} \notin A$. Let us find such a δ. For that, we will use the following two ideas:

- For every i from 1 to m, the rational function $f_{,i}(x_1, \ldots, x_m)$ is continuous on the compact set D_0. Therefore, the function $f_{,i}(x_1, \ldots, x_m)$ is uniformly continuous. In particular, there exists a $\delta_i > 0$ such that if $d(x, \tilde{x}) \le \delta_i$, then $|f_{,i}(x) - f_{,i}(\tilde{x})| \le 2^{-(k+2)}$.

- According to [336], the centered method has an accuracy $O(\max(\Delta_1, \ldots, \Delta_m))$. This means that for every i from 1 to m, there exists a constant C_i such that if all input intervals are absolutely δ-narrow, then the difference between the endpoints of the actual range $[\underline{a}_i, \overline{a}_i] = f_{,i}(\mathbf{x}_1, \ldots, \mathbf{x}_m)$ and the corresponding endpoints of the estimated range $\mathbf{e}_i = [\underline{e}_i, \overline{e}_i]$ does not exceed $C_i \cdot \delta$: $|\underline{e}_i - \underline{a}_i| \le C_i \cdot \delta$ and $|\overline{e}_i - \overline{a}_i| \le C_i \cdot \delta$.

Now, let us take

$$\delta = \min\left(\delta_0, \delta_1, \ldots, \delta_m, \frac{2^{-(k+2)}}{C_1}, \ldots, \frac{2^{-(k+2)}}{C_m}\right).$$

Let us show that for this δ, if the input intervals are absolutely δ-narrow, and $\tilde{x} \notin A$, then none of the intervals $\mathbf{e}_i$ contain zero and therefore, for these algorithms, the above-described algorithm returns the exact interval range.

Indeed, since $\tilde{x} \notin A$, by definition of the set A, we have $|f_{,i}(\tilde{x})| > 2^{-k}$ for all i. This means that either $f_{,i}(\tilde{x}) > 2^{-k}$, or $f_{,i}(\tilde{x}) < -2^{-k}$.

- Let us first consider the first case. In this case, for each j, if $x_j \in \mathbf{x}_j = [\tilde{x}_j - \Delta_j, \tilde{x}_j + \Delta_j]$, then (since the interval $\mathbf{x}_j$ is absolutely δ-narrow), we have $|x_j - \tilde{x}_j| \le \Delta_j \le \delta$. Hence, $d(x, \tilde{x}) = \max |x_j - \tilde{x}_j| \le \delta$. Due to our choice of δ as $\min(\ldots, \delta_i, \ldots)$, we have $\delta \le \delta_i$ and therefore, $d(x, \tilde{x}) \le \delta_i$. Due to our choice of δ_i this inequality, in turn, leads to the inequality $|f_{,i}(x) - f_{,i}(\tilde{x})| \le 2^{-(k+2)}$. In particular, we have $f_{,i}(x) \ge f_{,i}(\tilde{x}) - 2^{-(k+2)}$. Since $f_{,i}(\tilde{x}) > 2^{-k}$, we can conclude that $f_{,i}(x) > 2^{-k} - 2^{-(k+2)} = 3 \cdot 2^{-(k+2)}$. So, if $x_1 \in \mathbf{x}_1$, ..., and $x_m \in \mathbf{x}_m$, then $f_{,i}(x) \ge 3 \cdot 2^{-(k+2)}$. Therefore, the lower bound $\underline{a}_i$ of the actual range $f_{,i}(\mathbf{x}_1, \ldots, \mathbf{x}_m)$ satisfies the inequality $\underline{a}_i \ge 3 \cdot 2^{-(k+2)}$.

  Now, because of our choice of δ as $\delta = \min(\ldots, 2^{-(k+2)}/C_i, \ldots)$, we have $\delta \le 2^{-(k+2)}/C_i$. Hence, $C_i \cdot \delta \le 2^{-(k+2)}$. By the choice of C_i, this means that $|\underline{e}_i - \underline{a}_i| \le 2^{-(k+2)}$. Hence, $\underline{e}_i \ge \underline{a}_i - 2^{-(k+2)}$. We already know that $\underline{a}_i \ge 3 \cdot 2^{-(k+2)}$. Hence, $\underline{e}_i \ge 3 \cdot 2^{-(k+2)} - 2^{-(k+2)} = 2 \cdot 2^{-(k+2)} > 0$. The lower bound of the interval $\mathbf{e}_i$ is positive, hence, this interval can only contain positive numbers and cannot contain 0.

- If $f_{,i}(\tilde{x}) < -2^{-k}$, then we can similarly conclude that the upper bound $\overline{e}_i$ of the interval $\mathbf{e}_i$ is negative; hence, this interval contains only negative numbers and cannot contain 0.

So, for these intervals, the above algorithm gives the exact range. The theorem is proven.

Proof of Theorem 16.2. Theorem 16.2 follows from Theorem 16.1, if we take into consideration that the solution of a square $(n \times n)$ linear system is a *rational* function of the coefficients a_{ij} and b_i.

17

OPTIMIZATION – A FIRST EXAMPLE OF A NUMERICAL PROBLEM IN WHICH INTERVAL METHODS ARE USED: COMPUTATIONAL COMPLEXITY AND FEASIBILITY

In addition to the above-mentioned problems in which input is known with interval uncertainty, interval computations are also used to get guaranteed interval estimates for problems with purely numerical inputs. One such problem is optimization. In this chapter, we analyze computational complexity and feasibility of the corresponding optimization problems.

17.1. Interval methods are also used to solve numerical (non-interval) problems

So far, we have described interval methods for solving *interval* problems, i.e., problems in which the inputs are known with *interval* uncertainty. In addition to these problems, interval computations are also used to get *guaranteed* (interval) estimates for problems with purely *numerical* inputs. These are important applications of interval computations, and it is therefore desirable to analyze their computational complexity and feasibility.

A *number* x can be viewed as a particular case of an interval: namely, as a *degenerate interval* $\mathbf{x} = [x, x]$. Therefore, *numerical* computational problems can be viewed as particular cases of the corresponding *interval* problems. Hence:

- if a computational problem is *feasible* for interval inputs, it is feasible for numerical inputs as well;

- however, if a computational problem is *intractable* for *interval* inputs, its particular cases that correspond to *numerical* inputs may as well turn out to be feasible.

  For example, the problem of computing the value of a quadratic function $f(x_1, \ldots, x_n)$ is *NP-hard* for *interval* inputs $\mathbf{x}_1, \ldots, \mathbf{x}_n$, but straightforward and easy (and *feasible*) for *numerical* inputs.

It is therefore necessary, in addition to the above analysis of computational complexity and feasibility of *interval* problems, to analyze also the computational complexity and feasibility of the corresponding *numerical* problems. We will describe the results of this analysis in this chapter and in the few following chapters.

In this particular chapter, we analyze the class of numerical problems to which interval methods are most frequently used: namely, the *optimization* problems.

17.2. Constrained optimization: in brief

In most practical optimization problems, we know *a priori* bounds on the values of all variables and therefore, we have a *constrained* optimization problem.

In the basic problem of interval computations, we are interested in computing the endpoints $\underline{y}$ and $\overline{y}$ of the range $\mathbf{y} = f(\mathbf{x}_1, \ldots, \mathbf{x}_n)$ of the function $f(x_1, \ldots, x_n)$ for $x_i \in \mathbf{x}_i$. In other words, $\underline{y}$ is the *smallest* possible value of $f(x_1, \ldots, x_n)$ for $x_i \in \mathbf{x}_i$, and $\overline{y}$ is the *largest* possible value of $f(x_1, \ldots, x_n)$ under similar constraints. Therefore, the lower endpoint $\underline{y}$ of the desired interval $\mathbf{y}$ is the solution of the following (numerical) optimization problem:

$$f(x_1, \ldots, x_n) \to \min$$

under the constraints

$$\underline{x}_i \le x_i \le \overline{x}_i, \; 1 \le i \le n, \tag{17.1}$$

and the upper endpoint $\overline{y}$ is the solution of a similar optimization problem:

$$f(x_1, \ldots, x_n) \to \max$$

under the same constraints.

Hence, all our results about NP-hardness of interval computations can be reformulated as the results about NP-hardness of constrained optimization problems. In particular, we can conclude that *the constrained optimization problem for quadratic functions $f(x_1, \ldots, x_n)$ under interval constraints* (17.1) *is NP-hard*; that it is NP-hard even if we restrict ourselves to *bilinear* functions $f(x_1, \ldots, x_n)$, to quadratic functions $f(x_1, \ldots, x_n) = \sum a_{ij} \cdot x_i \cdot x_j + \sum a_i \cdot x_i + a_0$ with *sparse* matrices a_{ij}, etc.

Additional results on computational complexity and feasibility of constrained optimization problems can be found, e.g., in surveys Pardalos [324] and Horst *et al.* [157]. In particular, constrained optimization problems arising in *optimal control* turn out to be NP-hard (see, e.g., Abello *et al.* [2] and Smith *et al.* [399]).

17.2. Unconstrained optimization: finding the optimal value of the objective function $f(x_1, \ldots, x_n)$

In most practical optimization problems, we know *a priori* bounds on the values of all variables and therefore, we have a *constrained* optimization problem. In some practical cases, however, no *a priori* bounds are known: e.g., in *fundamental physics*, most equations are formulated in terms of *variational principles*: namely, the values $x_1, \ldots, x_n$ of physical quantities are such that a certain function $S(x_1, \ldots, x_n)$ (called *action* in physics) is optimized (see, e.g., Feynman [105]).

For example, a static configuration (e.g., a static configuration of electric charges) usually corresponds to the minimal *energy*, etc. In general, fundamental physical equations are usually formulated in the optimization form $S = \int L dV dt \to \min$ (here, dV denotes integration over a 3-D volume V, and dt denotes integration over time). This "variational principle" $S \to \min$ is the only condition on x_i, no additional *a priori* information is known; therefore, from the mathematical viewpoint, we have a problem of *unconstrained optimization.*

Since the unconstrained optimization problems are practically useful, it is important to analyze the computational complexity and feasibility of this class of problems.

Definition 17.1. *By the* ***precise unconstrained optimization problem,*** *we mean the following problem:*

GIVEN:

- *a positive integer* n*;*
- *a polynomial* $f(x_1, \ldots, x_n)$ *with rational coefficients.*

COMPUTE the value $\overline{y} = \sup f(x_1, \ldots, x_n)$*, where the supremum (least upper bound) is taken over all possible real numbers* $x_1, \ldots, x_n$ *(*$\overline{y}$ *is either a real number or a symbol* $+\infty$*).*

Comments.

- For infimum (greatest lower bound) $\underline{y} = \inf f(x_1, \ldots, x_n)$, a similar problem can be formulated, with $-\infty$ as a possible value.
- When we optimize a continuous (everywhere defined) function $f(x_1, \ldots, x_n)$ over a *bounded* (and closed) domain, then both $\sup f(x_1, \ldots, x_n)$ and $\inf f(x_1, \ldots, x_n)$ are real numbers. Over an unbounded domain, this is not always true: sup can be equal to $+\infty$, inf can be equal to $-\infty$.

Definition 17.2. *By the* ***$\varepsilon-$approximate unconstrained optimization problem,*** *we mean the following problem:*

GIVEN:

- *a positive integer* n*;*
- *a polynomial* $f(x_1, \ldots, x_n)$ *with rational coefficients;*
- *a rational number* $\varepsilon > 0$*.*

COMPUTE a real number $\tilde{y}$ *that is* $\varepsilon-$*close to* $\overline{y} = \sup f(x_1, \ldots, x_n)$*, where sup is taken over all possible real numbers* $x_1, \ldots, x_n$ *(i.e., either a real number* $\tilde{y}$ *for which* $|\tilde{y} - \overline{y}| \le \varepsilon$*, or* ∞ *is* $\overline{y} = +\infty$*).*

Our first comment is that this problem is algorithmically solvable:

Proposition 17.1. *There exists an algorithm that solves an arbitrary polynomial optimization problem.*

For example, we can use Tarski's algorithm (mentioned in Chapter 3) or one of its modern faster versions (mentioned in Chapter 4) to compute the desired minimum and maximum. Such algorithms have indeed been successfully applied to optimization (see, e.g., Weispfenning [425]). However, as we have mentioned in Chapter 3, such algorithms often requires *unrealistically long* time. So, it is desirable to know when a *feasible* algorithm is possible.

Optimization over an interval (i.e., over a *bounded* domain) is feasible (even linear time) for *linear* functions $f(x_1, \ldots, x_n)$ and *NP-hard* for *quadratic* (and higher order) polynomials. *Unconstrained* optimization turns out to be somewhat easier: it is *feasible* (polynomial time) for all *quadratic* and *cubic* polynomials as well:

Theorem 17.1.

- *There exists a polynomial-time algorithm that solves the unconstrained optimization problem for all polynomials of degree ≤ 3.*
- *For quartic polynomials, and for every $\varepsilon > 0$, the ε-approximate unconstrained optimization problem is NP-hard.*

The comparison between this result and the complexity of *constrained* optimization is given by the following table:

Objective function	Constrained optimization	Unconstrained optimization
Linear	Linear time	Linear time
Quadratic	NP-hard	Polynomial time
Cubic	NP-hard	Polynomial time
Quartic	NP-hard	NP-hard
5-th and higher degree	NP-hard	NP-hard

Comment. We want to attract the reader's attention to the fact that these comparison results may seem somewhat counterintuitive. Indeed:

- *Intuitively*, the complexity of the optimization problem depends on the *size* of the area in which a solution has to be found. This size describes the total number of objects that we need to analyze in order to find a solution; thus, the larger the size, the more complicated the problem. In general, this intuition is true: e.g., the more variables an optimization problem has, the more difficult it is to solve it. From this viewpoint, when we bind the variables x_i, then we drastically *decrease the size* of the area where the solution can be found. Therefore, it may seem at first glance that *constrained* optimization should be computationally *easier* (than unconstrained optimization).
- However, in reality, *unconstrained* optimization is computationally *easier*.

17.3. Unconstrained optimization: locating the values $x_1, \ldots, x_n$ for which the maximum is attained

In applications, we often want to know not only the *largest* (or the smallest) *value* of the objective function $f(x_1, \ldots, x_n)$, but also *where* exactly it is *attained*.

For *example*, in control applications, e.g., in designing the most fuel-efficient car, we want to know not only how much fuel can be saved, but also *how* to achieve these optimal savings.

In these cases, we get the following formulations:

Definition 17.3. *By the problem of precisely locating the optimizing values, we mean the following problem:*

GIVEN:

- *a positive integer n;*
- *a bounded-from-above polynomial $f(x_1, \ldots, x_n)$ with rational coefficients.*

COMPUTE the numbers $x_1, \ldots, x_n$ for which the objective function $f(x_1, \ldots, x_n)$ attains the largest possible value.

A similar problem can be formulated for computing the numbers for which the function $f(x_1, \ldots, x_n)$ attains the *smallest* possible value. In optimization, both types of *optimizing values* are also called the *optimal solution* to the original optimization problem.

Definition 17.4. *By the problem of ε-approximately locating the optimizing values, we mean the following problem:*

GIVEN:

- *a positive integer n;*
- *a bounded-from-above polynomial $f(x_1, \ldots, x_n)$ with rational coefficients;*
- *a rational number $\varepsilon > 0$.*

COMPUTE the rational numbers $\tilde{x}_1, \ldots, \tilde{x}_n$ that are ε-close to the numbers $x_1, \ldots, x_n$ for which the objective function $f(x_1, \ldots, x_n)$ attains the largest (correspondingly, smallest) possible value.

This problem seems to be somewhat more complicated than the problem of finding the largest value of f:

- If we know the vector $\vec{x} = (x_1, \ldots, x_n)$ where the maximum is attained, i.e., for which $\overline{y} = f(x_1, \ldots, x_n)$, then we can *easily compute* this maximum $\overline{y}$ by computing the value of the polynomial f for the known numbers x_i.
- On the other hand, even if we know the exact value of the maximum $\overline{y}$, it is still somewhat difficult to find the values $x_1, \ldots, x_n$ for which $f(x_1, \ldots, x_n)$ attains this maximum because this means solving a polynomial equation $f(x_1, \ldots, x_n) = \overline{y}$ in many variables, and no easy general algorithm is known for solving such equations (moreover, as we will see in the next chapter, the problem of solving such equations is NP-hard).

Our first comment is that this (somewhat more complicated) problem is still algorithmically *solvable*:

Proposition 17.2. *There exists an algorithm that locates the optimizing values for an arbitrary bounded-from-above polynomial objective function.*

For example, we can use Tarski's algorithm (mentioned in Chapter 3) or one of its modern faster versions to locate the optimizing values. Such algorithms have indeed been successfully applied to optimization (see, e.g., Weispfenning [425]). However, as we have mentioned in Chapter 3, such algorithms often require *unrealistically long* time. So, it is desirable to know when a *feasible* algorithm is possible.

Theorem 17.2.

- *There exists a polynomial-time algorithm that precisely locates the optimizing values for all polynomials of degree* ≤ 3.
- *For every* $\varepsilon > 0$*, an arbitrary algorithm that* ε*-approximately locates the optimizing values for all quadratic objective functions requires, for some instances, at least exponential time* ($O(2^n)$).

Comments.

- We will see from the proof that the result about exponential time holds even if we *know* the actual minimum value $\underline{y}$, and even if there is *only one* point $\vec{x} = (x_1, \ldots, x_n)$ at which this minimum is attained.
- This resulting computational complexity estimates are indeed somewhat harsher for this *location* problem than for the problem of *computing the minimum* (or *maximum*) *value*:

Objective function	Computing the maximum $\sup f(x_1, \ldots, x_n)$	Computing the values $x_1, \ldots, x_n$ for which the maximum is attained
Linear	Linear time	Linear time
Quadratic	Polynomial time	Polynomial time
Cubic	Polynomial time	Polynomial time
Quartic	NP-hard	Exponential time (or worse)
5-th and higher degree	NP-hard	Exponential time (or worse)

17.4. Optimization for polynomials with bounded coefficients

In Theorems 17.1 and 17.2, we considered polynomials with ***arbitrary coefficients***. It turns out that our computational complexity and feasibility results do not change if we impose ***a priori bounds*** on the values of these ***coefficients***:

Theorem 17.3.

- *For quartic polynomials with coefficients from the set* $\{0, 1, 2, 3\}$, *and for every* $\varepsilon > 0$, *the* ε*–approximate unconstrained optimization problem is NP-hard.*
- *For every* $\varepsilon > 0$, *an arbitrary algorithm that* ε*-approximately locates the optimizing values for all quadratic objective functions with coefficients from the set* $\{0, 1, 2\}$ *requires, for some instances, at least exponential time.*

Comment. We do not know whether each of these results holds for a smaller set of values, i.e.:

- whether NP-hardness results hold for values from the set $\{0, 1, 2\}$, and
- whether the exponential lower bound will stand if we only allow coefficients from the set $\{0, 1\}$ (i.e., coefficients that only take two values: 0 and 1).

17.5. Optimization problems with fixed number of variables

Theorems 17.1–17.3 show what happens if we restrict the *degrees* of the polynomials. If, instead, we restrict the *number of variables* n, then we get the following results:

Theorem 17.4. *For every* n, *there exists a polynomial-time algorithm that locates the optimizing values for all polynomials of* n *variables.*

Comment. This algorithm is similar to the one presented in Chapter 4: it is polynomial time, but it is not yet practical.

17.6. Stationary points

In some practical problems (including the problems from theoretical physics with which we started this chapter), we do not necessarily need the point $\vec{x} = (x_1, \ldots, x_n)$ where the *global minimum* or the *global maximum* is attained; it is often sufficient to find a *stationary* (extremal) point, in which all partial derivatives $\partial f/\partial x_i$ are equal to 0:

- In *theoretical physics*, traditional differential equations correspond exactly to the condition that the action S is *extremal*, not that we necessarily have a global optimum.
- Another example is when $y = f(x_1, \ldots, x_n)$ is potential energy. Then, a stationary point describes *equilibrium* (stable or unstable).

This problem turns out to be somewhat more complicated than the problem of finding global extrema:

Definition 17.5. *By the problem of precisely locating a stationary point, we mean the following problem:*

GIVEN:

- *a positive integer n;*
- *a polynomial $f(x_1, \ldots, x_n)$ with rational coefficients.*

COMPUTE the numbers $x_1, \ldots, x_n$ for which all n partial derivatives of the objective function $f(x_1, \ldots, x_n)$ equal 0.

Definition 17.6. *By the problem of $\varepsilon-$approximately locating a stationary point, we mean the following problem:*

GIVEN:

- *a positive integer n;*
- *a polynomial $f(x_1, \ldots, x_n)$ with rational coefficients;*
- *a rational number $\varepsilon > 0$.*

COMPUTE rational numbers $\tilde{x}_1, \ldots, \tilde{x}_n$ that are ε-close to the numbers $x_1, \ldots, x_n$ for which all n partial derivatives of the objective function $f(x_1, \ldots, x_n)$ equal 0.

Proposition 17.3. *There exists an algorithm that precisely locates stationary points of an arbitrary polynomial.*

(For example, we can use Tarski's algorithm.)

Theorem 17.5.

- *There exists a polynomial-time algorithm that precisely locates stationary points for all linear and quadratic polynomials.*
- *For every $\varepsilon > 0$, an arbitrary algorithm that ε-approximately locates the stationary points for all cubic objective functions requires, for some instances, at least exponential time.*

Comments.

- We will see from the proof that the result about exponential time holds even if there is *only one* stationary point.
- This exponential-time result is also true if we restrict ourselves to quartic polynomials in which each coefficient is equal to 0, 1, or 2.

Objective function	Computing the values $x_1, \ldots, x_n$ for which the maximum is attained	Computing stationary points
Linear	Linear time	Linear time
Quadratic	Polynomial time	Polynomial time
Cubic	Polynomial time	Exponential time (or worse)
Quartic	NP-hard	Exponential time (or worse)
5-th and higher degree	NP-hard	Exponential time (or worse)

17.7. Local optimization

In some practical problems, we need *more* than only *global* optima, but *less* than all the *stationary points*. For example, when $f(x_1, \ldots, x_n)$ is potential energy, then we may be looking only for *stable equilibria*, i.e., for *local minima*.

For computing *local minima* and *maxima*, we have the following complexity result:

Definition 17.7. *By the problem of precisely computing local maximum, we mean the following problem:*

GIVEN:

- *a positive integer* n;
- *a polynomial* $f(x_1, \ldots, x_n)$.

COMPUTE the numbers $x_1, \ldots, x_n$ *at which the objective function* $f(x_1, \ldots, x_n)$ *attains a local maximum.*

A similar problem can be formulated for computing local minima.

Definition 17.8. *By the problem of* ε*-approximately computing local maximum, we mean the following problem:*

GIVEN:

- *a positive integer* n;
- *a polynomial* $f(x_1, \ldots, x_n)$;
- *a rational number* $\varepsilon > 0$.

COMPUTE the rational numbers $\tilde{x}_1, \ldots, \tilde{x}_n$ *that are* ε*-close to the numbers* $x_1, \ldots, x_n$ *for which the objective function* $f(x_1, \ldots, x_n)$ *attains a local maximum.*

Proposition 17.4. *There exists an algorithm that precisely computes local maxima and local minima of an arbitrary polynomial.*

(For example, we can use Tarski's algorithm.)

Theorem 17.6.

- *There exists a polynomial-time algorithm that precisely locates local maxima and minima for all linear and quadratic polynomials.*
- *For every $\varepsilon > 0$, an arbitrary algorithm that ε-approximately locates local minima for all quartic objective functions requires, for some instances, at least exponential time.*

Comments.

- We will see from the proof that the result about exponential time holds even if there is *only one* local minimum.
- This exponential-time result is also true if we restrict ourselves to quartic polynomials in which each coefficient is equal to 0, 1, or 2.

Objective function	Global optimum	Stationary points	Local optimum
Linear	Linear time	Linear time	Linear time
Quadratic	Polynomial time	Polynomial time	Polynomial time
Cubic	Polynomial time	Exponential time time (or worse)	?
Quartic	NP-hard	Exponential time (or worse)	Exponential time (or worse)
5-th and higher degree	NP-hard	Exponential time (or worse)	Exponential time (or worse)

Proofs

Proof of Theorem 17.1. Let us first show that the unbounded optimization problem is easy for polynomials $f(x_1, \ldots, x_n)$ of degree ≤ 3.

Indeed, a polynomial of *0-th degree* is simply a constant: $f = a_0$. Therefore, $\sup f = \inf f = a_0$ (and this value is attained for an arbitrary tuple $\vec{x} = (x_1, \ldots, x_n)$).

If a polynomial of *first degree* $f(x_1, \ldots, x_n) = a_0 + a_1 x_1 + \ldots + a_n x_n$ is not a constant, this means that at least one of its coefficients a_i, $1 \le i \le n$, is different from 0. If we take $x_i \neq 0$ for this i and $x_j = 0$ for all $j \neq i$, then we can easily conclude that $\inf f = -\infty$ and $\sup f = +\infty$.

If a polynomial of *second degree*

$$f(x_1, \ldots, x_n) = a_0 + \sum_{i=1}^{n} a_i x_i + \sum_{i=1}^{n} \sum_{j=1}^{n} a_{ij} x_i x_j$$

is not a linear function, this means that at least one of the coefficients a_{ij} is different from 0. The maximum and minimum of the smooth function f are attained at a point where all the partial derivatives of f are equal to 0. Partial derivatives of a quadratic function are linear expressions, so, we get a system of linear equations to find the point where sup and inf are attained. There are known polynomial time algorithms (modified Gaussian elimination one of them) to solve systems of linear equations. Substituting these values into f, we get the desired inf and sup. (Degenerate case when the matrix a_{ij} is singular are also easy to handle: If the corresponding linear system has no solutions at all, this means that the quadratic function has no stationary points, so $\inf f = -\infty$ and $\sup f = +\infty$. If the system has infinitely many solutions, it is sufficient to take any of them.)

Let us now consider polynomials $f(x_1, \ldots, x_n)$ of *third degree* that are not quadratic. Each polynomial of this type can be represented as

$$f(x_1, \ldots, x_n) = f_0(x_1, \ldots, x_n) + f_1(x_1, \ldots, x_n) + f_2(x_1, \ldots, x_n) + f_3(x_1, \ldots, x_n),$$

where

$$f_0(x_1, \ldots, x_n) = a_0,$$

$$f_1(x_1, \ldots, x_n) = \sum_{i=1}^{n} a_i x_i,$$

$$f_2(x_1, \ldots, x_n) = \sum_{i=1}^{n} \sum_{j=1}^{n} a_{ij} x_i x_j,$$

$$f_3(x_1, \ldots, x_n) = \sum_{i=1}^{n} \sum_{j=1}^{n} \sum_{k=1}^{n} a_{ijk} x_i x_j x_k.$$

The fact that f is not a quadratic function means that $f_3(x_1, \ldots, x_n) \neq 0$ for some tuple $(x_1, \ldots, x_n)$. Then, for every real number λ, we can consider the value $g(\lambda) = f(\lambda x_1, \ldots, \lambda x_n)$. This new function g is a cubic function of λ:

$g(\lambda) = g_0 + \lambda g_1 + \lambda^2 g_2 + \lambda^3 g_3$, where we denoted $g_i = f_i(x_1, \ldots, x_n)$. This cubic polynomial attains all values from $-\infty$ to $+\infty$. Hence, for polynomials of third degree that are not quadratic, we have $\inf f = -\infty$ and $\sup f = +\infty$.

Let us now show that for every $\varepsilon > 0$, the ε–approximate unconstrained optimization problem for *quartic* polynomials is NP-hard.

Let us start with proving a slightly *weaker* result: that the *precise* unconstrained optimization problem is NP-hard for quartic polynomials. To prove this result, we will reduce the *PARTITION* problem (known to be NP-hard) to this problem. In the *PARTITION* problem, we are given a sequence of integers $s_1, \ldots, s_n$, and we must check whether there exist values $x_1, \ldots, x_n$ for which $x_i \in \{-1, 1\}$ and $s_1 \cdot x_1 + \ldots + s_n \cdot x_n = 0$.

For each instance of the *PARTITION* problem, we will construct the following quartic polynomial:

$$f(x_1, \ldots, x_n) =$$

$$(x_1 + 1)^2 \cdot (x_1 - 1)^2 + \ldots + (x_n + 1)^2 \cdot (x_n - 1)^2 + (s_1 \cdot x_1 + \ldots + s_n \cdot x_n)^2.$$

This polynomial is a sum of squares; therefore,

- it is always non-negative (i.e., greater than or equal to 0), and
- its value is equal to 0 if and only if all the squared terms are equal to 0, i.e., if $(x_i + 1) \cdot (x_i - 1) = 0$ for all i and $\sum s_i \cdot x_i = 0$.

Thus, if the infimum of this polynomial is equal to 0, this means that there exist values x_i for which $x_i \in \{-1, 1\}$ and $\sum s_i \cdot x_i = 0$, i.e., the answer to the given instance of the *PARTITION* problem is "yes". Vice versa, if the answer to the given instance is "yes", this means that there exist values $x_i \in \{-1, 1\}$ for which $\sum s_i \cdot x_i = 0$, and therefore, the infimum of the function $f(x_1, \ldots, x_n)$ is indeed equal to 0.

Thus, the infimum $\underline{y}$ of this function $f(x_1, \ldots, x_n)$ is equal to 0 if and only if the answer to the given instance of *PARTITION* problem is "yes". This reduction shows that the *precise* unconstrained optimization problem for quartic equations is indeed NP-hard.

To complete our proof, let us modify this reduction so that it will be applicable for *ε-approximate* unconstrained optimization. For this approximate problem, instead of the above function $f(x_1, \ldots, x_n)$, we will use a slightly different function $\tilde{f}(x_1, \ldots, x_n) = C \cdot f(x_1, \ldots, x_n)$; we will choose a constant $C > 0$ for which there is still a reduction, i.e., to be more precise, for which $\tilde{y} \le 2\varepsilon$ if and only if the answer to the given instance is "yes".

Indeed, if the answer is "yes", then the actual infimum $\underline{y}$ of the function $\tilde{f}(x_1, \ldots, x_n)$ is equal to 0, and therefore, its ε-approximation $\tilde{y}$ cannot exceed $\underline{y} + \varepsilon = 0 + \varepsilon = \varepsilon$; thus, $\tilde{y} \le \varepsilon < 2\varepsilon$.

Vice versa, if $\tilde{y} \le 2\varepsilon$, this means that the actual infimum $\underline{y}$ of the function $f(x_1, \ldots, x_n)$ is bounded by $\underline{y} \le \tilde{y} + \varepsilon \le 3\varepsilon$. Hence, there exist values $x_1, \ldots, x_n$ for which $\tilde{f}(x_1, \ldots, x_n) = C \cdot f(x_1, \ldots, x_n) \le 3\varepsilon$ and for which, therefore, $f(x_1, \ldots, x_n) \le \delta$, where we denoted $\delta = 3\varepsilon/C$. Since the function $f(x_1, \ldots, x_n)$ is the sum of several non-negative terms, this means that each of these terms is bounded by δ, i.e., $(x_i + 1)^2 \cdot (x_i - 1)^2 \le \delta$ and $(\sum s_i \cdot x_i)^2 \le \delta$.

Each of these $n + 1$ bounded terms is a square. Therefore, from the inequality on the square, we can extract two-sided inequalities on the squared expressions:

$$-\sqrt{\delta} \le (x_i + 1)(x_i - 1) \le \sqrt{\delta};$$

$$-\sqrt{\delta} \le s_1 \cdot x_1 + \ldots + s_n \cdot x_n \le \sqrt{\delta}.$$

The first inequality means that $-\sqrt{\delta} \le x_i^2 - 1 \le \sqrt{\delta}$, and $1 - \sqrt{\delta} \le x_i^2 \le 1 + \sqrt{\delta}$. From this inequality on x_i^2, we would like to extract an inequality for $|x_i| = \sqrt{x_i^2}$.

For $\delta < 1$, both bounds on x_i^2 are positive: $1 - \sqrt{\delta} < 1$ and $1 + \sqrt{\delta} > 1$. Thus, we can take square roots of all three sides of this double-sided inequality and conclude that $\sqrt{1 - \sqrt{\delta}} \le |x_i| \le \sqrt{1 + \sqrt{\delta}}$. This inequality is somewhat clumsy to use, so, we will try to deduce a slightly easier-to-handle inequality from it. To deduce this "easier" inequality, we will use the following two facts:

- For $z \ge 1$, we have $\sqrt{z} \le z$; therefore, $\sqrt{1 + \sqrt{\delta}} \le 1 + \sqrt{\delta}$.
- For $z \le 1$, we have $z \le \sqrt{z}$; therefore, $1 - \sqrt{\delta} \le \sqrt{1 - \sqrt{\delta}}$.

Therefore, from the above inequality on x_i^2, we can conclude that $1 - \sqrt{\delta} \le |x_i| \le 1 + \sqrt{\delta}$. Therefore:

- either $x_i \geq 0$ and $1 - \sqrt{\delta} \leq x_i \leq 1 + \sqrt{\delta}$,
- or $x_i \leq 0$ and $1 - \sqrt{\delta} \leq -x_i \leq 1 + \sqrt{\delta}$, in which case $-1 - \sqrt{\delta} \leq x_i \leq -1 + \sqrt{\delta}$.

In both cases, x_i is either $\sqrt{\delta}$-close to 1, or to -1. Since $\delta < 1$, we have $\sqrt{\delta} < 1$, and therefore, a number x_i cannot be $\sqrt{\delta}$-close to both 1 and -1. Let us denote by $\tilde{x}_i$ the number that is equal to 1 or -1 and that is $\sqrt{\delta}$-close to x_i: $|x_i - \tilde{x}_i| \leq \sqrt{\delta}$.

From the inequality $\sum s_i \cdot x_i \leq \sqrt{\delta}$, we can now conclude that

$$\sum s_i \cdot \tilde{x}_i = \sum s_i \cdot (x_i + (\tilde{x}_i - x_i)) = \sum s_i \cdot x_i + \sum s_i \cdot (\tilde{x}_i - x_i) \leq$$

$$\sqrt{\delta} + \sum |s_i| \cdot \sqrt{\delta} = \sqrt{\delta}(1 + \sum |s_i|).$$

So, $\sum s_i \cdot \tilde{x}_i \leq \Delta$, where we denoted $\Delta = \sqrt{\delta}(1 + \sum |s_i|)$. Similarly, $\sum s_i \cdot \tilde{x}_i \geq -\Delta$. Since s_i and $\tilde{x}_i$ are integers, the sum $S = \sum s_i \cdot \tilde{x}_i$ is also an integer. Therefore, if we choose δ in such a way that $\Delta = \sqrt{\delta}(1 + \sum |s_i|) < 1$, then from $-1 < -\Delta \leq S \leq \Delta < 1$, we conclude that $-1 < S < 1$, i.e., that $S = 0$. Thus, for such δ, we have n values $\tilde{x}_i \in \{-1, 1\}$ for which $\sum s_i \cdot \tilde{x}_i = 0$, and the answer to this instance of the *PARTITION* problem is indeed "yes".

To complete the proof, let us choose an appropriate $\varepsilon > 0$. The only condition on δ is

$$\Delta = \sqrt{\delta}(1 + \sum |s_i|) < 1,$$

i.e., $\sqrt{\delta} < 1/(1 + \sum |s_i|)$ (the original condition $\sqrt{\delta} < 1$ follows from this one). Therefore, we can, e.g., take δ for which $\sqrt{\delta} = 1/(2 \cdot (1 + \sum |s_i|))$, i.e., we can take $\delta = 1/(4 \cdot (1 + \sum |s_i|)^2)$. To achieve this value of $\delta = 3\varepsilon/C$, we must take $C = 3\varepsilon/\delta$. The theorem is proven.

Proof of Theorem 17.2. For *linear* and *cubic* (non-quadratic) polynomials, as we have mentioned in the proof of the previous theorem, the maximum is $+\infty$, the minimum is $-\infty$, so we can produce the answer "no such x_i" without even looking at the coefficients.

A *quadratic* (non-linear) objective function $f(x_1, \ldots, x_n) = a_0 + \sum a_i \cdot x_i + \sum a_{ij} \cdot x_i \cdot x_j$ has:

- a *finite* minimum (and *infinite* maximum) if the matrix a_{ij} is positive semi-definite;
- a *finite* maximum (and *infinite* minimum) if the matrix a_{ij} is negative semi-definite;
- *infinite* maximum and *infinite* minimum for all other matrices a_{ij}.

The maximum (minimum) is attained (if at all) at a point for which all n partial derivatives are equal to 0. All these derivatives are *linear* functions and therefore, to find the values x_i, we can solve a *system of linear equations* (which takes polynomial time).

Let us show that for *quartic* objective functions, we need exponential time to locate the optimizing values. Indeed, let us consider the following polynomial:

$$f(x_1, \ldots, x_n) = (x_1 - 2)^2 + (x_2 - x_1^2)^2 + (x_3 - x_2^2)^2 + \ldots + (x_n - x_{n-1}^2)^2.$$

This function is a sum of non-negative terms, and is, therefore, itself non-negative, and its minimum is non-negative.

Let us show that the minimum of this function is 0. For its value to be equal to 0, it is necessary for *all* the terms in the sum to be equal to 0, i.e., it is necessary that the following equations hold: $x_1 = 2$, $x_2 = x_1^2$, $x_3 = x_2^2$, ..., $x_n = x_{n-1}^2$. From this system, we can find x_1, x_2, ..., x_n, and get the (unique) solution $x_1 = 2$, $x_2 = 2^2$, $x_3 = (2^2)^2 = 2^{2^2}$, $x_4 = (2^{2^2})^2 = 2^{2^2 \cdot 2} = 2^{2^3}$, ..., $x_i = x_1^{2^{i-1}} = 2^{2^{i-1}}$. For these x_i, we indeed have $f(x_1, \ldots, x_n) = 0$. Thus, the minimum of this objective function is indeed 0, at it is attained for only one sequence of values $\vec{x} = (x_1, \ldots, x_n)$.

Let us now show that computing these values x_i is exponentially hard. Indeed, in binary representation, $x_n = 2^{2^{n-1}}$ is 1 followed by 2^{n-1} zeros. This number contains *exponentially many* bits, therefore, we need *exponential time* just to write this answer down. Similarly, if we are looking for a number that is ε-*close* to x_n, then, for large enough n, we will still need *exponentially many* bits. Thus, solving this particular systems of equation requires at least exponential time. The theorem is proven.

Proof of Theorem 17.3. Let us first prove the NP-hardness result. To prove it, we will use the same NP-hard problem *PARTITION* as in the proof of Theorem 17.1, but we will replace the polynomial

$$f(x_1, \ldots, x_n) = \sum (x_i^2 - 1)^2 + \left(\sum s_i \cdot x_i\right)^2$$

that was used in the proof of that theorem, by a polynomial whose coefficients only take values 0, 1, and 2.

The polynomial used in that proof is a sum of squares. Therefore, this polynomial is equal to 0 if and only if all the squared terms are equal to 0, i.e., if and only if the corresponding system of equations (consisting of equations $x_i^2 - 1 = 0$ and $\sum s_i \cdot x_i = 0$) is satisfied.

The new polynomial will also be equal to the sum of squares of linear and quadratic polynomials, so its value will be equal to 0 if and only if the corresponding system of equations is satisfied. (We will, therefore, interchangingly talk about the new polynomial $\tilde{f}$ and about the corresponding (new) system of equations.)

To get a *new* system of equations, let us describe, step-by-step, how each term in the left-hand side of the *original* system of equations is computed. We will introduce new variables to describe each intermediate step of these computations. Checking whether $x_i^2 = 1$ is done in two steps:

- First, we compute the square $q_i := x_i \cdot x_i$; this multiplication can be represented by the equation $q_i = x_i^2$.
- Then, we check whether $q_i = 1$.

Checking whether the sum $S = \sum s_i \cdot x_i$ is equal to 0, is done as follows: we start with the partial sum equal to 0 ($S_0 = 0$), and then, for i from 1 to n, we do the following two elementary operations:

- First, we multiply s_i and x_i: $q'_i := s_i \cdot x_i$ (equation $q'_i = s_i \cdot x_i$).
- Then, we add the result to the existing partial sum, thus getting the new partial sum: $S_i := S_{i-1} + q'_i$ (equation $S_i = S_{i-1} + q'_i$).

Finally, we check whether $S_n = 0$.

In this description, we assumed that the constants s_i are already given. If we want to be more precise, in the computer, the binary representations for these constants will be obtained bit-by-bit from the decimal values that we enter into the program. We can represent the process of generating these constants as the following computation process:

- First, we generate the powers of 2 that are needed to represent all the digits in all the binary expansions of all the values s_i: we start with $d_0 = 1$, and get $d_1 := d_0 + d_0$, $d_2 := d_1 + d_1$, ..., $d_i := d_{i-1} + d_{i-1}$, ... (equations $d_i = d_{i-1} + d_{i-1}$, $i = 1, 2, \ldots$).

  The total number of necessary powers of two does not exceed the length of the binary code of s_i, and therefore, does not exceed the length of the input.

- Then, each *positive* values of s_i can be represented as a sum of powers of two (e.g., $9_2 = 1001_2 = 2^3 + 2^0 = d_3 + d_0$), and this sum can be, in its turn, represent in the same step-by-step manner as the sum $\sum s_i \cdot x_i$.

- each *negative* coefficient s_i can be represented as $s_i = -|s_i|$, where $|s_i|$ is represented as before.

As a result, we get a system of simple equations, of the type $a = b \cdot c$, $a = b + c$, $a = b - c$, $a = -b$, $a = 0$, and $a = 1$, that is equivalent to the original system of equations.

For example, if $n = 2$, $s_1 = 9 = 2^3 + 2^0$, and $s_2 = -5 = -(2^2 = 2^0)$, we get the following systems of equations:

- $q_1 = x_1^2$, $q_1 = 1$, $q_2 = x_2^2$, $q_2 = 1$;
- $S_0 = 0$, $q_1' = s_1 \cdot x_1$, $S_1 = S_0 + q_1'$, $q_2' = s_2 \cdot x_2$, $S_2 = S_1 + q_2'$, $S_2 = 0$;
- $d_0 = 1$, $d_1 = d_0 + d_0$, $d_2 = d_1 + d_1$, $d_3 = d_2 + d_2$;
- (to compute $s_1 = 9$) $S_{10} = 0$, $S_{11} = S_{10} + d_0$, $s_1 = S_{11} + d_3$;
- (to compute $s_2 = -5$) $S_{20} = 0$, $S_{21} = S_{20} + d_0$, $S_{22} = S_{21} + d_2$, $s_2 = -S_{22}$.

In general, according to this system of equations, we start with 0 and 1, and perform elementary arithmetic operations and comparisons. Let us order all the constants and variables in the order of their appearance in this computation process, and let us correspondingly rename them by $r_0, r_1, r_2, \ldots$ Since we start with 0 and 1, we can always assume that $r_0 = 0$ and $r_1 = 1$.

In the above *example*, if we first compute the power of two (d_i), then the values s_i, and then check the equations, we will get the following new notations for the constant and variables:

- $r_0 = 0$, $r_1 = 1$, $r_1 = d_0$, $r_2 = d_1$, $r_3 = d_2$, $r_4 = d_3$;
- $r_5 = S_{10}$, $r_6 = S_{11}$, $r_7 = s_1$;
- $r_8 = S_{20}$, $r_9 = S_{21}$, $r_{10} = s_2$;
- $r_{11} = S_0$, $r_{12} = q'_1$, $r_{13} = S_1$, $r_{14} = q'_2$, $r_{15} = S_2$;
- $r_{16} = q_1$, $r_{17} = q_2$.

In terms of these new variables, on each computational step, we express each term r_i in terms of the previous terms r_j, $j < i$.

In principle, one and the same term r_j can be used in computing several terms r_i, and in each of these computations, it can be used several times: e.g., in $d_1 = d_0 + d_0$, the same value d_0 is used twice: as the first and as the second argument of addition. Inside the computer, the value r_j is *copied* as many times as necessary. To describe this "copying" process, let us introduce different variables $r_j^{(1)}, r_j^{(2)}, \ldots$, for different uses of r_j. For each of these variables, we will also add a new variable $s_j^{(k)}$ with the intended meaning $-r_j^{(k)}$. Then the equality between different copies of the same variable r_j can be described by the following equations: $r_j^{(1)} + s_j^{(1)} = 0$, $s_j^{(1)} + r_j^{(2)} = 0$, $r_j^{(2)} + s_j^{(2)} = 0$, $s_j^{(2)} + r_j^{(3)} = 0$, $\ldots$, $r_j^{(k)} + s_j^{(k)} = 0$, $s_j^{(k)} + r_j^{(k+1)} = 0$. From the first equation, we conclude that $s_j^{(1)} = -r_j^{(1)}$, then the second leads to $r_j^{(2)} = -s_j^{(1)} = r_j^{(1)}$, etc.

We will now use these "negative" variables s_j to re-write the elementary equations in a way that avoids negative coefficients:

- An equation $r_i = r_j + r_k$ is rewritten as $r_j + r_k + s_i = 0$.
- An equation $r_i = r_j - r_k$ is rewritten as $r_j + s_k + s_i = 0$.
- An equation $r_i = r_j \cdot r_k$ is rewritten as $r_j \cdot r_k + s_i = 0$.
- An equation $r_i = -r_j$ is rewritten as $r_j + r_i = 0$.
- An equation $r_i = 0$ stays.
- An equation $r_i = 1$ is rewritten as $s_i + 1 = 0$.

As a result, we get a system of equations with coefficients 0 and 1. As the desired function $\tilde{f}$, we will take the sum of the squares of all these equations.

After our renaming, each variable ($r_j^{(k)}$ or $s_j^{(k)}$) occurs at most in three equations: when it is first introduced, when it is copied to the next copy, and when it is used in some computations or checking. Thus, the coefficient at the variable's square is equal to 0, 1, 2, or 3. One can check that all other (non-square) terms in $\widetilde{f}$ occur only once, therefore, the coefficient at each of these terms is equal to 1 or to 2.

Since each equation is quadratic, the sum of their squares is a quartic polynomial. Thus, we get a quartic polynomial whose coefficients belong to the set $\{0, 1, 2, 3\}$ and whose minimum is equal to 0 if and only if the answer to the original instance of the *PARTITION* problem is "yes". Therefore, we have reduced *PARTITION* to the problem of computing the minimum and hence, the problem of computing the minimum is indeed NP-hard. The first statement is proven.

Let us now prove the exponential bound. For that, we will consider the following function of $n+3$ variables:

$$f(z^-, z^+, z_0, y_1, \ldots, y_n) =$$

$$(z^- + 1)^2 + (z^+ + 1)^2 + (z_0 + z^- + z^+)^2 + (y_1 + z_0)^2 +$$
$$(y_2 + y_1^2)^2 + (y_3 + y_2^2)^2 + \ldots + (y_n + y_{n-1}^2)^2.$$

If we open all the parentheses, we will easily see that every coefficient in this polynomial is indeed equal to 0, 1, or 2.

This objective function is non-negative, and it is equal to 0 if and only if all the squared terms in the sum (that forms this function) are equal to 0, i.e., if $z^- + 1 = 0$, $z^+ + 1 = 0$, $z_0 + z^- + z^+ = 0$, $y_1 + z_0 = 0$, $y_2 + y_1^2 = 0$, $y_3 + y_2^2 = 0$, ..., $y_n + y_{n-1}^2 = 0$. From the first four equations, we conclude that $z^- = z^+ = -1$, $z_0 = -(z^- + z^+) = 2$, and $y_1 = -2$. From the following equations, we conclude that $y_2 = -2^{2^1}$, $y_3 = -2^{2^2}$, ..., $y_i = -2^{2^{i-1}}$, ..., $y_n = -2^{2^{n-1}}$. Thus, similarly to Theorem 17.2, computing y_n requires exponentially many steps. The theorem is proven.

Proof of Theorem 17.4 is similar to the proofs from Chapter 4.

Proof of Theorem 17.5. Non-constant *linear* functions do not have stationary points, and stationary points of *quadratic* objective functions can be obtained as solutions of a system of linear equations (see the proof of Theorem 17.2).

Let us show that finding stationary points for *cubic* polynomials requires *exponential time*. Indeed, let us consider the following cubic polynomial with $n+1$ variables $x_0, x_1, \ldots, x_n$:

$$f(x_0, x_1, \ldots, x_n) = x_0 \cdot (x_1 - 2)^2 + x_0 \cdot (x_2 - x_1^2)^2 + x_0 \cdot (x_3 - x_2^2)^2 + \ldots +$$

$$x_0 \cdot (x_n - x_{n-1}^2)^2 + x_0^3.$$

What are the stationary points of this polynomial? Differentiating with respect to x_0 and equating the resulting partial derivative to 0, we conclude that

$$(x_1 - 2)^2 + (x_2 - x_1^2)^2 + (x_3 - x_2^2)^2 + \ldots + (x_n - x_{n-1}^2)^2 + 3x_0^2 = 0.$$

Since the sum of several non-negative terms is equal to 0, it means that each of the terms is equal to 0, i.e., $x_1 = 2$, $x_2 = x_1^2$, $x_3 = x_2^2$, ..., $x_n = x_{n-1}^2$.

Differentiation with respect to any other variable x_i leads to an expression that is proportional to x_0. Since, from $\partial f / \partial x_0 = 0$, we already know that $x_0 = 0$, it follows that all other partial derivatives are also equal to 0. Thus, stationary points of our cubic polynomial coincide with the solutions of a system of equations, the same system of equations whose solving, as we have already noticed in the proof of Theorem 17.2, requires at least exponential time. The theorem is proven.

Proof of the comment after Theorem 17.5. In our proof of Theorem 17.5, we started with a function from the proof of Theorem 17.2. If, instead of this starting function, we start with a function used in the proof of Theorem 17.3, we get an objective function

$$f(x_0, z^-, z^+, z_0, y_1, \ldots, y_n) = x_0 \cdot (z^- + 1)^2 + x_0 \cdot (z^+ + 1)^2 + x_0 \cdot (z_0 + z^- + z^+)^2 +$$

$$x_0 \cdot (y_1 + z_0)^2 + x_0 \cdot (y_2 + y_1^2)^2 + x_0 \cdot (y_3 + y_2^2)^2 + \ldots + x_0 \cdot (y_n + y_{n-1}^2)^2 + x_0^3,$$

we will be able to show exponential lower bound for cubic polynomials whose coefficients can only take values 0, 1, or 2. Comment is proven.

Proof of Theorem 17.6. For *linear* and *quadratic* objective functions, local maxima and minima coincide with global ones, so the result follows from Theorem 17.2.

Let us show that for *quartic* polynomials, the problem of computing local optima is *exponentially hard.* We already know, from Theorem 17.2, that for such polynomials, computing the values at which the *global maximum* is attained is *exponentially hard.* Let us show that for the function considered in Theorem 17.2, there are *no local minima except for the global minimum*, and thus, computing a local minimum is as hard as computing a global minimum, i.e., exponentially hard. We will even show that this local minimum is *the only* stationary point of this objective function.

Indeed, a stationary point of a function

$$f(x_1, \ldots, x_n) = (x_1 - 2)^2 + (x_2 - x_1^2)^2 + \ldots + (x_{n-1} - x_{n-2}^2)^2 + (x_n - x_{n-1}^2)^2$$

is a point where all its partial derivatives are equal to 0. Let us apply this condition to derivatives with respect to x_n, x_{n-1}, ..., x_1 (in this order).

- Differentiating with respect to x_n and equating the result to 0, we get $2(x_n - x_{n-1}^2) = 0$, i.e., $x_n = x_{n-1}^2$.
- Differentiating with respect to x_{n-1}, we get
$$2(x_{n-1} - x_{n-2}^2) + 2(x_n - x_{n-1}^2) \cdot (-2x_{n-1}) = 0.$$
Since we already know that $x_n - x_{n-1}^2 = 0$, we conclude that $2(x_{n-1} - x_{n-2}^2) = 0$ and $x_{n-1} = x_{n-2}^2$.
- Similarly, by equating $\partial f / \partial x_{n-2}$ to 0, and using the already proven equation $x_{n-1} = x_{n-2}^2$, we can conclude that $x_{n-2} = x_{n-3}^2$, etc.
- ...
- Finally, by equating $\partial f / \partial x_1$ to 0, we conclude that $x_1 = 2$.

So, at every stationary point $\vec{x} = (x_1, \ldots, x_n)$ of the objective function $f(x_1, \ldots, x_n)$ all the terms that form f are equal to 0, and therefore, $f(x_1, \ldots, x_n) = 0$, i.e., f indeed attains the global minimum. We already know that computing this global minimum requires exponential time. The theorem is proven.

18

SOLVING SYSTEMS OF EQUATIONS

In this chapter, we analyze the computational complexity and feasibility of yet another computational problem in which interval methods are often used: solving systems of equations. It turns out that already for systems of quadratic equations, solving these systems is NP-hard.

18.1. Solving systems of equations

Definition 18.1. *By precisely solving systems of polynomial equations, we mean the following problem:*

GIVEN:

- *an integer* n, *and*
- *a finite sequence of polynomials* $f_1(x_1, \ldots, x_n)$, $\ldots$, $f_k(x_1, \ldots, x_n)$ *with rational coefficients.*

COMPUTE: a tuple $x_1, \ldots, x_n$ *that satisfies all equations from the following system:*

$$\begin{aligned} f_1(x_1, \ldots, x_n) &= 0, \\ &\ldots \\ f_k(x_1, \ldots, x_n) &= 0. \end{aligned} \tag{18.1}$$

Definition 18.2. *By ε-approximately solving systems of polynomial equations, we mean the following problem:*

GIVEN:

- *an integer n,*
- *a finite sequence of polynomials $f_1(x_1, \ldots, x_n)$, ..., $f_k(x_1, \ldots, x_n)$ with rational coefficients, and*
- *a rational number $\varepsilon > 0$.*

COMPUTE: rational numbers $x_1, \ldots, x_n$ that are $\varepsilon-$close to a tuple that satisfies the system (18.1).

The first natural question is: is this problem algorithmically solvable at all? The answer to this question is given by the following proposition:

Proposition 18.1. *There exists an algorithm that solves an arbitrary system of polynomial equations.*

For example, we can use Tarski's algorithm (mentioned in Chapter 3) to compute the solutions of every system of polynomial equations. However, as we have mentioned, this algorithm often requires *unrealistically long* time. So, it is desirable to know when a *feasible* algorithm is possible. The result is as follows:

Theorem 18.1.

- *There exists a polynomial-time algorithm that precisely solves systems of linear equations.*
- *For every $\varepsilon > 0$, an arbitrary algorithm that ε-approximately solves all systems of quadratic equations requires, for some systems, at least exponential time.*

Comment. We will see from the proof that the result about exponential time holds even if we only consider systems of quadratic equations that have a *unique* solution.

Since numerical equations are particular (degenerate) case of interval equations, solving systems of *interval* quadratic equations is also at least exponentially hard. Thus, we can describe the following comparison between the computational complexity and feasibility of solving *numerical* and *interval* equations:

Polynomials	Solving numerical systems of equations	Solving interval systems of equations
Linear f_i	Polynomial time	NP-hard
Quadratic f_i	Exponential time (or worse)	Exponential time (or worse)
Cubic f_i (or higher order)	Exponential time (or worse)	Exponential time (or worse)

18.2. Checking whether a system of equations is solvable

Definition 18.3. *By checking solvability of systems of polynomial equations, we mean the following problem:*

GIVEN:

- *an integer n, and*
- *a finite sequence of polynomials $f_1(x_1, \ldots, x_n)$, ..., $f_k(x_1, \ldots, x_n)$ with rational coefficients.*

CHECK: whether the system of polynomial equations (18.1) *has a solution.*

Theorem 18.2.

- *There exists a polynomial-time algorithm that checks whether a system of linear equations is solvable.*
- *The problem of checking solvability of a system of quadratic equations is NP-hard.*

Theorems 18.1 and 18.2 can be represented as a table:

Polynomials	Solving systems of equations	Checking solvability of systems of equations
Linear f_i	Polynomial time	Polynomial time
Quadratic f_i	Exponential time (or worse)	NP-hard
Cubic (or higher order) f_i	Exponential time (or worse)	NP-hard

18.3. Systems of equations: other possible restrictions

Theorems 18.1 and 18.2 show what happens if we restrict the *degrees* of the polynomials. In these theorems, we did not restrict the *size* of the coefficients of the polynomials $f_j(x_1, \ldots, x_n)$, the *number of variables* n, or *number of equations* k. If we restrict one of these parameters, we get the following results.

If we impose *a priori* bounds on the coefficients of the polynomials f_k, the above computational complexity and feasibility results do not change, even if we require that all the coefficients of these polynomials are equal either to 0 or to 1. We will call such polynomials *0-1-polynomials* and corresponding equations *0-1-polynomial* equations. For example, linear equations with linear 0-1-polynomials will be called *0-1-linear* equations; quadratic equations with quadratic 0-1-polynomials will be called *0-1-quadratic*, etc.

Theorem 18.3.

- *For every $\varepsilon > 0$, an arbitrary algorithm that ε-approximately solves all systems of 0-1-quadratic equations requires, for some systems, at least exponential time.*
- *The problem of checking solvability of a system of 0-1-quadratic equations is NP-hard.*

If we restrict the *number of variables* n, the results change:

Theorem 18.4.

- *For every n, there exists a polynomial-time algorithm that ε-approximately solves an arbitrary system of polynomial equations with n unknowns.*
- *For every n, there exists a polynomial-time algorithm that checks solvability of an arbitrary system of polynomial equations with n unknowns.*

Comments.

- This algorithm is similar to the one presented in Chapter 4: it is polynomial-time, but it is not yet practical.
- If we fix the number of *equations* instead of the number of *variables*, the results are different. Namely, for the case $k = 1$, when we have a *single equation* instead of a system of equations, we get the following result:

Theorem 18.5.

- *There exists a linear-time algorithm that solves each linear equation.*
- *There exists a polynomial-time algorithm that checks solvability of each quadratic or cubic equation.*
- *The problem of checking solvability of an arbitrary quartic (4-th order) equation is NP-hard.*

Comments.

- From the proof of Theorem 17.3, it follows that for *quartic* polynomials, the problem if NP-hard even if we only allow polynomials whose coefficients come from the set $\{0, 1, 2, 3\}$.
- The results about checking solvability can be represented in the following table:

	Single equation	Systems of equations
Linear f_i	Linear time	Polynomial time
Quadratic f_i	Polynomial time	NP-hard
Cubic f_i	Polynomial time	NP-hard
Quartic f_i	NP-hard	NP-hard
5-th or higher order	NP-hard	NP-hard

Proofs

Proof of Theorem 18.1. It is well known that systems of *linear* equations can be solved in polynomial time (see, e.g., Cormen *et al.* [76]).

Let us show that solving systems of *quadratic* equations requires exponential time. Indeed, for a system $x_1 = 2$, $x_2 = x_1^2$, $x_3 = x_2^2$, ..., $x_n = x_{n-1}^2$, the only possible solution is $x_i = x_1^{2^{i-1}} = 2^{2^{i-1}}$. In binary representation, x_n is 1 followed by 2^{n-1} zeros. This number contains exponentially many bits, therefore, we need exponential time just to write this answer down. Similarly, if we are looking for a number that is ε-close to x_n, we will still need exponentially many bits. Thus, solving this particular system of equations requires at least exponential time. The theorem is proven.

Proof of Theorem 18.2. For systems of *linear* equations, known methods (modified Gaussian elimination for one) check solvability in polynomial time (see, e.g., Cormen *et al.* [76]).

Let us show that checking solvability of systems of *quadratic* equations is NP-hard. To prove this result, we will reduce the *PARTITION* problem (known to be NP-hard) to this problem. In the *PARTITION* problem, we are given a sequence of integers $s_1, \ldots, s_n$, and we must check whether there exist values $x_1, \ldots, x_n$ for which $x_i \in \{-1, 1\}$ and $s_1 \cdot x_1 + \ldots + s_n \cdot x_n = 0$.

For each instance of the *PARTITION* problem, let us consider the following system of $n+1$ quadratic equations: $(x_i + 1) \cdot (x_i - 1) = 0$, $1 \le i \le n$, and $s_1 \cdot x_1 + \ldots + s_n \cdot x_n = 0$.

- If real numbers x_i satisfy this system, then from the first n equations, we conclude that for each i, either $x_i = -1$, or $x_i = 1$, and from the $(n+1)$-st equation, that $\sum s_i \cdot x_i = 0$.

- Vice versa, if the values x_i satisfy the conditions of the *PARTITION* problem, then all $n+1$ equations hold.

So, our system of equations is solvable if and only if the answer to the given instance of the *PARTITION* problem is "yes". Hence, the *PARTITION* problem (that is known to be NP-hard) is reduced to our problem, and therefore, our problem is NP-hard too. The theorem is proven.

Proof of Theorem 18.3. Let us first show that computing the solutions of systems of 0-1-quadratic equations indeed requires exponential time. Indeed, let us consider the following system of equations with $2(n+1)$ unknowns $x_0, x_1, \ldots, x_n$, and $y_0, y_1, \ldots, y_n$: $x_0 + 1 = 0$, $y_0 + 1 = 0$, $x_1 + x_0 + y_0 = 0$, $x_i + y_i = 0$ $(1 \le i \le n)$, and $y_i + x_{i-1}^2 = 0$ $(2 \le i \le n)$. All the coefficients of all the polynomials in the left-hand side are equal to 0 or to 1.

If x_i and y_i satisfy this system of equations, then $x_0 = y_0 = -1$,

$$x_1 = -(x_0 + y_0) = 2,$$

and $x_i = -y_i = x_{i-1}^2$. Thus, for x_i, $1 \le i \le n$, we get the same system of equations that we have considered in the proof of Theorem 18.1 (and x_0 and y_i are easily computed from these x_i). For this system, $x_n = 2^{2^{n-1}}$ takes exponential time to compute. The statement is proven.

Let us now prove the *NP-hardness* result. To prove it, we will re-formulate the system from the proof of Theorem 18.2 in the desired 0-1-form. The equation $x_i^2 - 1 = 0$ is the easiest to represent in this form: we can take a *new variable* x_0, add a *new equation* $x_0 + 1 = 0$, and replace each equation $x_i^2 = 1$ by an equivalent equation $x_i^2 + x_0 = 0$.

To represent the equation $s_1 \cdot x_1 + \ldots + s_n \cdot x_n = 0$ in the 0-1-form, we replace each constant s_i by a *new variable* v_i, and add *new equations* that guarantee that $v_i = s_i$. How can we do that? In the computer, each constant s_i is represented as a binary number: e.g., if $s_1 = 5$, then $s_1 = 5_{10} = 111_2 = 1 \cdot 2^2 + 1 \cdot 2^1 + 1 \cdot 2^0$. The maximal power of 2 in these representations of s_i does not exceed the length of each of these numbers. Therefore, to describe these variables, we will introduce auxiliary variables $d_0^+, d_1^+, \ldots, \widetilde{d_0^+}, \widetilde{d_1^+}, \ldots, d_0^-, d_1^-, \ldots$, and $\widetilde{d_0^-}, \widetilde{d_1^-}, \ldots$ with the intended meaning $d_i^+ = \widetilde{d_i^+} = 2^i$ and $d_i^- = \widetilde{d_i^-} = -2^i$. The number of these auxiliary variables is equal to the largest length of the numbers $s_1, \ldots, s_n$. The following equations guarantee the desired values for x_i:

- First, the equations

$$d_0^- + 1 = 0,\ \tilde{d}_0^- + 1 = 0,\ d_0^+ + d_0^- = 0, \text{ and } \tilde{d}_0^+ + \tilde{d}_0^- = 0$$

guarantee that $d_0^- = \tilde{d}_0^- = -1$ and $d_0^+ = \tilde{d}_0^+ = -(-1) = 1$.

- Then, for every $i \geq 1$, four equations

$$d_i^+ + d_{i-1}^- + \tilde{d}_{i-1}^- = 0,\ \tilde{d}_i^+ + d_{i-1}^- + \tilde{d}_{i-1}^- = 0,$$

$$d_i^- + d_{i-1}^+ + \tilde{d}_{i-1}^+ = 0, \text{ and } \tilde{d}_i^+ + d_{i-1}^+ + \tilde{d}_{i-1}^+ = 0,$$

guarantee that if $d_{i-1}^+ = \tilde{d}_{i-1}^+ = 2^{i-1}$ and $d_{i-1}^- = \tilde{d}_{i-1}^- = -2^{i-1}$, then $d_i^+ = \tilde{d}_i^+ = 2^i$ and $d_i^- = \tilde{d}_i^- = -2^i$.

Then, for every i from 1 to n, to guarantee that $v_i = s_i$, we add the following equation:

- If $s_i \geq 0$, then we take the binary representation of s_i:

$$s_i = \varepsilon_0 \cdot 2^0 + \varepsilon_1 \cdot 2^1 + \ldots + \varepsilon_i \cdot 2^i + \ldots \quad (\varepsilon_i \in \{0,1\}),$$

and add an equation

$$v_i + \varepsilon_0 \cdot d_0^- + \varepsilon_1 \cdot d_1^- + \ldots + \varepsilon_i \cdot d_i^- + \ldots = 0$$

- If $s_i < 0$, then we take the binary representation of $-s_i$:

$$-s_i = \varepsilon_0 \cdot 2^0 + \varepsilon_1 \cdot 2^1 + \ldots + \varepsilon_k \cdot 2^i + \ldots \quad (\varepsilon_i \in \{0,1\}),$$

and add an equation

$$v_i + \varepsilon_0 \cdot d_0^+ + \varepsilon_1 \cdot d_1^+ + \ldots + \varepsilon_i \cdot d_i^+ + \ldots = 0$$

In the resulting system of equations, we have $v_i = s_i$ and therefore, the equation $v_1 \cdot x_1 + \ldots + v_n \cdot x_n = 0$ is equivalent to $s_1 \cdot x_1 + \ldots + s_n \cdot x_n = 0$.

Thus, the *new system* of quadratic equations, in which all polynomials are 0-1-polynomials, is *equivalent* to the *original one*; hence, solving an instance of the *PARTITION* problem is equivalent to this problem. Therefore, we have reduced the *PARTITION* problem to the problem of solving systems of 0-1-quadratic equations and so, the problem of solving these equations is indeed NP-hard. The theorem is proven.

Proof of Theorem 18.4 is similar to the proofs from Chapter 4.

Proof of Theorem 18.5. A *linear* equation $a_0 + a_1 \cdot x_1 + \ldots + a_n \cdot x_n = 0$ is solvable if either one of the coefficients $a_1, \ldots, a_n$ is different from 0, or all the values $a_1, \ldots, a_n, a_0$ are equal to 0.

- In the first case, if $a_i \neq 0$, we can take $x_i = -a_0/a_i$, and $x_j = 0$ for all $j \neq i$.
- In the second case, arbitrary values $x_1, \ldots, x_n$ form a solution, so we can take, e.g., $x_1 = \ldots = x_n = 0$.

This algorithm takes *linear time*.

The possibility of checking solvability of quadratic and cubic equations $f(x_1, \ldots, x_n) = 0$ follows from Theorem 17.2: Indeed, due to continuity of a a polynomial $f(x_1, \ldots, x_n)$, solvability of this equation is equivalent to the condition $0 \in [\inf f, \sup f]$. According to Theorem 17.2, for a quadratic or cubic polynomial, we can compute the values $\inf f$ and $\sup f$ in polynomial time and thus, check, in polynomial time, whether the corresponding polynomial equation is solvable.

Let us prove that for *quartic* polynomials, solving a polynomial equation is NP-hard. We will prove it by reducing the same *PARTITION* problem as we did in the proof of Theorem 18.2. Namely, for each instance of the *PARTITION* problem, we can consider the equation $f(x_1, \ldots, x_n) = 0$, where $f(x_1, \ldots, x_n) = (x_1 + 1)^2 \cdot (x_1 - 1)^2 + \ldots + (x_n + 1)^2 \cdot (x_n - 1)^2 + (s_1 \cdot x_1 + \ldots + s_n \cdot x_n)^2$. Since this quartic polynomial is a sum of squares, it is equal to 0 if and only if all the squared terms are equal to 0, i.e., if $(x_i + 1) \cdot (x_i - 1) = 0$ for all i and $\sum s_i \cdot x_i = 0$. We already know (from the proof of Theorem 18.2) that the solvability of this system is equivalent to the "yes" answer to the given instance of the *PARTITION* problem. Thus, the equation $f(x_1, \ldots, x_n) = 0$ is solvable if and only if the answer to this instance is "yes". This reduction shows that solvability of quartic equations is NP-hard. The theorem is proven.

19

APPROXIMATION OF INTERVAL FUNCTIONS

Another problem where interval computations are used is a problem of approximating functions with simpler ones. In this chapter, we show that already the problem of optimal (narrowest) approximation of a quadratic interval function $\mathbf{f}(x_1, \ldots, x_n)$ by a linear one is NP-hard. For a practically important 1D case ($n = 1$), an efficient approximation is possible.

This chapter was written in collaboration with M. Koshelev and L. Longpré.

19.1. Introduction: why approximation?

In many practical problems, we know that a quantity y depends on the quantities $x_1, \ldots, x_n$ (i.e., that $y = f(x_1, \ldots, x_n)$ for some function f), but we do not know the exact form of this dependence; instead, for every $x = (x_1, \ldots, x_n)$, we know an *interval* $\mathbf{y} = [\underline{y}, \overline{y}]$ of possible values of y.

For example, to determine the desired dependence, we may:

- measure the value of y at certain points $x^{(1)}, \ldots, x^{(M)}$, thus getting the intervals $\mathbf{f}(x^{(1)}), \ldots, \mathbf{f}(x^{(M)})$, and then
- use the *a priori* known bounds D_i on the *rate of change* of f, i.e., on the partial derivatives:
$$\left|\frac{\partial f}{\partial x_i}\right| \le D_i,$$
to estimate the values of $f(x)$ for all other x.

If we make only one measurement, then we get a *piece-wise linear* estimate

$$\mathbf{f}(x_1,\ldots,x_n) = [\underline{f}(x_1,\ldots,x_n), \overline{f}(x_1,\ldots,x_n)],$$

where

$$\underline{f}(x_1,\ldots,x_n) = \underline{f}(x_1^{(1)},\ldots,x_n^{(1)}) - D_1 \cdot |x_1 - x_1^{(1)}| - \ldots - D_n \cdot |x_n - x_n^{(1)}|;$$

$$\overline{f}(x_1,\ldots,x_n) = \overline{f}(x_1^{(1)},\ldots,x_n^{(1)}) + D_1 \cdot |x_1 - x_1^{(1)}| + \ldots + D_n \cdot |x_n - x_n^{(1)}|.$$

For several measurements ($M > 1$), we get a slightly *more complicated* but still *piece-wise linear* bounds on $f(x_1,\ldots,x_n)$.

The case $n = 1$ is especially practically important, because it represents the dependence on time $x_1 = t$:

- we measure the values of physical quantities at different moments of time and then
- we use *a priori* bounds on the time derivative to estimate the values of these quantities at intermediate moments of time.

In particular, in Lhomme *et al.* [252, 253] and in Loiez *et al.* [256, 257], such piece-wise linear interval functions were effectively used in *electrical and electronic engineering*: namely, in analysis of circuits with linear elements.

In many real-life applications, we need to *process* the resulting values y, and the data processing algorithms are often *non-linear*. As a result:

- if we *start* with the intervals for $y(x)$ that are *piece-wise linear* in x,
- we *end up* with intervals for the result r of data processing whose dependence on x is much *more complicated*.

For example:

- if we *add* or *subtract* two interval functions whose endpoints are piece-wise linear in x, we still get the result that is piece-wise linear in x; but
- if we *multiply* two intervals that are linear in x, then their endpoints also get multiplied, and, as a result, we get an interval function in which the endpoints are *quadratic* functions of x (i.e., we get *quadratic interval functions*).

If we multiply more, we get cubic, quartic, etc. functions; see, e.g., Loiez *et al.* [257]. In principle, there is nothing wrong with this complexity, except for the fact that, e.g., while a linear interval function of one variable requires only 4 numbers to store (2 coefficients of the lower endpoint and 2 coefficients of the upper endpoint), quadratic functions require 6 coefficients, cubic functions require 8, etc. The more complicated the function becomes, the more memory we need to store these coefficients, and the longer it takes to process these functions.

Thus, since we are often limited both in processing time and in memory (especially if the processing is done in an on-board computer), we must ***approximate*** the given complicated interval function by a simpler one.

In other words, if we have an interval function $\mathbf{y}(x) = [\underline{y}(x), \overline{y}(x)]$ that is known to contain the actual value $y(x)$, we want to be able to find a simpler interval function $\mathbf{z}(x) = [\underline{z}(x), \overline{z}(x)]$ that, for each x, contains the entire interval $\mathbf{y}(x)$ and is, thus, guaranteed to contain the actual value $y(x)$.

Of course, this approximation comes at a trade-off: we simplify the expression, but we make the interval wider (and therefore, lose some information). Therefore, the narrower the approximating interval function, the better.

The simplest approximation problem of this type is the problem of approximating a quadratic interval function by linear ones. Due to the practical importance, this problem has been analyzed in several papers; see, e.g., Schmitgen [379], Oelschlägel *et al.* [314, 315, 313, 316], Rokne [363], Beaumont [22].

In this chapter, following Koshelev *et al.* [183, 182]:

- we show that, *in general*, this problem is *computationally intractable* (NP-hard); and
- we will present an *efficient* optimal solution to the practically important *1D case* of this problem.

19.2. When is an approximation optimal? Mathematical formulation of the problem

Definition 19.1. *Let a box* $\mathbf{X} = [\underline{x}_1, \overline{x}_1] \times \ldots \times [\underline{x}_n, \overline{x}_n]$ *be fixed. By an interval function, we mean a mapping* $\mathbf{y}$ *that puts into correspondence to each element* $x \in \mathbf{X}$ *an interval* $\mathbf{y}(x) = [\underline{y}(x), \overline{y}(x)]$.

Definition 19.2. *If both functions* $\underline{y}(x)$ *and* $\overline{y}(x)$ *are linear functions of* x*, i.e.:*

- $\underline{y}(x) = \underline{y}_0 + \sum \underline{y}_i \cdot x_i$ *for some constants* $\underline{y}_0$ *and* $\underline{y}_i$, *and*
- $\overline{y}(x) = \overline{y}_0 + \sum \overline{y}_i \cdot x_i$,

we say that the interval function $\mathbf{y}(x) = [\underline{y}(x), \overline{y}(x)]$ *is linear.*

Definition 19.3. *If both functions* $\underline{y}(x)$ *and* $\overline{y}(x)$ *are quadratic functions of* x*, i.e.,*

- $\underline{y}(x) = \underline{y}_0 + \sum \underline{y}_i \cdot x_i + \sum \underline{y}_{ij} \cdot x_i \cdot x_j$ *for some constants* $\underline{y}_0$, $\underline{y}_i$, *and* $\underline{y}_{ij}$, *and*
- $\overline{y}(x) = \overline{y}_0 + \sum \overline{y}_i \cdot x_i + \sum \overline{y}_{ij} \cdot x_i \cdot x_j$,

we say that the interval function is quadratic.

Definition 19.4. *We say that a linear interval function* $\mathbf{z}(x) = [\underline{z}(x), \overline{z}(x)]$ *approximates a quadratic interval function* $\mathbf{y}(x) = [\underline{y}(x), \overline{y}(x)]$ *(or is an approximation of* $\mathbf{y}$*) if* $\mathbf{y}(x) \subseteq \mathbf{z}(x)$ *for all* x.

Comment. The narrower the intervals, the better. So, our goal is to minimize the worst-case width of the approximating interval, i.e., the value $W(\mathbf{z}) = \max_x(\overline{z}(x) - \underline{z}(x))$.

We will see that in some cases, for a given quadratic interval function $\mathbf{y}(x)$, there are several approximating linear interval functions $\mathbf{z}$ with the same value of $W(\mathbf{z})$. If two different approximating functions have the same worst-case widths, then it is reasonable to choose the one for which the best-case width $w(\mathbf{z}) = \min_x(\overline{z}(x) - \underline{z}(x))$ is the smallest. Thus, we arrive at the following definition:

Definition 19.5.

- *For every interval function* $\mathbf{z}(x)$*:*
  - *by its worst-case width, we mean the value*

$$W(\mathbf{z}) = \max_{x \in \mathbf{X}} (\overline{z}(x) - \underline{z}(x)).$$

  - *by its best-case width, we mean the value*

$$w(\mathbf{z}) = \min_{x \in \mathbf{X}} (\overline{z}(z) - \underline{z}(x)).$$

- *Let a quadratic interval function* $\mathbf{y}$ *be fixed on a box* $\mathbf{X}$*. We say that a linear function* $\mathbf{z}$ *is an optimal approximation of* $\mathbf{y}$ *if the following conditions are satisfied:*
  - *first,* $\mathbf{z}$ *approximates* $\mathbf{y}$*;*
  - *second, among all linear approximations to* $\mathbf{y}$*, the function* $\mathbf{z}$ *has the smallest value of the worst-case width* $W(\mathbf{z})$*;*
  - *third, if there exist several linear approximations* $\mathbf{u}$ *to* $\mathbf{y}$*, with the same smallest value of the worst-case width* $W(\mathbf{u})$*, the function* $\mathbf{z}$ *has the smallest value of the best-case width* $w(\mathbf{z})$*.*

19.3. Results

In general, this problem is computationally intractable (NP-hard):

Theorem 19.1. (Koshelev *et al.* [182]) *The problem of computing the optimal linear approximation to a given quadratic interval function is NP-hard.*

For a practically important 1D case ($n = 1$), an efficient algorithm is possible:

Proposition 19.1. ($n = 1$; Koshelev *et al.* [183, 182]) *The following Algorithm 19.1 computes the optimal linear approximation to a given quadratic interval function of one variable.*

The resulting complexity of the approximation problem can be represented by the following table:

$n=1$	n fixed and finite	general case
Feasible	?	NP-hard

Comment. If we are given a *piece-wise* quadratic function, then we can apply this algorithm to each quadratic piece and thus, get the optimal piece-wise linear approximation.

Algorithm 19.1. (Koshelev *et al.* [183, 182]) *We start with a quadratic interval function*

$$\mathbf{y}(t) = [\underline{y}_0 + \underline{y}_1 \cdot t + \underline{y}_2 \cdot t^2, \overline{y}_0 + \overline{y}_1 \cdot t + \overline{y}_2 \cdot t^2]$$

defined on an interval $[\underline{t}, \overline{t}]$. *The formulas for the optimal linear approximation* $\mathbf{z}(t) = [\underline{z}(t), \overline{z}(t)]$ *depend on the signs of the coefficients* ($\underline{y}_2$ *and* $\overline{y}_2$) *at the quadratic term* t^2.

- *Case 1:* $\underline{y}_2 \le 0$, $\overline{y}_2 \ge 0$.
  In this case, the function $\underline{z}(t)$ *is the secant of* $\underline{y}(t)$, *i.e., a straight line whose endpoints are the endpoints of the quadratic function* $\underline{y}(t)$ *on this interval. Similarly,* $\overline{z}(t)$ *is a secant of* $\overline{y}(t)$, *i.e.,*

$$\underline{z}(t) = \underline{y}(\underline{t}) + \frac{\underline{y}(\overline{t}) - \underline{y}(\underline{t})}{\overline{t} - \underline{t}} \cdot (t - \underline{t}), \tag{2}$$

$$\underline{z}(t) = \overline{y}(\underline{t}) + \frac{\overline{y}(\overline{t}) - \overline{y}(\underline{t})}{\overline{t} - \underline{t}} \cdot (t - \underline{t}). \tag{3}$$

- *Case 2:* $\underline{y}_2 \le 0$, $\overline{y}_2 < 0$.
  In this case, the lower line $\underline{z}(t)$ *is a secant (2). To determine the upper line* $\overline{z}(t)$, *we apply the formula*

$$t_m = -(\overline{y}_1 - \underline{z}_1)/(2\overline{y}_2). \tag{4}$$

to compute the value t_m. *Then:*

  - *If* $t_m \in [\underline{t}, \overline{t}]$, *we take* $\overline{z}(t) = \underline{z}(t) + (\overline{y}(t_m) - \underline{z}(t_m))$.
  - *If* $t_m > \overline{t}$, *then* $\overline{z}$ *is the tangent to* $\overline{y}$ *at* $\overline{t}$*:*

$$\overline{z}(t) = \overline{y}(\overline{t}) + (\overline{y}_1 + 2\overline{y}_2 \cdot \overline{t})(t - \overline{t}). \tag{5}$$

  - *If* $t_m < \underline{t}$, *then* $\overline{z}$ *is the tangent to* $\overline{y}$ *at* $\underline{t}$*:*

$$\overline{z}(t) = \overline{y}(\underline{t}) + (\overline{y}_1 + 2\overline{y}_2 \cdot \underline{t})(t - \underline{t}). \tag{6}$$

- *Case 3:* $\underline{y}_2 > 0$, $\overline{y}_2 \geq 0$.
  In this case, the upper line $\overline{z}(t)$ *is a secant* (3). *To determine the lower line* $\underline{z}(t)$, *we compute the value* $t_m = -(\underline{y}_1 - \overline{z}_1)/(2\underline{y}_2)$. *Then:*
  - *If* $t_m \in [\underline{t}, \overline{t}]$, *we take* $\underline{z}(t) = \overline{z}(t) - (\overline{z}(t_m) - \underline{y}(t_m))$.
  - *If* $t_m > \overline{t}$, *then* $\underline{z}$ *is the tangent to* $\underline{y}$ *at* $\overline{t}$*:*
  $$\underline{z}(t) = \underline{y}(\overline{t}) + (\underline{y}_1 + 2\underline{y}_2\overline{t})(t - \overline{t}). \tag{7}$$
  - *If* $t_m < \underline{t}$, *then* $\underline{z}$ *is the tangent to* $\underline{y}$ *at* $\underline{t}$*:*
  $$\underline{z}(t) = \underline{y}(\underline{t}) + (\underline{y}_1 + 2\underline{y}_2\underline{t})(t - \underline{t}). \tag{8}$$
- *Case 4:* $\underline{y}_2 > 0$, $\overline{y}_2 < 0$.
  In this case, we first compute $t_M = -(\overline{y}_1 - \underline{y}_1)/[2(\overline{y}_2 - \underline{y}_2)]$. *Then:*
  - *If* $t_M \in [\underline{t}, \overline{t}]$, *then* $\underline{z}$ *is the tangent to* $\underline{y}$ *at* t_M, *and* $\overline{z}$ *is the tangent to* $\overline{y}$ *at* t_M*:* $\underline{z}(t) = \underline{y}(t_M) + (\underline{y}_1 + 2\underline{y}_2 \cdot t_M)(t - t_M)$; $\overline{z}(t) = \overline{y}(t_M) + (\overline{y}_1 + 2\overline{y}_2 \cdot t_M)(t - t_M)$.
  - *If* $t_M > \overline{t}$, *then* $\underline{z}$ *is the tangent* (7) *to* $\underline{y}$ *at* $\overline{t}$, *and* $\overline{z}$ *is the tangent* (5) *to* $\overline{y}$ *at* $\overline{t}$*;*
  - *If* $t_M < \underline{t}$, *then* $\underline{z}$ *is the tangent* (8) *to* $\underline{y}$ *at* $\underline{t}$ *and* $\overline{z}$ *is the tangent* (6) *to* $\overline{y}$ *at* $\underline{t}$.

Proofs

Proof of Theorem 19.1. To show that our problem $\mathcal{P}$ is NP-hard, we will reduce a problem $\mathcal{V}$ already known to be NP-hard to $\mathcal{P}$. As such a problem $\mathcal{V}$, we will take the problem of *minimizing a given non-negative quadratic function* $y(x)$ *on a given box* $\mathbf{X}$. (The fact that the problem $\mathcal{V}$ is NP-hard was shown in Chapter 3.)

Indeed, let us assume that an algorithm $\mathcal{U}$ solves all particular cases of our problem $\mathcal{P}$ in polynomial time. Then, we can find the minimum m of a given non-negative quadratic function $y(x)$ as follows:

1. We compute an upper bound B for the quadratic function $y(x)$ on a box $\mathbf{X}$. This upper bound can be obtained, e.g., by using naive interval computations.

2. We apply the algorithm $\mathcal{U}$ to a quadratic interval function

$$\mathbf{y}(x) = [0, B - y(x)].$$

As a result of applying this algorithm, we get the linear interval function $\mathbf{z}(x) = [\underline{z}(x), \overline{z}(x)]$ that is the optimal approximation of $\mathbf{y}(x)$.

3. We compute the worst-case width $W = W(\mathbf{z})$ of the linear interval function $\mathbf{z}(x)$, i.e., the maximum of a linear function

$$z(x) = \overline{z}(x) - \underline{z}(x) = z_0 + z_1 \cdot x_1 + \ldots + z_n \cdot x_n$$

on a box $\mathbf{X} = [\underline{x}_1, \overline{x}_1] \times \ldots \times [\underline{x}_n, \overline{x}_n]$. This maximum is easy to compute:

$$W = z(\text{mid}(\mathbf{X})) + |z_1| \cdot \text{half}(\mathbf{x}_1) + \ldots + |z_n| \cdot \text{half}(\mathbf{x}_n),$$

where

$$\text{mid}(\mathbf{X}) = \left(\frac{\underline{x}_1 + \overline{x}_1}{2}, \ldots, \frac{\underline{x}_n + \overline{x}_n}{2}\right)$$

is the midpoint of the box, and

$$\text{half}([\underline{x}_i, \overline{x}_i]) = \frac{\overline{x}_i - \underline{x}_i}{2}$$

is the radius (half-width) of the corresponding interval.

4. Finally, we take the difference $B - W$ as the desired value of the minimum m.

Let us show that the value $B - W$ computed by our algorithm is indeed the minimum of the original function $y(x)$:

- First, we will show that $m \geq B - W$.

  Indeed, since the function $\mathbf{z}(x)$ is an approximation to $\mathbf{y}(x)$, we have $\mathbf{y}(x) \subseteq \mathbf{z}(x)$; hence, for every x, the interval $\mathbf{z}(x)$ is wider than or equal to the interval $\mathbf{y}(x)$. Therefore, the worst-case width $W(\mathbf{z})$ is greater than or equal to the worst-case width $W(\mathbf{y})$ of the quadratic interval function $\mathbf{y}(x)$: $W = W(\mathbf{z}) \geq W(\mathbf{y})$. To use this inequality, we must find the value $W(\mathbf{y})$.

  For every x, the width $\overline{y}(x) - \underline{y}(x)$ of the interval $\mathbf{y}(x) = [0, B - y(x)]$ is equal to $B - y(x)$. Therefore, the largest possible value $W(\mathbf{y})$ of this width is attained when $B - y(x)$ is the largest possible, i.e., when $y(x)$ is the smallest possible. Hence, $W(\mathbf{y}) = \max(B - y(x)) = B - \min y(x) = B - m$.

  So, $W = W(\mathbf{z}) \geq W(\mathbf{y}) = B - m$, and therefore, $m \geq B - W$.

- On the other hand, since $\mathbf{z}$ is the *optimal* approximation to $\mathbf{y}$, its worst-case width $W = W(\mathbf{z})$ must be the smallest possible of the worst-case widths $W(\mathbf{u})$ of all linear interval approximations $\mathbf{u}(x)$ to the function $\mathbf{y}(x)$.

  In particular, since $y(x) \geq m$ for all x, we have $B - y(x) \leq B - m$ and therefore, a linear interval function $\mathbf{u}(x) = [0, B - m]$ is an approximation for $\mathbf{y}(x) = [0, B - y(x)]$. The worst-case width $W(\mathbf{u})$ is equal to

  $$W(\mathbf{u}) = \max(B - m) = B - m.$$

  Therefore, $W = W(\mathbf{z}) \leq W(\mathbf{u}) = B - m$, i.e., $W \leq B - m$ and thence, $m \leq B - W$.

So, $m \geq B - W$ and $m \leq B - W$ and hence, $m = B - W$. The reduction is proven, so, our problem is indeed NP-hard. The theorem is proven.

Proof of Proposition 19.1. Let us present the proofs for all 4 cases.

Case 1: $\underline{y}_2 \leq 0$, $\overline{y}_2 \geq 0$. In this case, both functions $\underline{y}(x)$ and $\overline{y}(x)$ are "pointing inward".

If $\mathbf{z}(t)$ is a linear interval approximation to the given quadratic interval function, then, from $\mathbf{y}(t) \subseteq \mathbf{z}(t)$, it follows, in particular, that $\underline{z}(t) \leq \underline{y}(t)$ for all $t \in [\underline{t}, \overline{t}]$; therefore, $\underline{z}(\underline{t}) \leq \underline{y}(\underline{t})$ and $\underline{z}(\overline{t}) \leq \underline{y}(\overline{t})$.

Since $\underline{y}_2 \leq 0$, the quadratic function $\underline{y}(t)$ is concave and hence, from $\underline{z}(\underline{t}) \leq \underline{y}(\underline{t})$ and $\underline{z}(\overline{t}) \leq \underline{y}(\overline{t})$, i.e., from the fact that the line $\underline{z}(t)$ lies below the two endpoints of the graph of $\underline{y}(t)$, it automatically follows that the entire graph of $\underline{y}(t)$ is above the line $\underline{z}(t)$. So, it is sufficient to guarantee that at the endpoints, the values $\underline{z}(\underline{t})$ and $\underline{z}(\overline{t})$ do not exceed the corresponding values of $\underline{y}(\underline{t})$ and $\underline{y}(\overline{t})$.

For fixed $\overline{z}(t)$, the resulting intervals are the narrowest when the line $\underline{z}(t)$ is at the highest possible location. Thus, to minimize the widths of the intervals, we must move both points $\underline{z}(\underline{t})$ and $\underline{z}(\overline{t})$ up as much as possible. The highest possible location for $\underline{z}(\underline{t})$ is $\underline{y}(\underline{t})$, and the highest possible location for $\underline{z}(\overline{t})$ is $\underline{y}(\overline{t})$. Thus, $\underline{y}(t)$ is a a straight line going through $\underline{y}(\underline{t})$ and $\underline{y}(\overline{t})$, i.e., a *secant.*

Similarly, $\overline{z}(t)$ is a secant of $\overline{y}(t)$.

Case 2: $\underline{y}_2 \leq 0$, $\overline{y}_2 < 0$. The arguments given for Case 1 show that in this case, the function $\underline{z}(t)$ is still the secant. With $\underline{z}(t)$ fixed, it is sufficient to find the upper function $\overline{z}(t)$ for which the resulting approximation is optimal.

Let us first guarantee that the worst-case width is indeed the smallest possible. Let us denote the worst-case width by W_0. By definition, for every t, we have $\overline{z}(t) \leq \underline{z}(t) + W_0$, and therefore, $\overline{y}(t) \leq \overline{z}(t) \leq \underline{z}(t) + W_0$. Thus, to guarantee that W_0 takes the smallest possible values, we must choose W_0 as the smallest possible value for which $\overline{y}(t) \leq \underline{z}(t) + W_0$ for all $t \in [\underline{t}, \overline{t}]$. In other words, as W_0, we take the maximum of the function $\overline{y}(t) - \underline{z}(t)$ on the interval $[\underline{t}, \overline{t}]$.

The maximum of the concave quadratic function

$$\overline{y}(t) - \underline{z}(t) = (\overline{y}_0 - \underline{z}_0) + (\overline{y}_1 - \underline{z}_1) \cdot t + \underline{y}_2 \cdot t^2$$

is attained at the point where its derivative is equal to 0, i.e., at the point t_m determined by the formula (4). If this maximum is attained at the internal point t_m of this interval, then at t_m, the line $\underline{z}(t) + W_0$ is a tangent to $\overline{y}(t)$, and therefore, there is no way to find a lower line without increasing the worst-case width $W(\mathbf{z})$. So, in this case, $\overline{z}(t) = \underline{z}(t) + W_0$.

If the maximum is attained in one of the endpoints, e.g., at $\overline{t}$, then, we can, keeping the straight line $\overline{z}(t)$ at the point $(\overline{t}, \overline{y}(\overline{t}))$, lower its other end and still get the same worst-case width. The lowest value of the best-case width is attained when we lower the other end to the lowest possible position in which it is still above $\overline{y}(t)$, i.e., to the position of a *tangent* to $\overline{y}(t)$.

Case 3: $\underline{y}_2 > 0$, $\overline{y}_2 \geq 0$. This case is similar to case 2.

Case 4: $\underline{y}_2 > 0$, $\overline{y}_2 < 0$. In this case, both functions $\underline{y}(x)$ and $\overline{y}(x)$ are "pointing outward".

As a first step of constructing the optimal linear approximation $\mathbf{z}(t)$, let us first make sure that we have the smallest possible value of the worst-case width. For every $t \in [\underline{t}, \overline{t}]$, from $\mathbf{y}(t) \subseteq \mathbf{z}(t)$, we can conclude that the width of $\mathbf{z}(t)$ is at least as large as the width of the interval $\mathbf{y}(t)$. Thus, the worst-case width $W(\mathbf{z})$ of the approximating linear function $\mathbf{z}(t)$ cannot be smaller than the worst-case width $W(\mathbf{y})$ of the original (quadratic) interval function $\mathbf{y}$. Since we are minimizing $W(\mathbf{z})$, it is therefore desirable to choose $\mathbf{z}$ in such a way that its worst-case width is *exactly equal* to $W(\mathbf{y})$.

The worst-case width $W(\mathbf{y})$ is a maximum of the quadratic width function $\overline{y}(t) - \underline{y}(t)$ on the interval $[\underline{t}, \overline{t}]$. By differentiating this difference, one can easily get an explicit expression for this maximum point t_M (this expression is given in the formulation of the algorithm).

If this maximum point t_M is inside the interval $[\underline{t}, \overline{t}]$, then at this point t_M, both approximating lines $\underline{z}(t)$ and $\overline{z}(t)$ must be tangent to the corresponding functions $\underline{y}(t)$ and $\overline{y}(t)$, because otherwise, in at least one of the directions, the width will increase.

If this maximum t_M is attained at one the endpoints, e.g., for $\overline{t}$, then we must have $\underline{z}(\overline{t}) = \underline{y}(\overline{t})$ and $\overline{z}(\overline{t}) = \overline{y}(\overline{t})$. In this case, to guarantee the smallest possible *best-case* width, we must place $\overline{z}(t)$ as low as possible (i.e., along the tangent to $\overline{y}$), and $\underline{z}(t)$ as high as possible, i.e., similarly, along the tangent to $\underline{y}(t)$.

As a result, we arrive at the formulas given in the algorithm. Proposition is proven.

20

SOLVING DIFFERENTIAL EQUATIONS

Yet another problem where interval computations are used is a problem of solving differential equations. In this chapter, we show that in general, this problem requires at least exponential time, and briefly describe the heuristics that are used to solve important particular classes of differential equations in polynomial time.

This chapter was written in collaboration with M. Berz.

20.1 In General, Solving Differential Equations is Exponentially Hard

In the previous chapters, we have shown that even simple problems, such as solving a system of polynomial equations or optimizing a polynomial objective function, are, in general, NP-hard or even exponentially hard (i.e., require at least exponential running time for some instances). It is therefore natural to expect that if, instead of the simple *static* problems, we consider more realistic *dynamic* problems (e.g., if we solve *differential* equations instead of simpler algebraic equations), then the required computation time will only increase. Indeed, solving even the simplest *linear* differential equations is, in general, exponentially hard:

Definition 20.1. *By ε-approximately solving systems of polynomial differential equations, we mean the following problem:*

GIVEN:

- *an integer n;*
- *a finite sequence of polynomials $f_1(x_1,\ldots,x_n),\ldots,f_n(x_1,\ldots,x_n)$ with rational coefficients;*
- *n rational numbers $x_1^{(0)},\ldots,x_n^{(0)}$;*
- *rational numbers $t_0 < T$; and*
- *a rational number $\varepsilon > 0$.*

COMPUTE: rational numbers $\widetilde{x}_1,\ldots,\widetilde{x}_n$ that are ε–close to the values of $x_i(T)$, where $x_1(t),\ldots,x_n(t)$ are a solution to the system of differential equations

$$\frac{dx_i}{dt} = f_i(x_1,\ldots,x_n) \tag{20.1}$$

for which $x_i(t_0) = x_i^{(0)}$ $(i = 1,\ldots,n)$.

Comment. In particular, when $n = 1$, i.e., when the system consists of only *one* differential equation, we will talk about *ε-approximately solving polynomial differential equations.* Already this problem is exponentially hard:

Theorem 20.1. *For every $\varepsilon > 0$, an arbitrary algorithm that ε-approximately solves all linear differential equations requires, for some equations, at least exponential time.*

This is indeed much worse than for systems of *algebraic* equations:

	Systems of algebraic equations	Systems of differential equations
Linear $f_i(x_1,\ldots,x_n)$	Polynomial time	Exponential time (or worse)
Quadratic $f_i(x_1,\ldots,x_n)$	Exponential time (or worse)	Exponential time (or worse)

20.2. Important Heuristics: Traditional Numerical Methods and Taylor Series Methods

It may be possible to have a feasible heuristic. As we can see from the proof, the exponential running time appears if we allow arbitrarily large values of integration time $T - t_0$ and of the right-hand sides $f_i(x_1, \ldots, x_n)$. It is, therefore, reasonable to fix some T_0 and B and consider only systems of equations for which $|T - t_0| \leq T_0$ and $|f_i(x_1, \ldots, x_n)| \leq B$. With this restriction, we can hope to have reasonable algorithms or at least reasonable heuristics.

Traditional numerical methods, and why, unfortunately, their worst-case computational complexity is still exponential. People have been solving differential equations for more than three centuries, and many good practical methods have been designed. So, the first natural idea is to check the computational complexity of the known methods. Alas, it turns out that although these methods are very *good in many practical problems*, their *worst-case* computational complexity is still *exponential.*

Indeed, most of these methods are based on a step-by-step integration of the system (20.1). The exponential complexity can be illustrated on the simplest example of Euler integration, in which we choose some computation step Δt and sequentially compute, for $t_1 = t_0 + \Delta t$, $t_2 = t_0 + 2\Delta t$, etc., the values

$$x_i(t_{k+1}) = x_i(t_k) + \Delta t \cdot f_i(x_1(t_k), \ldots, x_n(t_k))$$

until we reach $t_k \approx T$. This method requires $T/\Delta t$ iterations.

The accuracy of such an integration scheme is proportional to Δt, so to attain the desired accuracy ε, we need to make $\approx 1/\varepsilon$ computation steps. For k-digit values, e.g., for $\varepsilon = 2^{-k}$, this means that we need a number of computational steps that exceed an exponent 2^k of the length of the input. In other words, these methods require *exponential time*.

The same exponential lower bound for computation time holds for *more sophisticated* methods, for which the accuracy is proportional to $(\Delta t)^a$ for some a: to achieve the desired accuracy $\varepsilon > 0$, we need to take $\Delta t \approx \varepsilon^{1/a}$, i.e., for $\varepsilon = 2^{-k}$, still *exponential time*.

This worst-case behavior of traditional methods is a real phenomenon. The exponential-time behavior of traditional methods is, unfortunately, not just a purely theoretical phenomenon. It is indeed happening for some real-life differential equations. Let us give two examples.

When we design a *particle accelerator*, we must guarantee that the particles will not get out of the camera after billions of cycles. To do this, we must solve systems of differential equations. With particles moving at approximately the speed of light, and the resulting fast changes, traditional methods of solving differential equations would require millions of years to compute.

Another example is *geophysics*, when we want to simulate long-time behavior of complex interacting geophysical system. In this case, too, traditional methods do not work if we want to predict what will happen after (a geologically meaningful period of) millions of years.

How to avoid exponential time: Taylor series methods. If the right-hand sides $f_i(x_1, \ldots, x_n)$ of the differential equations (20.1) are polynomials, or, more generally, *analytical functions*, then the well-known Cauchy theorem says that the solution is also analytical. We can, therefore, represent each solution as a Taylor series. If we take only first d terms of this expansion, we get an expression $x_i(t) = a_{i0} + a_{i1} \cdot t + a_{i2} \cdot t^2 + \ldots + a_{id} \cdot t^d$ with unknown coefficients a_{ij}. We can find these coefficients if we substitute these expressions into the original system (20.1) (and apply known automatic differentiation techniques). As a result, we get a system of $n \times (d+1)$ algebraic equations with $n \times (d+1)$ unknowns, a system which is often easy to solve (and which, at least, is not doomed to exponential computation time).

How many terms do we need to achieve the given accuracy? If we restrict ourselves to d-th order terms in Taylor series, then the remainder is, crudely speaking, proportional to $(T/R)^{d+1}$, where R is the radius of convergence. Therefore, to achieve an accuracy $\varepsilon = 10^{-k}$, we must choose d for which $(T/R)^{d+1} \approx 10^{-k}$, i.e., for which $d \approx \text{const} \cdot k$. Thus, the number of terms is bounded by a *polynomial* of the length of the input. If we are in a situation where solving the resulting system of algebraic equations is feasible (i.e., polynomial time), we have thus reduced the original *exponential* time to *polynomial* time.

Comment. From the computational complexity viewpoint, these methods were first described in Longpré *et al.* [258] and Kreinovich *et al.* [230].

Taylor series methods indeed lead to practically feasible algorithms. The Taylor series methods indeed work well for many practical problems, in particular, for the particle accelerator example; see, e.g., Berz *et al.* [38, 39, 40, 42, 41], Makino *et al.* [267, 268].

Proofs

Proof of Theorem 20.1. Indeed, for $t_0 = 0$ and $x_1^{(0)} = 1$, a system $dx/dt = C \cdot x$ has a solution $x(t) = \exp(C \cdot t)$ for which $x(T) = \exp(C \cdot T)$. Thus, if we are given k-digit numbers C and T, the resulting value requires an order of 2^k digits to describe. Thus, we need exponentially many computational steps just to write down the solution. Similarly to the previous chapters, requiring that the solution be ε-approximate does not eliminate this exponential time. The theorem is proven.

21

PROPERTIES OF INTERVAL MATRICES I: MAIN RESULTS

In the previous chapters, we analyzed the computational complexity and feasibility of the problems in which the main goal was to compute a number (or an interval). In many practical situations, however, we are not interested in the exact value of this number; all we need to know is whether a certain property is true or not: e.g., whether a given controlled system is stable, etc. It turns out that the most interesting practical problems of this type relate to numerical and interval-values matrices: to check whether a given matrix is regular, positive definite, stable, etc.

In this chapter, we describe the main results related to computational complexity and feasibility of such problems. Proofs and several important auxiliary results are presented in the next chapter.

21.1. Properties of Interval Objects: Informal Introduction

"Checking" problems and why they are practically important. In the previous chapters, we analyzed the computational complexity and feasibility of the problems in which the main goal was to *compute* a number $y = f(x_1, \ldots, x_n)$, or, to be more precise, to find the set of all possible values of the function $y = f(x_1, \ldots, x_n)$ when its inputs x_i take values from given intervals $\mathbf{x}_i$. In many *practical* situations, however, we are not interested in the exact value of this number; all we need to know is whether a certain property is true or not.

For *example*, in *environmental* applications, we may want to know whether the pollution level exceeds the legal limit or not; in *control* applications, we may want to know whether the given controlled system is stable or not, etc.

"Checking" problems are (somewhat) simpler than the corresponding computation problems. In most problems of this type, the condition that we want to check can be re-formulated as an inequality of the type $f(x_1, \ldots, x_n) \geq y_0$ for a given value y_0 (or as several inequalities of this type). For example, the problem of checking the pollution level is already given in this form; the problem of checking stability of a controlled system is formulated in a much more complicated form, but, as we will see later in this chapter, it can also be reduced to checking inequalities of this type.

Since we only know *intervals* $\mathbf{x}_j$ of possible values of x_j, we can only get a *interval* $\mathbf{y} = [\underline{y}, \overline{y}]$ of possible values of y. Depending on the values $\underline{y}$ and $\overline{y}$, we can have three different possibilities:

- If $\underline{y} \geq y_0$, this means that *all* possible values of $y = f(x_1, \ldots, x_n)$ (i.e., all values y from the interval $[\underline{y}, \overline{y}]$) satisfy the desired inequality, so the desired property is *guaranteed to be true.*
- If $\overline{y} < y_0$, this means that *none* of possible values of $y = f(x_1, \ldots, x_n)$ satisfy the desired inequality, so the desired property is *guaranteed to be false.*
- If $\underline{y} < y_0 \leq \overline{y}$, this means that *some* possible values of $y = f(x_1, \ldots, x_n)$ satisfy the desired inequality, while some other possible values do not satisfy this property. In this case, based on our information (as described by the intervals $\mathbf{x}_1, \ldots, \mathbf{x}_n$), we cannot decide whether the desired property is satisfied or not, so, *additional information is needed.*

When there exists a *feasible* algorithm for *computing* the interval $\mathbf{y} = f(\mathbf{x}_1, \ldots, \mathbf{x}_n)$, we can thus check the desired property in reasonable time. However, as all the previous chapters show, in many cases, the problem of *computing* the interval $\mathbf{y}$ is *computationally intractable* (NP-hard). In these cases, we *cannot* simply compute the interval $\mathbf{y}$ and then check whether elements of this interval are larger than or equal to y_0. The fact that we cannot *compute* the endpoints $\underline{y}$ and $\overline{y}$ of the interval $\mathbf{y}$ in reasonable time does not necessarily mean that we cannot *check* the desired property in reasonable time by using some feasible algorithm: for this checking, we do not need the *exact* values of $\underline{y}$ and $\overline{y}$, we do not even need ε-approximations to these endpoints: all we need to know is whether $\underline{y} \geq y_0$ and whether $\overline{y} \geq y_0$. Since the desired output consists of a single bit of information (yes or no), this problem is clearly *simpler* than the corresponding *computing* problem in which the desired output (a number) consist of several bits.

Due to this relative simplicity, in some cases, this *simpler* "checking" problem becomes *feasible* even when the corresponding computational problem is NP-hard.

> For *example*, if $f(x_1, \ldots, x_n)$ is a non-linear polynomial, then, in general, computing the exact range $\mathbf{y}$ is NP-hard. On the other hand, we can apply naive interval computations or some more sophisticated interval technique and get a reasonable *enclosure* $\mathbf{Y} \supset \mathbf{y}$. If *all* the values y from this enclosure are greater than or equal to y_0, then the desired property is *guaranteed to be true*; if all the values from $\mathbf{Y}$ are smaller than y_0, then we can *guarantee* that the desired property is *false*.

In some other cases, the "checking" problem is still NP-hard. However, since checking problem is, in general, *simpler* than the computing problem, we cannot simply deduce this NP-hardness from the NP-hardness of the corresponding computational problem: we have to prove it anew.

The main goal of this chapter is to analyze computational complexity and feasibility of the "checking" problems.

21.2. Practically Important Properties of Interval Objects: Informal Introduction Continued

Three basic stages of solving practical problems: a brief reminder. To describe practically important properties of interval objects, let us recall the basic stages of solving a generic practical problem:

- At first, we *determine the current state* of the system and, ideally, its *dynamics*.
- Then, we use the known *dynamics* to *predict* the future states of the system.
- Finally, if we are not satisfied with this prediction, we try to find the *control* that leads to, say, the largest possible value of the objective function.

Let us briefly recall the computational problems related to each of these three stages, and describe natural *"checking"* problems that appear on each of these stages.

First stage: Determining the current state. To determine the current state of the system, we must undertake some *measurements*. In some situations, we can *directly* measure the values of the variables $x_1, \ldots, x_n$ that describe the system's state. In other situations, we *cannot directly* measure the values x_j (or at least cannot *easily* directly measure these values). In such situations, we measure some *other* quantities $y_1, \ldots, y_m$ (that are related to the desired quantities x_j), and use the results $\widetilde{y}_i$ of measuring y_i and the known relations between x_j and y_i to reconstruct the values of the desired quantities x_j.

In some cases, we have *explicit* formulas that describe x_j in terms of y_i: $x_j = f_j(y_1, \ldots, y_m)$. In such cases, from the *computational* viewpoint, all we need to do to compute estimates $\widetilde{x}_j$ is to perform some explicit computations: $\widetilde{x}_j = f_j(\widetilde{y}_1, \ldots, \widetilde{y}_m)$. If we take *interval* uncertainty into consideration, i.e., if we take into consideration that we only know the *intervals* $\mathbf{y}_j$ of possible values of y_i, then we get the above basic problem of interval computations: given the intervals $\mathbf{y}_i$, compute the range $f_i(\mathbf{y}_1, \ldots, \mathbf{y}_m)$. In these cases, this problem is a purely *computational* problem, with no significant *"checking"* component.

In other cases, we only have *implicit* formulas that relate x_j and y_i:

$$F_k(x_1, \ldots, x_n, y_1, \ldots, y_m) = 0, \; 1 \le k \le K. \tag{21.1}$$

In such cases, to find x_j, we must actually *solve* the system of equations (21.1). The corresponding functions F_k can be arbitrarily complicated. Usually, we know the *approximate* values $x_j^{(0)}$ of the measured quantities x_j. If this knowledge is accurate enough, i.e., if the differences $\delta x_j = x_j - x_j^{(0)}$ are *small*, we can *simplify* these equations if we expand the functions

$$F_k(x_1, \ldots, x_n, y_1, \ldots, y_m) = F_k(x_1^{(0)} + \delta x_1, \ldots, x_n^{(0)} + \delta x_n, y_1, \ldots, y_m)$$

into Taylor series in δx_j and retain only the first few main terms in this expansion. In particular, if we only retain *linear* terms, we get a system of *linear* equations

$$\sum a_{kj} x_j = b_k, \; 1 \le k \le K, \tag{21.2}$$

(where

$$a_{kj} = \frac{\partial F_k}{\partial x_j}(x_1^{(0)}, \ldots, x_n^{(0)}, y_1, \ldots, y_m)$$

and $b_k = \sum a_{kj} \cdot x_j^{(0)} - F_k(x_1^{(0)}, \ldots, x_n^{(0)}, y_1, \ldots, y_m)$.) If we take interval uncertainty into consideration, then, as we have shown in the previous chapters, even in this *simplest* case, the problem of estimating x_j is NP-hard.

"Checking" problems emerging from the first stage. If the relations (21.1) or (21.2) are *absolutely* true (and the measuring instruments that measure y_i are functioning correctly), then we have to solve the system (21.1) or (21.2) (or its interval analogue) anyway, because this is how we determine x_j. In this case, there is no checking problem. However, in reality, it happens sometimes that the relations (21.1) that we assumed to be (absolutely) true are only *approximately* true or, even worse, these relations are only true for *some* values x_j and y_i (including all the values for which they have been tested before) and not true for the values x_j and y_i that we are currently measuring. It can also happen that one of the measuring instruments is broken, and the values y_i that we get from this instrument are way off. In these cases, the system (21.1) may have no solution at all. If this is the case, we must *re-analyze* the situation: e.g., replace the sensors, make new measurements of y_i, and try again to solve the system (21.1).

Since solving a system of nonlinear equations is a *computationally hard* problem, the algorithms that solve such systems are often very *time-consuming.* Since there is a risk that the original system is not solvable at all (and thus, the time spent on trying to solve this system is wasted), it is desirable, before starting this time-consuming algorithm, to *check* whether the system (21.1) has a solution. Thus, on the first stage of practical problem solving, we have an important *"checking"* problem: *to check whether a given system of equations* (21.1) *has a solution.*

In the *ideal* (numerical) case when all the intervals are *degenerate* (i.e., numbers) and all the measurements are *precise* (i.e., when the measured values $\widetilde{y}_i$ coincide with the actual values y_i of the directly measured quantities), we should expect that not only the system has a solution, but that there should be a *unique* solution. In this ideal case, in order to determine n values $x_1, \ldots, x_n$, we need only n equations, i.e., n relations (21.1); every other relation would be redundant and extra measurements used to establish this relation unnecessary. So, we have a system of n equations with n unknowns.

For the simplest case of a *linear* system, the requirement that the resulting $n \times n$ system of linear equations (21.2) has a unique solution can be easily reformulated in purely algebraic terms: it is equivalent to the requirement that the matrix A formed by the coefficients a_{kj} is *regular* (i.e., *invertible*).

Thus, the first stage of problem solving leads to the following important "checking" problem: *to check whether a given matrix is regular* (non-singular).

Comment. To be more precise: the general problem is to check solvability of a system of equations. Regularity of a matrix is only the *simplest* case of the general problem; more complicated (and more general) cases are also practically interesting. However, as we will show, in the presence of interval uncertainty, already this simplest case is NP-hard; so there is no need to consider computational complexity of other, more general and more complicated cases: all these more general and more complicated cases are NP-hard too.

Second stage: Predicting the future states. On the second stage of practical problem solving, we use the known *initial state* of the system and its known *dynamics* to *predict* the future states of the system. There are two basic ways to describe the dynamics:

- Traditionally, in *fundamental physics*, we describe the dynamics in such a way so as to be able to predict what is happening after an *arbitrary* period of time Δt. In particular, we must be able, knowing the state $x_i(t)$ at the initial moment of time t, to predict the state $x_i(t+\Delta t)$ for *arbitrarily small* values Δt. For small Δt, we have $x_i(t+\Delta t) \approx x_i(t)+\Delta t \cdot \dot{x}_i(t)$ (where $\dot{x}_i(t)$ denotes the time derivative). Therefore, being able to predict the values $x_i(t + \Delta t)$ is equivalent to being able, knowing the *values* $x_1, \ldots, x_n$, to predict the values of the *derivatives* $\dot{x}_i(t)$. Hence, we describe the dynamics in terms of a system of *differential equations*:
$$\frac{dx_i(t)}{dt} = f_i(x_1(t), \ldots, x_n(t)). \tag{21.3}$$
- In many *practical problems*, we are only interested in the state $x_i(t)$ for *discrete* moments of time: e.g., in daily temperatures, yearly crop, etc. In other words, we know the values $x_i(t)$ at a certain initial moment of time t, and we are interested in the values $x_i(t + \Delta t)$, $x_i(x + 2\Delta t)$, ..., where $\Delta t > 0$ is a given positive constant (*observation period*).
  - In *principle*, we can still describe the dynamics in terms of differential equation (21.3), solve this differential equations (i.e., find the values $x_i(t)$ for all t), and then extract the desired values of $x_i(t)$ from the resulting solutions.
  - However, from the *computational* viewpoint, this is a *waste* of computing resources, because, in addition to the states $x_i(t+\Delta t)$, $x_i(t+2\Delta t)$, etc., in which we are really interested, we compute all intermediate states $x_i(t)$ as well.

To make computations more efficient, it is desirable to directly describe the transition from the state $x_i(t)$ in the initial moment of time to the state $x_i(t+\Delta)$ in the moment of time $t+\Delta t$:

$$x_i(t+\Delta t) = F_i(x_1(t), \ldots, x_n(t)). \tag{21.4}$$

If we know the *exact* initial values of the parameters $x_i(t)$ that characterize the initial state and the *exact* functions f_i that describe the ***dynamics***, then we can explicitly ***solve*** the corresponding system of equations ((21.3) or (21.4)) and get the desired prediction.

These prediction problems are doable, although often they are time-consuming, especially if n is large, and we want to predict a reasonably distant future. For *example*, *weather prediction* requires a significant amount of time on modern supercomputers that perform 10^9 to 10^{12} operations per second, and still the results of modern computer weather prediction are still far from being perfect.

Similarly to the first stage, we can ***simplify*** the equations (21.3) and (21.4) by retaining only the first few terms in the Taylor expansion of the functions f_i or F_i. In particular, in the simplest case when we only retain *linear* terms (and thus, approximate the original functions f_i or F_i by linear functions), we get the following *linear* equations:

$$\frac{dx_i(t)}{dt} = a_0 + \sum_{j=1}^{n} a_{ij} \cdot x_j(t); \tag{21.5}$$

$$x_i(t+\Delta t) = a_0 + \sum_{j=1}^{n} a_{ij} \cdot x_j(t). \tag{21.6}$$

"Checking" problems emerging from the second stage. In real life, there are inevitable measurement errors in the initial values of $x_i(t)$: the *actual* values $x_i(t)$ are slightly different from the *measured* values $\widetilde{x}_i(t)$. As a result, sometimes, the prediction process becomes ***unstable***: as we try to predict the states at the times $t \to \infty$, even small errors $\Delta x_i(t_0) = \widetilde{x}_i(t_0) - x_i(t_0)$ at the initial moment of time t_0 exponentially increase as $t \to \infty$. In this case, due to the possibility of large prediction errors $\Delta x_i(t) = \widetilde{x}_i(t) - x_i(t)$, the results $\widetilde{x}_i(t)$ of integrating the equations (21.3) or (21.4) do not lead to any meaningful information about the desired *actual* future values $x_i(t)$. Since prediction is often very time-consuming, it is desirable to avoid unnecessary computations and *check* beforehand whether the given prediction problem is stable or not.

From this *computational* viewpoint, stability means that small initial deviations $\Delta x_i(t)$ do not grow exponentially to ∞ as $t \to \infty$, but remain reasonably small.

Stability checking is also very important for *control* applications, where the equations (21.3) or (21.4) characterize the dynamics of the controlled system. The typical goal of control is to *stabilize* the system, i.e., to make sure that an arbitrary initial deviation $x_i(t_0) \neq x_i^{\rm des}(t_0)$ of the actual trajectory $x_i(t)$ from the ideal (desired) trajectory $x_i^{\rm des}(t)$ will eventually get corrected, i.e., the deviations $\Delta x_i(t) = x_i(t) - x_i^{\rm des}(t)$ from the desired trajectory tend to 0 as $t \to \infty$.

Notice that *control* applications motivate a *stronger* stability requirement than *computational* applications:

- In computational applications, we only required that the deviations do not grow too fast.
- In *control* applications, we want the deviations to *decrease* and tend to 0.

For *non-linear* systems, checking each of these stability properties can be very complicated, but for *linear* systems, the problem is much easier: there exist *explicit formulas* for solving the systems (21.5) and (21.6), and these formulas lead to *explicit formulas* for checking stability. Let us briefly describe how this is done; a detailed description can be found, e.g., in Chorlton [67] and Bellman [25].

Indeed, in both problems, we are interested in the *difference* $\Delta x_i(t)$ between the two solutions of the corresponding equation (21.5) or (21.6):

- In the *computational* applications, $\Delta x_i(t)$ is the difference between the trajectory $\widetilde{x}_i(t)$ based on the *approximate* initial state $\widetilde{x}_i(t_0)$ and the trajectory $x_i(t)$ based on the exact (unknown) initial state $x_i(t_0)$.
- In the *control* applications, $\Delta x_i(t)$ is the difference between the *actual* trajectory $x_i(t)$ and the *ideal* (desired) trajectory $x_i^{\rm des}(t)$.

From the fact that both solutions satisfy the equation (21.5) (or (21.6)), it follows that the difference $\Delta x_i(t)$ between these two solutions satisfies one of the following equations:

$$\frac{d\Delta x_i(t)}{dt} = \sum_{j=1}^{n} a_{ij} \cdot \Delta x_j(t); \tag{21.7}$$

$$\Delta x_i(t + \Delta t) = \sum_{j=1}^{n} a_{ij} \cdot \Delta x_j(t). \tag{21.8}$$

These equations are the easiest to solve when $n = 1$, then the first equation leads to $\Delta x_1(t) = \Delta x_1(t_0) \cdot \exp(a_{11} \cdot (t - t_0))$, and the second to $\Delta x_1(t_0 + k \cdot \Delta t) = \Delta x_1(t_0) \cdot a_{11}^k$. For $n > 1$, we can try to find *partial* solutions of the similar type $x_i(t) = C_i \cdot \exp(\lambda \cdot t)$ or $x_i(t) = C_i \cdot \lambda^t$, and then use the fact that the equations are linear to describe the *general* solution as an arbitrary linear combination of these partial solutions. If we substitute the desired partial solution into the equations (21.7) or (21.8), we get the system $\sum_j a_{ij} C_j = \lambda \cdot C_i$. In other words, λ is an *eigenvalue* of the matrix a_{ij}, while C_i is the corresponding *eigenvector*. So, if a matrix a_{ij} has n *different* eigenvalues, then we have n linearly independent partial solutions of this type, and an arbitrary solution can be represented as a linear combination of these partial ones. The *degenerate* case when two or more eigenvalues are equal ($\lambda' = \lambda$) can be treated as the limit case of the non-degenerate case $\lambda' = \lambda + \varepsilon$ when $\varepsilon \to 0$:

- For the system of *differential* equations (21.7), in the non-degenerate case, we have an arbitrary linear combination of $\exp(\lambda \cdot t)$ and $\exp((\lambda + \varepsilon) \cdot t)$. In the limit $\varepsilon \to 0$, in addition to $\exp(\lambda \cdot t)$, we also have a solution

$$\lim_{\varepsilon \to 0} \frac{\exp((\lambda + \varepsilon) \cdot t) - \exp(\lambda \cdot t)}{\varepsilon} = \frac{d}{d\lambda} \exp(\lambda \cdot t) = t \cdot \exp(\lambda \cdot t).$$

  If we have *three* coinciding eigenvalues, then we, similarly, get solutions proportional to $t^2 \cdot \exp(\lambda \cdot t)$, etc.

- Similarly, for the equation (21.8), we get solutions of the type $t \cdot \lambda^k$, $t^2 \cdot \lambda^t$, etc.

In general, an arbitrary solution $\Delta x_i(t)$ of the system (21.7) can be represented as a linear combination of the terms $x^p \cdot \exp(\lambda \cdot t)$, where p is a non-negative integer and λ is an eigenvalue, and an arbitrary solution of the system (21.8) can be represented as a linear combination of the terms $x^p \cdot \lambda^k$ for similar p and λ. This explicit description of all possible solutions of these systems leads to explicit conditions for stability; in deriving these conditions, we must take into consideration that eigenvalues λ are, in general, complex numbers:

- To describe stability conditions for the system (21.7), we can use the fact that
$$t^p \cdot \exp(\lambda \cdot t) = t^p \cdot \exp((\mathrm{Re}\lambda + \mathrm{i} \cdot \mathrm{Im}\lambda) \cdot t) =$$
$$t^p \cdot \exp(\mathrm{Re}\lambda \cdot t) \cdot (\cos(\mathrm{Im}\lambda \cdot t) + \mathrm{i} \cdot \sin(\mathrm{Im}\lambda \cdot t)).$$
This term tends to 0 if and only if $\mathrm{Re}\lambda < 0$, and it does not increase exponentially if and only if $\mathrm{Re}\lambda \le 0$. Thus:
  - Stability needed for *control* applications is equivalent to the requirements that $\mathrm{Re}\lambda < 0$ for each eigenvalue λ of the matrix a_{ij}. Matrices that satisfy this property are called *stable*, or *Hurwitz stable*, after the mathematician who first analyzed this type of stability.
  - Stability needed for *computational* applications is equivalent to the requirements that $\mathrm{Re}\lambda \le 0$ for each eigenvalue λ of the matrix a_{ij}. Matrices that satisfy this property are called *semi-stable*, or *Hurwitz semi-stable*.

- To describe stability condition for the system (21.7), we used a standard representation of a complex number in terms of two real numbers: $\lambda = \mathrm{Re}\lambda + \mathrm{i} \cdot \mathrm{Im}\lambda$; in the standard geometric representation of complex numbers as points on a plane, this representation corresponds to Cartesian coordinates. For the system (21.8), it is more appropriate to use an alternative (polar-coordinate) representation $\lambda = |\lambda| \cdot \exp(\mathrm{i} \cdot \varphi)$. We can now use the fact that
$$t^p \cdot \lambda^k = t^p \cdot |\lambda|^k \cdot \exp(\mathrm{i} \cdot k \cdot \varphi) = t^p \cdot |\lambda|^k \cdot (\cos(k \cdot \varphi) + \mathrm{i} \cdot \sin(k \cdot \varphi)).$$
This term tends to 0 when $k \to \infty$ if and only if $|\lambda| < 1$, and it does not increase exponentially if and only if $|\lambda| \le 1$. Thus:
  - Stability needed for *control* applications is equivalent to the requirements that $|\lambda| < 1$ for each eigenvalue λ of the matrix a_{ij}. Matrices that satisfy this property are called *Schur stable*, after the mathematician who first analyzed this type of stability.
  - Stability needed for *computational* applications is equivalent to the requirements that $|\lambda| \le 1$ for each eigenvalue λ of the matrix a_{ij}. Matrices that satisfy this property are called *Schur semi-stable*.

Thus, the second stage of problem solving leads to the following important "checking" problem: *to check whether a given matrix is stable* (see, e.g., Barmish [16]).

Third stage: Optimization. Finally, if we are not satisfied with this prediction, we try to find the *alternative* that leads to the smallest (or the largest) possible value of some objective function $y = f(x_1, \ldots, x_n)$:

$$f(x_1, \ldots, x_n) \to \min \tag{21.9}$$

The objective function can be very complicated, but if we know the approximate values $x_i^{(0)}$ of the desired parameters x_i, we can expand the objective function into Taylor series in $\delta x_i = x_i - x_i^{(0)}$ and keep only a few main terms in the expansion.

In our analysis of the first two stages of problem solving, we kept only linear terms in the corresponding Taylor expansions. For unconstrained optimization, keeping *only linear* terms in the expansion of $f(x_1, \ldots, x_n)$ does not make sense, because a linear function does not have any minima or maxima at all. Thus, we need to keep at least *quadratic* terms as well. If we keep quadratic terms, we get the following (approximate) optimization problem:

$$a_0 + \sum_{i=1}^{n} a_i \cdot x_i + \sum_{i=1}^{n} \sum_{j=1}^{n} a_{ij} \cdot x_i \cdot x_j \to \min. \tag{21.10}$$

"Checking" problems emerging from the third stage. In many real-life situations, we only have an *approximate* information about how the desired objective (e.g., cost or time) depend on the controlled parameters x_i. Due to this approximate character of the objective function, it *may* happen (and it *does* happen sometimes) that the resulting optimization problem (21.9) does not have a solution at all. If this turns out to be the case, this means that we need to make a more accurate description of the objective function and repeat the computations.

We already know, from the previous chapters, that optimization is, in general, a computationally difficult problem. Thus, it is desirable, before we start the actual optimization, to *check whether this optimization problem has a solution at all.*

Optimization problems also lead to another, related "checking" problem: If the original optimization problem has a *single* solution, i.e., a *unique* combination of parameters $x_1, \ldots, x_n$ that minimizes the objective function (21.9), then we simply choose these values x_i. But what if the same smallest value of the objective function is attained for *several* different combinations of parameters? There exist two possible approaches to this situation:

- A *pragmatic* approach is to pick one of these combinations and use it.
- On the other hand, if the practical problem is really important, the existence of several solutions to the optimization problems means that we can do some further optimization.

  For *example*, let us assume that we are designing a super-computer, and our initial goal is to achieve the fastest *speed*. If it turns out that several different designs guarantee the best possible speed, then we can use this non-uniqueness to select, e.g., the *least costly* design.

  In other words, in this second approach, we change the objective function and re-solve the optimization problem.

Since we are solving the second optimization problem anyway, it makes sense to avoid the (often time-consuming) process of solving the original optimization problem. In other words, it is desirable to be able, ***given an optimization problem, to check whether it has a unique solution or not.***

For quadratic objective functions (21.10), these two "checking" problems can be easily reformulated in terms of the corresponding matrix a_{ij}:

- the (unconstrained) minimization problem (21.10) ***has a solution*** if and only if the matrix a_{ij} is *positive semi-definite*, i.e., if $\sum a_{ij} \cdot x_i \cdot x_j \geq 0$ for every vector $\vec{x} = (x_1, \ldots, x_n)$;
- the (unconstrained) minimization problem (21.10) ***has a unique solution*** if and only if the matrix a_{ij} is *positive definite*, i.e., if $\sum a_{ij} \cdot x_i \cdot x_j > 0$ for every vector $\vec{x} = (x_1, \ldots, x_n) \neq \vec{0} = (0, \ldots, 0)$.

Auxiliary "checking" problems. We already know, from the previous chapters, that many problems of interval computations are NP-hard. This means, crudely speaking, that no algorithm is likely to exist that would solve *all* instances of these problems in feasible time. There are, however, feasible algorithms that solve *some* instances. For many such algorithms, we know *conditions* that guarantee the algorithm's usefulness. Often, these conditions are formulated in *theoretical* terms, *without a* ready-made *algorithm* for checking. In such cases, we have an important practical problem of checking whether these conditions are true. Let us give two important examples of such properties:

- In interval computations, many computational problems are difficult to solve. One practically useful case when they are relatively easy to solve is the case of *monotonic* functions: if we know that a function $f(x_1, \ldots, x_n)$ is non-decreasing in each of its variables x_i, then its range $\mathbf{y} = f(\mathbf{x}_1, \ldots, \mathbf{x}_n)$ over the intervals $\mathbf{x}_i = [\underline{x}_i, \overline{x}_i]$ is simply equal to $\mathbf{y} = [f(\underline{x}_1, \ldots, \underline{x}_n), f(\overline{x}_1, \ldots, \overline{x}_n)]$. (Examples of successful applications of monotonicity were given and cited in Chapter 16.) In particular, for the problem of solving a system of linear equations $\sum_j a_{ij} x_j = b_i$, we have $x_j = \sum_i m_{ji} b_i$, where m_{ji} is an inverse matrix to the matrix a_{ij}. Each of the resulting linear functions $b_i \rightarrow x_j$ is non-decreasing in each of the variables b_j if and only if all the coefficients m_{ji} of the inverse matrix are non-negative. Matrix a_{ij} with this property are called ***non-negative invertible***. So, we arrive at the problem of checking whether a given matrix is non-negative invertible.

  Real good algorithms for solving linear interval systems are known for the case when an ***additional*** condition is satisfied: that $a_{ij} \leq 0$ for all $i \neq j$; such matrices are called *M-matrices* (see, e.g., Neumaier [303]). So, it is also reasonable to check whether a given matrix is an M-matrix.

- Since even unconstrained optimization problems are difficult to solve, constrained optimization problems are often even more difficult. One class of constrained optimization problems for which reasonable algorithms exist is a class of ***linear complementarity*** problems (see, e.g., Murty [297]), i.e., problem of finding a solution to the system of equations $x_i - \sum a_{ij} y_j = b_i$ for given b_i and a_{ij} under the conditions that for every i, both values x_i and y_i are non-negative and one of them is equal to 0. This algorithm is useful, e.g., for solving ***linear programming*** problems (which can be reformulated in this form). It is known that this problem is guaranteed to have exactly one solution for all b_i if and only if all the principal minors of the matrix a_{ij} are positive; such matrices were introduced by Fiedler and Pták [107] under the name of *P-matrices*. Thus, we get another "checking" problem: ***check whether a given matrix is a P-matrix.***

Summarizing: practically important "checking" problems. Let us summarize the important "checking" problems that we have described so far: given a matrix, check whether this matrix is: regular, stable, semi-stable, Schur stable, Schur semi-stable, positive semi-definite, positive definite, nonnegative invertible, an M-matrix, or a P-matrix. These are the problems whose computational complexity and feasibility we will analyze in this chapter.

We will show that for *numerical* matrices, almost all these problems are feasible, while for *interval* matrices, almost all these problems are NP-hard. For such NP-hard problems, we also describe finitely verifiable necessary and sufficient conditions that help in solving these problems for small n (but whose worst-case checking time grows exponentially with n). For some of these problems, we also describe feasibly verifiable *sufficient* conditions that may help in practical applications.

A comment about eigenvalues. Most of the matrix properties in which we are interested are *invariant* with respect to arbitrary linear change of variables x_i. Therefore, they can be re-formulated in terms of the characteristics of the matrix that are invariant with respect to such transformations. Since every matrix can be represented in a standard (Jordan) form, that is determined only by its eigenvalues and their multiplicity, we can, therefore, re-formulate the desired properties in terms of eigenvalues. We have already done that for stability properties; other properties can also be re-formulated in this form: e.g., regularity means that there is no vector x_i for which $\sum a_{ij}x_j = 0$, i.e., that 0 is not an eigenvalue.

In view of this importance of eigenvalue-related properties of matrices, we will also analyze the computational complexity and feasibility of *computing* eigenvalues and of checking whether all eigenvalues of a given matrix satisfy a certain property.

Let us now summarize definitions of the matrix properties in which we are interested:

21.3. For numerical matrices, almost all "checking" problems are feasible

Definition 21.1. *An $n \times n$ matrix A with elements a_{ij} is called:*

- *regular if this matrix has an inverse;*
- *stable* (or *Hurwitz stable) if* $\text{Re}\lambda < 0$ *for all its eigenvalues* λ;
- *semi-stable* (or *Hurwitz semi-stable) if* $\text{Re}\lambda \leq 0$ *for all its eigenvalues* λ;
- *Schur stable if* $|\lambda| < 1$ *for all its eigenvalues* λ;
- *Schur semi-stable if* $|\lambda| \leq 1$ *for all its eigenvalues* λ;
- *positive semi-definite if* $\sum a_{ij} \cdot x_i \cdot x_j \geq 0$ *for all* $\vec{x} = (x_1, \ldots, x_n)$;
- *positive definite if* $\sum a_{ij} \cdot x_i \cdot x_j > 0$ *for all* $\vec{x} = (x_1, \ldots, x_n) \neq \vec{0} = (0, \ldots, 0)$;
- *nonnegative invertible if there exists an inverse matrix* m_{ij}*, and all the elements of this inverse matrix are non-negative;*
- an *M-matrix if it is nonnegative invertible and* $a_{ij} \leq 0$ *for all* $i \neq j$;
- a *P-matrix if all its principal minors (i.e., determinants of square submatrices formed from rows and columns with the same indices) are positive.*

How complicated is it to check these properties?

For numerical matrices, there exist a polynomial-time algorithm for computing its eigenvalues with an arbitrary accuracy (see, e.g., Schrijver [382], Cormen [76]); therefore, we can check, in polynomial time, whether a matrix is stable, semi-stable, Schur stable, or Schur semi-stable. Several other matrix properties can be easily reformulated in terms of eigenvalues:

- A matrix is *regular* if an only if 0 is not its eigenvalue.
- A matrix a_{ij} is *positive semi-definite* if all eigenvalues of its symmetric part $a_{ij}^{\text{sym}} = (1/2) \cdot (a_{ij} + a_{ji})$ are non-negative.
- A matrix a_{ij} is *positive definite* if all eigenvalues of its symmetric part are positive.

Therefore, the corresponding three matrix properties are also easy to check.

Computing the inverse matrix can be done, e.g., by (modified) Gaussian elimination, in polynomial time (see, e.g., Edmonds [98], Schrijver [382], Cormen [76]). After computing this inverse matrix, we can check whether each of the resulting elements is non-negative. Thus, checking whether a given numerical matrix is nonnegative invertible is also a feasible problem. Thus, checking whether a given matrix is an M-matrix is also feasible.

The only remaining property is the property of being a P-matrix.

- A *symmetric* matrix A is a P-matrix if and only if it is positive definite (Wilkinson [429]), hence for symmetric matrices, P-property can be checked in polynomial time.
- For *generic* matrices, this problem was shown by Coxson [77] to be NP-hard (the proof will be given in the Proofs section).

Summarizing, we arrive at the following result:

Theorem 21.1.

- *There exist polynomial-time algorithms that check, given a numerical matrix A with rational elements a_{ij}, whether this matrix is regular, stable, semi-stable, Schur stable, Schur semi-stable, positive semi-definite, positive definite, nonnegative invertible, or an M-matrix.*
- *There exists a polynomial-time algorithm that checks, given a symmetric matrix a_{ij}, where this matrix is a P-matrix.*
- *The problem of checking whether an arbitrary matrix a_{ij} with rational elements is a P-matrix is NP-hard.*

In the previous section, we promised to try to describe, for all NP-hard matrix "checking" problems, how to check the corresponding properties for small n. For numerical matrices, the only such problem is whether a given matrix is indeed a P-matrix. By definition, P-property means that all principal minors are positive. Since there exist feasible algorithms for computing the determinant of a numerical matrix, we can check this property as follows: For every non-empty set $S \subseteq \{1, 2, \ldots, n\}$, we take a matrix formed by the elements a_{ij} with

$i, j \in S$, and check whether the determinant of this matrix is indeed positive. Since there are $2^n - 1$ possible non-empty subsets, we thus need to check $2^n - 1$ submatrices.

Comment. For other examples of NP-hard matrix problems, see, e.g., Brimkov *et al.* [59] and references therein.

21.4. For interval matrices, almost all "checking" problems are NP-hard

Let us start with the necessary definitions (the first definition was already given in Chapter 11):

Definition 21.2.

- *A collection of intervals* $\mathbf{a}_{ij} = [\underline{a}_{ij}, \overline{a}_{ij}]$ *is called an interval matrix. We will denote interval matrices by bold capital letters, e.g.,* $\mathbf{A} = [\underline{A}, \overline{A}]$.
- *We say that a numerical matrix* A *with elements* a_{ij} *belongs to an interval matrix* $\mathbf{A}$ *with elements* $\mathbf{a}_{ij}$ *(and denote it by* $A \in \mathbf{A}$*) if* $a_{ij} \in \mathbf{a}_{ij}$ *for all* i *and* j.
- *We say that an interval matrix* $\mathbf{A}$ *satisfies a property* $\mathcal{P}$ *if all (numerical) matrices* A *that belong to* $\mathbf{A}$ *satisfy this property.*

For example, an interval matrix $\mathbf{A}$ is called *regular* if all matrices $A \in \mathbf{A}$ are regular (i.e., non-singular); an interval matrix $\mathbf{A}$ is called *stable* if all matrices $A \in \mathbf{A}$ are stable, etc.

Terminological comment. Non-regular *numerical* matrices are also called *singular*. Similarly, an *interval* matrix that is not regular is also sometimes called a *singular* interval matrix. To avoid potential confusion, we want to mention that although the resulting definition of a singular interval matrix seems quite natural, it does *not* follow the general pattern established by Definition 21.2: namely, contrary to Definition 21.2, a singular interval matrix is defined *not* as an interval matrix for which *all* numerical matrices $A \in \mathbf{A}$ are singular, but as an interval matrix for which *some* numerical matrix $A \in \mathbf{A}$ is singular.

Comment. Some interval algorithms originate from *measurement* applications, where an interval $\mathbf{x}$ is of the form $[\widetilde{x} - \Delta, \widetilde{x} + \Delta]$, where $\widetilde{x}$ is the measurement result and Δ is the upper bound on the measurement errors. To apply these algorithms to a general interval $\mathbf{x} = [\underline{x}, \overline{x}]$, we must represent it in this form. This is easily done by taking the *center* $\widetilde{x} = (1/2) \cdot (\underline{x} + \overline{x})$ and the *radius* $\Delta = (1/2) \cdot (\overline{x} - \underline{x})$. Similarly, for an arbitrary interval *matrix* $\mathbf{A} = [\underline{A}, \overline{A}]$, it is often useful to represent it in the form $[\widetilde{A} - \Delta, \widetilde{A} + \Delta]$, where:

- $\widetilde{A} = (1/2) \cdot (\underline{A} + \overline{A})$ is the *center* matrix, with elements $\widetilde{a}_{ij} = (1/2) \cdot (\underline{a}_{ij} + \overline{a}_{ij})$; and
- $\Delta = (1/2) \cdot (\overline{A} - \underline{A})$ is the *radius* matrix, with elements $\Delta_{ij} = (1/2) \cdot (\overline{a}_{ij} - \underline{a}_{ij})$.

Theorem 21.2.

- *For each of the following six properties, the problem of checking whether a given interval matrix satisfies the property is NP-hard: regularity, stability, semi-stability, positive semi-definiteness, positive definiteness, and P-matrix property.*
- *There exist polynomial-time algorithms that check, given a interval matrix* $\mathbf{A}$, *whether* $\mathbf{A}$ *is nonnegative invertible, and whether* $\mathbf{A}$ *is an M-matrix.*

Comments. We will see from the proof that all six NP-hardness results hold even if we restrict ourselves to *symmetric* interval matrices (i.e., matrices for which $\mathbf{a}_{ij} = \mathbf{a}_{ji}$ for all i and j), and to matrices formed by *narrow* intervals, i.e., matrices for which, for a given rational number $\delta > 0$, $\overline{a}_{ij} - \underline{a}_{ij} \leq \delta$ for all i and j.

Historical comments.

- The result about NP-hardness of *regularity* of interval matrices was published by Poljak and Rohn in a report form [328] in 1988 and in a journal form [329] in 1993. This result was proved independently (and also published in 1993) by Nemirovskii [300].
- NP-hardness of checking stability was proven in Nemirovskii [300] for general (not necessarily symmetric) interval matrices and in Rohn [351] for symmetric ones.

- NP-hardness of checking *positive semi-definiteness* was proven by Nemirovskii [300].
- NP-hardness of checking *positive definiteness* was proven by Rohn [351].
- NP-hardness of checking P-property was proven in Rohn and Rex [361];
- Feasibility of checking nonnegative invertibility and M-matrix property easily follows from a result of Kuttler [242].

In some practical situations, we know that the (unknown) matrix a_{ij} is *symmetric.* In such situations, we may want to check a property not for *all* matrices $A \in \mathbf{A}$, but only for *all symmetric* ones:

Definition 21.3. *We say that a symmetric interval matrix* $\mathbf{A}$ *s-satisfies a property* $\mathcal{P}$ *if all symmetric numerical matrices* A *that belong to* $\mathbf{A}$ *satisfy this property.*

Theorem 21.3.

- *For each of the following eight properties* P*, the problem of checking whether a given interval matrix s-satisfies the property is NP-hard: regularity, stability, semi-stability, Schur stability, Schur semi-stability, positive semi-definiteness, positive definiteness, and P-matrix property.*
- *There exist polynomial-time algorithms that check, given an interval matrix* $\mathbf{A}$*, whether* $\mathbf{A}$ *s-satisfies the nonnegative invertibility property, and whether* $\mathbf{A}$ *s-satisfies the M-matrix property.*

Historical comment. NP-hardness of checking s-Schur stability was proven in Rohn [351].

The results on computational complexity of "checking" problems can be represented by the following table:

	Numerical matrices (general)	Numerical matrices (symmetric)	Interval matrices (general)	Interval matrices (symmetric)
Regular	Polynomial time	Polynomial time	NP-hard	NP-hard
Stable	Polynomial time	Polynomial time	NP-hard	NP-hard
Semi-stable	Polynomial time	Polynomial time	NP-hard	NP-hard
Schur stable	Polynomial time	Polynomial time	?	NP-hard
Schur semi-stable	Polynomial time	Polynomial time	?	NP-hard
Positive semi-definite	Polynomial time	Polynomial time	NP-hard	NP-hard
Positive definite	Polynomial time	Polynomial time	NP-hard	NP-hard
Nonnegative invertible	Polynomial time	Polynomial time	Polynomial time	Polynomial time
M-matrix	Polynomial time	Polynomial time	Polynomial time	Polynomial time
P-matrix	NP-hard	Polynomial time	NP-hard	NP-hard

21.5. How to Check Properties of Interval Matrices of Small Size

Let us describe, for all NP-hard matrix "checking" problems, how to check the corresponding properties for small n.

21.5.1. *Regularity*

In view of the NP–hardness result of Theorem 21.2, no easily verifiable necessary and sufficient regularity conditions may be expected to exist. Indeed, 13 such conditions are proved in Theorem 5.1 in Rohn [347], all of which exhibit exponential behavior. Probably the most easily implementable criterion is that one by Baumann [20] (Theorem 21.4 below) which employs matrices $A^{(y,z)}$, defined for an $n \times n$ interval matrix $\mathbf{A} = [\widetilde{A} - \Delta, \widetilde{A} + \Delta] = [\underline{A}, \overline{A}]$ and for vectors

$y = (y_1, \ldots, y_n) \in \{-1, 1\}^n$ and $z = (z_1, \ldots, z_n) \in \{-1, 1\}^n$ by the formula

$$a_{ij}^{(y,z)} = \widetilde{a}_{ij} - \Delta_{ij} \cdot y_i \cdot z_j \tag{21.11}$$

It is easy to check that

$$a_{ij}^{(y,z)} = \begin{cases} \overline{a}_{ij} & \text{if } y_i \cdot z_j = -1, \\ \underline{a}_{ij} & \text{if } y_i \cdot z_j = 1 \end{cases} \tag{21.12}$$

for each i, j, hence $A^{(y,z)} \in \mathbf{A}$. Baumann's criterion employs a finite set of test matrices A_{yz} for $y, z \in \{-1, 1\}^n$ of cardinality at most 2^{2n-1}. (We have 2^n possible vectors $y \in \{-1, 1\}^n$ and 2^n possible vectors $z \in \{-1, 1\}^n$, so, totally, we have $2^n \cdot 2^n = 2^{2n}$ pairs, but due to $A^{(-y,-z)} = A^{(y,z)}$, it is sufficient to check only half of these pairs.)

Theorem 21.4. *An interval matrix* $\mathbf{A}$ *is regular if and only if determinants of all the matrices* $A^{(y,z)}$, $y, z \in \{-1, 1\}^n$ *are nonzero and of the same sign.*

In view of the exponentiality inherent in the necessary and sufficient conditions, in practical computations we must resort to verifiable *sufficient* conditions (that are not necessary conditions). These conditions are usually formulated in terms of the *spectral radius* $\varrho(A)$, minimal and maximal *singular values* $\sigma_{\min}(A)$ and $\sigma_{\max}(A)$, and similar characteristics of a matrix A. The most useful sufficient conditions are given by the following theorem (in this theorem, $|A|$ is a matrix with elements $|a|_{ij} = |a_{ij}|$, and $A \geq 0$ means that all elements a_{ij} of the matrix A are non-negative):

Theorem 21.5. *Each of the following two conditions implies regularity of the interval matrix* $[\widetilde{A} - \Delta, \widetilde{A} + \Delta]$*:*

(i) $\varrho(|\widetilde{A}^{-1}|\Delta) < 1$,

(ii) $\sigma_{\max}(\Delta) < \sigma_{\min}(\widetilde{A})$.

Each of the following two conditions implies that the interval matrix $[\widetilde{A} - \Delta, \widetilde{A} + \Delta]$ *is not regular:*

(iii) $\max_j(\Delta|\widetilde{A}^{-1}|)_{jj} \geq 1$,

(iv) $(\Delta - |\widetilde{A}|)^{-1} \geq 0$.

Historical comment. The condition (i), which is most frequently used, is due to Beeck [23]; an interval matrix satisfying this condition (i) is called *strongly regular* (Neumaier [303]). The second condition is due to Rump [370]. The condition (iii) is proved in Rohn [347], and (iv) comes from Rohn [359].

21.5.2. *Stability*

The following theorem establishes a link with another matrix property:

Theorem 21.6. (Rohn [350]) *A symmetric interval matrix* $\mathbf{A} = [\underline{A}, \overline{A}]$ *is stable if and only if the symmetric interval matrix* $-\mathbf{A} = [-\overline{A}, -\underline{A}]$ *is positive definite.*

Consider now the matrices $A^{(y,z)}$ defined by the formula (21.12) with $y = -z$, i.e. the matrices for which

$$a_{ij}^{(-z,z)} = \begin{cases} \overline{a}_{ij} & \text{if } z_i \cdot z_j = 1, \\ \underline{a}_{ij} & \text{if } z_i \cdot z_j = -1 \end{cases}$$

$(i, j = 1, \ldots, n)$. Each of these matrices $A^{(-z,z)}$ is symmetric for a symmetric $\mathbf{A}$.

Theorem 21.7. *A symmetric interval matrix* $\mathbf{A}$ *is stable if and only if each of the matrices* $A^{(-z,z)}, z \in \{-1,1\}^n$ *is stable.*

Historical comment. Each matrix $A^{(-z,z)}, z \in \{-1,1\}^n$ is a so-called *vertex matrix*, i.e., a matrix for which for all i and j, the corresponding value $a_{ij}^{(-z,z)}$ is either equal to $\underline{a}_{ij}$, or equal to $\overline{a}_{ij}$. The first attempt to use vertex matrices for characterization of stability was made by Białas [45] who conjectured that a general interval matrix $\mathbf{A}$ is stable if and only if all the vertex matrices are stable. His conjecture, however, was shown to be erroneous by Karl, Greschak and Verghese [168] and by Barmish and Hollot [18], see also Barmish, Fu, and Saleh [17]. Soh proved later [400] that for *symmetric* interval matrices this result is true: namely, a symmetric interval matrix is stable if and only if all the $2^{n(n+1)/2}$ symmetric vertex matrices are stable. Theorem 21.7, where the number of vertex matrices to be tested is reduced to 2^{n-1} (since $A^{(-z,z)} = A^{(z,-z)}$), was proven (in a slightly different form) by Hertz [149] and by Wang and Michel [424], and in the present form by Rohn [350]. A branch-and-bound algorithm for checking stability of symmetric interval matrices, that is based on Theorem 21.7, was given in Rohn [361].

For practical purposes we may use the following *sufficient* condition that is formulated in terms of the largest eigenvalue $\lambda_{\max}$ of a related symmetric matrix; this condition is valid both for symmetric and non-symmetric interval matrices

(Rohn [350], Delgado-Romero *et al.* [88]). It is formulated in terms of a *symmetrization* $\mathbf{A}^{\text{sym}}$ of an interval matrix $\mathbf{A}$ that is defined as an interval matrix with elements

$$\mathbf{a}_{ij}^{\text{sym}} = \left[\frac{1}{2}\cdot(\underline{a}_{ij} + \underline{a}_{ji}), \frac{1}{2}\cdot(\overline{a}_{ij} + \overline{a}_{ji})\right]. \tag{21.13}$$

For the symmetrization, the center matrix $\widetilde{A}^{\text{sym}}$ and the radius matrix Δ^{sym} are defined, correspondingly, by the formulas

$$\widetilde{a}_{ij}^{\text{sym}} = \frac{1}{2}\cdot(\widetilde{a}_{ij} + \widetilde{a}_{ji});$$

$$\Delta_{ij}^{\text{sym}} = \frac{1}{2}\cdot(\Delta_{ij} + \Delta_{ji}).$$

Theorem 21.8. *An interval matrix* $[\widetilde{A} - \Delta, \widetilde{A} + \Delta]$ *is stable if*

$$\lambda_{\max}(\widetilde{A}^{\text{sym}}) + \varrho(\Delta^{\text{sym}}) < 0. \tag{21.14}$$

21.5.3. *Positive definiteness*

Positive definiteness of interval matrices is closely related to regularity:

Theorem 21.9. (Rohn [350]) *An interval matrix* $\mathbf{A}$ *is positive definite if and only if its symmetrization* $\mathbf{A}^{\text{sym}}$ *is regular and contains at least one positive definite matrix.*

It is known that a numerical matrix A is positive definite if and only if its symmetrization A^{sym} is positive definite. Theorem 21.9 now implies that the same relationship holds for interval matrices:

Theorem 21.10. *An interval matrix* $\mathbf{A}$ *is positive definite if and only if its symmetrization* $\mathbf{A}^{\text{sym}}$ *is positive definite.*

A finite characterization of positive definiteness of interval matrices was first given by Shi and Gao [395] who proved that a symmetric interval matrix $\mathbf{A} = [\underline{A}, \overline{A}]$ is positive definite if and only if each symmetric vertex matrix $A \in \mathbf{A}$ of the form $a_{ii} = \underline{a}_{ii}, a_{ij} \in \{\underline{a}_{ij}, \overline{a}_{ij}\}$ for $i \neq j$, is positive definite. There are $2^{n(n-1)/2}$ such matrices. In [350] it was shown that the number of test matrices may be reduced down to 2^{n-1} if we employ instead the set of matrices $A^{(z,z)}$ defined for $z \in \{-1,1\}^n$ by

$$a_{ij}^{(z,z)} = \begin{cases} \overline{a}_{ij} & \text{if } z_i z_j = -1, \\ \underline{a}_{ij} & \text{if } z_i z_j = 1 \end{cases} \tag{21.15}$$

$(i, j = 1, \ldots, n)$. These are exactly the matrices $A^{(y,z)}$ (see (21.12)) used in the Baumann regularity criterion (Theorem 21.4), with $y = z$. Each matrix $A^{(z,z)}$ is symmetric if $\mathbf{A}$ is symmetric.

Theorem 21.11. *An interval matrix* $\mathbf{A}$ *is positive definite if and only if each of the matrices* $A^{(z,z)}, z \in \{-1,1\}^n$ *is positive definite.*

In practical computations we may use the following sufficient condition (where $\lambda_{\min}$ denotes the *minimal eigenvalue* of a symmetric matrix and ϱ is the spectral radius):

Theorem 21.12. (Rohn [350]) *An interval matrix* $\mathbf{A} = [\widetilde{A} - \Delta, \widetilde{A} + \Delta]$ *is positive definite if* $\varrho(\Delta^{\text{sym}}) < \lambda_{\min}(\widetilde{A}^{\text{sym}})$.

21.5.4. *P-property*

The following theorem is due to Białas and Garloff [46], reformulation using matrices $A^{(z,z)}$ comes from Rohn and Rex [361].

Theorem 21.13. *An interval matrix* $\mathbf{A}$ *is a P-matrix if and only if each matrix* $A^{(z,z)}, z \in \{-1,1\}^n$ *is a P-matrix.*

For symmetric matrices we also have the following useful reduction:

Theorem 21.14. *A symmetric interval matrix* $\mathbf{A}$ *is a P-matrix if and only if it is positive definite.*

21.6. Related computational problems are also NP-hard

If an interval matrix does not satisfy a certain property, it is desirable to calculate how close it is to matrices that do. In this section, we will show that the corresponding computational problems are also NP-hard.

21.6.1. *Problems related to regularity I: Computing determinants*

In linear algebra, several criteria are known that determine when a given (numerical) matrix is regular. The most well known criterion is that a matrix A is regular if and only if its determinant $\det A$ is not equal to 0. Similarly, an *interval* matrix $\mathbf{A}$ is regular if and only if the range of the determinant function $\det \mathbf{A} = \{\det A | A \in \mathbf{A}\}$ does not contain 0. Thus, one way to check regularity of an interval matrix is to compute the endpoints of this interval $\det \mathbf{A} = [\underline{\det}(\mathbf{A}), \overline{\det}(\mathbf{A})]$ and to check whether 0 is indeed located between these endpoints. It turns out that the problem of computing these endpoints is NP-hard:

Theorem 21.15. *The problem of computing the endpoints* $\underline{\det}(\mathbf{A})$ *and* $\overline{\det}(\mathbf{A})$ *of the determinant interval* $\det \mathbf{A}$ *of a given rational-valued interval matrix* $\mathbf{A}$ *is NP-hard.*

For small n, we can use the following result to compute these endpoints:

Theorem 21.16. (Rohn [344]) *For every interval matrix* $\mathbf{A} = [\underline{A}, \overline{A}]$, *each of the endpoints* $\underline{\det}(\mathbf{A})$ *and* $\overline{\det}(\mathbf{A})$ *of the determinant interval* $\det \mathbf{A}$ *is attained at one of the vertex matrices.*

Thus, to find the desired endpoints, it is sufficient to compute $\det A$ for all 2^{n^2} vertex matrices A, for which, for every i and j, $a_{ij} = \underline{a}_{ij}$ or $a_{ij} = \overline{a}_{ij}$.

21.6.2. *Problems related to regularity II: Computing eigenvalues*

We have already mentioned that most checking problems can be re-formulated in terms of *eigenvalues*; in particular, a numerical matrix is *non-regular* if and only if 0 is its eigenvalue, and an *interval* matrix is non-regular if and only if 0 belongs to its set of possible eigenvalues. Therefore, from the fact that checking regularity of an interval matrix is NP-hard, it follows that:

Theorem 21.17. *The problem of checking, for a given interval matrix* $\mathbf{A}$ *and a given rational number* λ, *whether* λ *is an eigenvalue of one of the matrices* $A \in \mathbf{A}$, *is NP-hard.*

Comment. It is interesting that, in contrast to eigen*values*, rational eigen*vectors* can be checked in polynomial time (Rohn [348]).

In view of the relation between regularity and eigenvalues, an alternative way to check whether an interval matrix is regular or not is to find its possible eigenvalues. Let us show that this problem is also NP-hard.

For an interval matrix $\mathbf{A}$ define

$$\overline{\lambda}(\mathbf{A}) = \max\{\mathrm{Re}\lambda | \lambda \text{ is an eigenvalue of some } A \in \mathbf{A}\}.$$

We will show that computing $\overline{\lambda}(\mathbf{A})$ approximately with relative error less than 1 is NP-hard:

Theorem 21.18. *Suppose there exists a polynomial-time algorithm which for each symmetric interval matrix* $\mathbf{A}$ *computes a rational number* $\widetilde{\lambda}(\mathbf{A})$ *for which*

$$\left| \frac{\widetilde{\lambda}(\mathbf{A}) - \overline{\lambda}(\mathbf{A})}{\overline{\lambda}(\mathbf{A})} \right| < 1$$

if $\overline{\lambda}(\mathbf{A}) \neq 0$ *and* $\widetilde{\lambda}(\mathbf{A}) \geq 0$ *otherwise. Then* P=NP.

In general, an eigenvalue can be a complex number. Let us show that the problem of computing eigenvalues is NP-hard even if we restrict ourselves to *symmetric* numerical matrices, for which all eigenvalues are *real* numbers. Before formulating the result, let us show that the corresponding set of possible values is indeed an interval:

Theorem 21.19. *For a symmetric interval matrix* $\mathbf{A}$*, the set*

$$\lambda_{\max}(\mathbf{A}) = \{\lambda_{\max}(A) | A \text{ symmetric, } A \in \mathbf{A}\}$$

is a (finite and closed) interval.

It is not only NP-hard to compute the endpoints of this interval, but it is also NP-hard to check whether a given interval $(\underline{a}, \overline{a})$ is indeed an enclosure for the interval $\lambda_{\max}(\mathbf{A})$:

Theorem 21.20. *The following problem is NP-hard: given a symmetric interval matrix* $\mathbf{A}$ *and an interval* $(\underline{a}, \overline{a})$*, check whether* $\lambda_{\max}(\mathbf{A}) \subset (\underline{a}, \overline{a})$.

How can we actually compute the endpoints of the interval $\lambda_{\max}(\mathbf{A})$, at least for small n? By definition, the desired interval is a set of all values $\lambda_{\max}(A)$

for *all* symmetric matrices $A \in \mathbf{A}$. Some simplification comes from the fact that to compute theses endpoints, it is not necessary to consider *all* symmetric matrices $A \in \mathbf{A}$, only so-called *edge matrices*:

Definition 21.4. *Let $\mathbf{A} = [\underline{A}, \overline{A}]$ be an interval matrix. A numerical matrix $A \in \mathbf{A}$ is called its edge matrix if for all pairs (i, j) except for maybe one pair, either $a_{ij} = \underline{a}_{ij}$ or $a_{ij} = \overline{a}_{ij}$.*

This name comes from the following geometric representation: Every numerical matrix can be represented by a point in n^2-dimensional space; an *interval* matrix $\mathbf{A}$ is then represented as a *box* (parallelepiped) in R^{n^2}, and "edge matrices" form *edges* of this box.

Theorem 21.21. (Rohn [347]) *If a real number λ is an eigenvalue of some $A \in \mathbf{A}$, then it is also an eigenvalue of one of $\mathbf{A}$'s edge matrices.*

This result is not always true for *complex* eigenvalues: there exists an interval matrix $\mathbf{A}$ for which a complex eigenvalue λ of one of the matrices $A \in \mathbf{A}$ is different from all eigenvalues of all its edge matrices (Barmish, Fu, and Saleh [17]); an appropriate modification of the "edge theorem" is, however, true for complex eignevalues as well (Hollot and Bartlett in [153]).

21.6.3. *Radius of non-singularity (and subordinate matrix norms)*

We can also describe a slightly different computational problem: assuming that a given numerical matrix A is regular, how accurately do we need to describe it so that the resulting matrix will still be guaranteed to be regular? In other words, for each "direction", how wide, in this "direction", can an interval matrix centered in A be that it will still be regular?

Formally, given an $n \times n$ matrix A and a nonnegative "directional" $n \times n$ matrix Δ, the *radius of non-singularity* is defined by

$$d(A, \Delta) = \inf\{\varepsilon \geq 0 | [A - \varepsilon\Delta, A + \varepsilon\Delta] \text{ is not a regular interval matrix}\} \quad (21.16)$$

($d(A, \Delta) = \infty$ if no such ε exists; if $d(A, \Delta) < \infty$, then the infimum is attained as minimum). This notion was, to the best of our knowledge, first formulated by Neumaier [302] and was since studied by Poljak and Rohn [328, 329], Demmel [89], Rohn [348] and Rump [372], [371] (Demmel and Rump use the term "componentwise distance to the nearest singular matrix").

This problem is NP-hard even for the special case when all the values of Δ_{ij} are equal to 1, i.e., when Δ coincides with the matrix E all of whose elements are equal to 1. We will denote the radius $d(A,E)$ by $d(A)$. This radius $d(A)$ is always finite and $d(A)=0$ if and only if A is singular. For regular matrices, the radius $d(A)$ can be explicitly described in terms of one of the known matrix norms. To describe this result, let us recall how matrix norms are defined.

Given two vector-space norms $||x||_\alpha$ in R^n and $||x||_\beta$ in R^m, a *subordinate matrix norm* in $R^{m\times n}$ is defined as

$$||A||_{\alpha,\beta} = \max_{||x||_\alpha=1} ||Ax||_\beta$$

(see Golub and van Loan [130] or Higham [151]). $||A||_{\alpha,\beta}$ is a matrix norm, i.e., it possesses the three usual properties:

1) $||A||_{\alpha,\beta} \geq 0$ and $||A||_{\alpha,\beta}=0$ if and only if $A=0$,
2) $||A+B||_{\alpha,\beta} \leq ||A||_{\alpha,\beta} + ||B||_{\alpha,\beta}$,
3) $||\lambda A||_{\alpha,\beta} = |\lambda|\cdot||A||_{\alpha,\beta}$.

It is worth mentioning, however, that generally, such a norm does not satisfy the property $||AB||_{\alpha,\beta} \leq ||A||_{\alpha,\beta}||B||_{\alpha,\beta}$ (it does satisfy this property, e.g., if $\alpha=\beta$).

By combining the three most frequently used norms

$$||x||_1 = \sum_i |x_i|, \quad ||x||_2 = \sqrt{\sum_i x_i^2}, \quad ||x||_\infty = \max_i |x_i|,$$

we get nine subordinate norms, including the three usual norms

$$||A||_1 = ||A||_{1,1} = \max_j \sum_i |a_{ij}|,$$

$$||A||_2 = ||A||_{2,2} = \sqrt{\lambda_{\max}(A^TA)},$$

$$||A||_\infty = ||A||_{\infty,\infty} = \max_i \sum_j |a_{ij}|.$$

It turns out that the radius $d(A)$ is related to one of these nine norms, namely, to the norm

$$||A||_{\infty,1} = \max_{||x||_\infty=1} ||Ax||_1.$$

Theorem 21.22. (Poljak *et al.* [329]) *For each non-singular matrix A,*

$$d(A) = \frac{1}{\|A^{-1}\|_{\infty,1}}. \quad (21.17)$$

This norm has an exceptional behavior in the sense that it is much more difficult to compute than the other ones:

Theorem 21.23. *Computing $\|A\|_{\infty,1}$ is NP-hard.*

Comment. In sharp contrast to this result, the norm $\|A\|_{1,\infty}$ (with indices swapped) can be computed in polynomial time (see Higham [151]):

$$\|A\|_{1,\infty} = \max_{i,j} |a_{ij}| \quad (21.18)$$

As an immediate consequence of Theorem 21.23, we obtain the following result Poljak *et al.* [329]:

Theorem 21.24. *Computing the radius of non-singularity $d(A, \Delta)$ is NP-hard (even in the special case $\Delta = E$).*

Not only *computing* this norm, but even computing an *approximation* to it or checking whether a given interval is an *enclosure* for the norm are NP-hard problems:

Theorem 21.25.

- *Checking whether $\|A\|_{\infty,1} \geq 1$ for a given matrix A is NP-hard.*
- *Checking whether $d(A) \leq 1$ for a given matrix A is NP-hard.*

Theorem 21.26.

- *For every $\delta > 0$, computing a rational number that is δ-close to $||A||_{\infty,1}$ is NP-hard.*
- *Suppose there exists a polynomial-time algorithm which for each nonnegative symmetric positive definite rational matrix A computes a rational approximation $\tilde{d}(A)$ to $d(A)$ satisfying*

$$\left|\frac{\tilde{d}(A) - d(A)}{d(A)}\right| \leq \frac{1}{4n^2},$$

where n is the size of A. Then P=NP.

How can we actually compute the radius of non-singularity for small n? A general formula for $d(A, \Delta)$ was given in Poljak and Rohn [329]:

$$d(A, \Delta) = \frac{1}{\max\{\varrho_0(A^{-1}T_1\Delta T_2) \mid |T_1| = |T_2| = I\}}, \tag{21.19}$$

where I is the unit matrix, ϱ_0 denotes the real spectral radius defined as $\varrho_0(A) = \max\{|\lambda| \mid \lambda$ is a real eigenvalue of $A\}$ (and as $\varrho_0(A) = 0$ if a matrix A has no real eigenvalues). A matrix T satisfying $|T| = I$ is obviously a diagonal matrix with ± 1 entries on the diagonal. There are 2^n such matrices, hence the formula (21.19) enables us, for small n, to compute the radius of non-singularity in finitely many steps.

For $\Delta = E$, we can compute the radius $d(A) = d(A, E)$ as follows:

- first, we compute the matrix norm of a matrix A by trying all 2^n vectors $z \in \{-1, 1\}^n$ and using the following theorem;
- then, we use the formula (21.17) that relates the desired radius to this matrix norm.

Theorem 21.27. *For each $A \in R^{m \times n}$ we have*

$$||A||_{\infty,1} = \max_{z \in \{-1,1\}^n} ||Az||_1. \tag{21.20}$$

If A is symmetric positive semidefinite, then

$$||A||_{\infty,1} = \max_{z \in \{-1,1\}^n} z^T A z. \tag{21.21}$$

The above method can be used for small n, when it is still computationally feasible to analyze all 2^n possible vectors $z \in \{-1,1\}^n$. For larger n, this is not possible; for such n, *bounds* on the radius of non-singularity can be derived from sufficient regularity or singularity conditions. E.g., from Theorem 21.5, we conclude that

$$\frac{1}{\varrho(|A^{-1}|\Delta)} \le d(A,\Delta) \le \frac{1}{\max_j(\Delta|A^{-1}|)_{jj}}.$$

Using a more sophisticated reasoning, Rump [372], [371] recently proved a "symmetric" estimation

$$\frac{1}{\varrho(|A^{-1}|\Delta)} \le d(A,\Delta) \le \frac{6n}{\varrho(|A^{-1}|\Delta)}.$$

Comment. Related to the radius of non-singularity is the ***structured singular value*** introduced by Doyle [97]. The NP-hardness of its computation was proved by Braatz, Young, Doyle, and Morari [56] and independently by Coxson and DeMarco [79, 80].

21.6.4. *Radius of stability*

Similarly to the radius of *non-singularity* $d(A,\Delta)$, we may define ***radius of stability*** as

$$s(A,\Delta) = \inf\{\varepsilon \ge 0 | [A - \varepsilon\Delta, A + \varepsilon\Delta] \text{ is unstable}\}.$$

Hence, $[A-\varepsilon\Delta, A+\varepsilon\Delta]$ is stable if $0 \le \varepsilon < s(A,\Delta)$ and unstable if $\varepsilon \ge s(A,\Delta)$. These two radii are related:

Theorem 21.28. *If A is a symmetric stable matrix, and Δ is a symmetric matrix, then $s(A,\Delta) = d(A,\Delta)$.*

Hence, we may apply the results of the previous subsection to the radius of stability. In particular: for a symmetric stable matrix A we have

$$s(A,E) = \frac{1}{||A^{-1}||_{\infty,1}},$$

and computing $s(A,E)$ is NP-hard (even approximately).

22

PROPERTIES OF INTERVAL MATRICES II: PROOFS AND AUXILIARY RESULTS

In this chapter, we prove the results about computational complexity and feasibility of properties of interval matrices that were formulated in the previous chapter. Along the way, we also describe some important auxiliary results.

22.1. Notations

In the proofs, we will use the following notations.

- For two matrices A, B of the same size, inequalities like $A \leq B$ or $A < B$ are understood componentwise. A is called ***nonnegative*** if $A \geq 0$ and ***symmetric*** if $A^T = A$ (A^T is the transpose of A). The ***absolute value*** of a matrix $A = (a_{ij})$ is defined by $|A| = (|a_{ij}|)$; properties like $|A + B| \leq |A| + |B|$ or $|AB| \leq |A||B|$ are easy to prove.
- The same notations also apply to ***vectors*** that are treated as one-column matrices. In particular, for vectors $a = (a_i)$ and $b = (b_i)$, $a^T b = \sum_i a_i b_i$ is the ***scalar product***, whereas ab^T is the matrix $(a_i b_j)$.
- $\lambda_{\min}(A)$, $\lambda_{\max}(A)$ denote, correspondingly, the smallest and the largest ***eigenvalue*** of a symmetric matrix A. As is well known, $\lambda_{\min}(A) = \min_{\|x\|_2=1} x^T Ax$ and $\lambda_{\max}(A) = \max_{\|x\|_2=1} x^T Ax$.
- $\sigma_{\min}(A), \sigma_{\max}(A)$ denote the minimal and maximal ***singular value*** of a matrix A, and $\varrho(A)$ is the ***spectral radius*** of A.
- I denotes the unit matrix, e_j is the jth column of I and $e = (1, \ldots, 1)^T$ is the vector of all ones, $E = ee^T$ is the matrix of all ones.

22.2. The norm $\|A\|_{\infty,1}$

22.2.1. *Properties of the norm*

Proof of Theorem 21.27. Many results will use properties of the norm $\|A\|_{\infty,1}$. So, let us start the proofs with proving Theorem 21.27 that provides an explicitly computable (although not feasibly computable) expression for this matrix norm: $\|A\|_{\infty,1} = \max_z \|Az\|_1$, where max is taken over all $z \in \{-1,1\}^n$ (i.e., over all ± 1-vectors), and $\|A\|_{\infty,1} = \max_z z^T Az$ for symmetric positive semidefinite matrices A.

If $\|x\|_\infty = 1$, then x belongs to the unit cube $\{x| -e \le x \le e\}$, which is a convex polyhedron, therefore x can be expressed as a convex combination of its vertices which are exactly the points in $\{-1,1\}^n$:

$$x = \sum_{z\in\{-1,1\}^n} \lambda_z z, \tag{22.1}$$

where $\lambda_z \ge 0$ for each $z \in \{-1,1\}^n$ and $\sum_{z\in\{-1,1\}^n} \lambda_z = 1$. From (22.1), we get

$$\|Ax\|_1 = \|\sum_{z\in\{-1,1\}^n} \lambda_z Az\|_1 \le \max_{z\in\{-1,1\}^n} \|Az\|_1,$$

hence

$$\max_{\|x\|_\infty=1} \|Ax\|_1 \le \max_{z\in\{-1,1\}^n} \|Az\|_1 \le \max_{\|x\|_\infty=1} \|Ax\|_1$$

(since $\|z\|_\infty = 1$ for each $z \in \{-1,1\}^n$) and (21.20) follows.

Let now A be symmetric positive semidefinite and let $z \in \{-1,1\}^n$. Define $y \in \{-1,1\}^n$ by $y_j = 1$ if $(Az)_j \ge 0$ and $y_j = -1$ if $(Az)_j < 0$ $(j = 1,\dots,n)$, then $\|Az\|_1 = y^T Az$. Since A is symmetric positive semidefinite, we have $(y - z)^T A(y-z) \ge 0$, which implies

$$2y^T Az \le y^T Ay + z^T Az \le 2 \max_{z\in\{-1,1\}^n} z^T Az,$$

hence

$$\|Az\|_1 = y^T Az \le \max_{z\in\{-1,1\}^n} z^T Az$$

and

$$\|A\|_{\infty,1} = \max_{z\in\{-1,1\}^n} \|Az\|_1 \le \max_{z\in\{-1,1\}^n} z^T Az. \tag{22.2}$$

Conversely, for each $z \in \{-1,1\}^n$ we have

$$z^T Az \leq |z|^T \cdot |Az| = \|Az\|_1 \leq \max_{z\in Z} \|Az\|_1 = \|A\|_{\infty,1},$$

hence

$$\max_{z\in\{-1,1\}^n} z^T Az \leq \|A\|_{\infty,1},$$

which together with (22.2) gives (21.21). The theorem is proven.

22.2.2. *Computing* $\|A\|_{\infty,1}$ *is NP-hard*

In order to prove the NP-hardness for a possibly narrow class of matrices, we introduce the following concept (first formulated in Rohn [351]):

Definition 22.1. *A numerical symmetric $n \times n$ matrix $A = (a_{ij})$ is called an MC-matrix if it is of the form*

$$a_{ij} \begin{cases} = n & \text{if } i = j \\ \in \{0,-1\} & \text{if } i \neq j \end{cases}$$

$(i, j = 1, \ldots, n)$.

Comment. MC comes from "maximum cut" (see the proof of Theorem 22.2 below)

Since an MC-matrix is symmetric by definition, there are altogether $2^{n(n-1)/2}$ MC-matrices of size n. The basic properties of MC-matrices are summed up in the following proposition:

Theorem 22.1. *An MC-matrix $A \in R^{n\times n}$ is symmetric positive definite, nonnegative invertible and satisfies*

$$\|A\|_{\infty,1} = \max_{z\in\{-1,1\}^n} z^T Az, \tag{22.3}$$

$$n \leq \|A\|_{\infty,1} \leq n(2n-1) \tag{22.4}$$

and

$$\|A^{-1}\|_1 \leq 1.$$

Proof. A is symmetric by definition; it is positive definite since for $x \neq 0$,

$$x^T Ax \geq n\|x\|_2^2 - \sum_{i \neq j} |x_i x_j| = (n+1)\|x\|_2^2 - \|x\|_1^2 \geq \|x\|_2^2 > 0$$

($\|x\|_1 \leq \sqrt{n}\|x\|_2$ by Cauchy-Schwartz inequality; see, e.g., Golub and van Loan [130]). Hence (22.3) holds by Theorem 21.27. Since $|a_{ij}| \leq 1$ for $i \neq j$, for each $z \in \{-1,1\}^n$ and $i \in \{1,\ldots,n\}$ we have

$$z_i(Az)_i = n + \sum_{j \neq i} a_{ij} z_i z_j \in [1, 2n-1],$$

hence

$$n \leq z^T Az \leq n(2n-1)$$

for each $z \in \{-1,1\}^n$, and (22.3) implies (22.4). By definition, A is of the form

$$A = nI - A_0 = n(I - \frac{1}{n}A_0)$$

where $A_0 = nI - A \geq 0$ and $\|\frac{1}{n}A_0\|_1 \leq \frac{n-1}{n} < 1$, hence

$$A^{-1} = \frac{1}{n}\sum_0^\infty \left(\frac{1}{n}A_0\right)^j \geq 0$$

and

$$\|A^{-1}\|_1 \leq \frac{1}{n - \|A_0\|_1} \leq 1.$$

The theorem is proven.

The following basic result is due to Poljak and Rohn [329] (given there in a slightly different formulation without using the concept of an MC-matrix).

Theorem 22.2. *The following decision problem is NP-complete:*
Instance. *An MC-matrix A and a positive integer ℓ.*
Question. *Is $z^T Az \geq \ell$ for some $z \in \{-1,1\}^n$?*

Proof. Let n be a positive integer, and let $(\mathcal{V}, \mathcal{E})$ be an arbitrary graph whose vertices are integers 1 through n; in this proof, we will denote its set of vertices by $\mathcal{V}$ (i.e., take $\mathcal{V} = \{1,\ldots,n\}$) and its set of edges by $\mathcal{E}$. Let $A = (a_{ij})$ be given by $a_{ij} = n$ if $i = j$, $a_{ij} = -1$ if $i \neq j$ and the nodes i, j are connected by an edge (i.e., if $(i,j) \in \mathcal{E}$), and $a_{ij} = 0$ if $i \neq j$ and i, j are not connected. Then A is an MC-matrix. For an arbitrary set $S \subseteq \mathcal{V}$, define the cut $c(S)$ as the number of edges in $\mathcal{E}$ whose one endpoint belongs to the set S and the other one does not belong to this set. We will prove that

$$\|A\|_{\infty,1} = 4 \max_{S \subseteq \mathcal{V}} c(S) - 2\mathrm{Card}(\mathcal{E}) + n^2 \tag{22.5}$$

holds. Given a $S \subseteq \mathcal{V}$, define a $z \in \{-1,1\}^n$ by

$$z_i = \begin{cases} 1 & \text{if } i \in S \\ -1 & \text{if } i \notin S. \end{cases}$$

Then we have

$$\begin{aligned} z^T A z &= \sum_{i,j} a_{ij} z_i z_j = \sum_{i \neq j} a_{ij} z_i z_j + n^2 \\ &= \sum_{i \neq j} [-\frac{1}{2} a_{ij} (z_i - z_j)^2 + a_{ij}] + n^2 \\ &= -\frac{1}{2} \sum_{z_i z_j = -1} a_{ij} (z_i - z_j)^2 + \sum_{i \neq j} a_{ij} + n^2 \\ &= -\frac{1}{2} \sum_{z_i z_j = -1} 4 a_{ij} + \sum_{i \neq j} a_{ij} + n^2, \end{aligned}$$

hence

$$z^T A z = 4c(S) - 2\mathrm{Card}(\mathcal{E}) + n^2. \tag{22.6}$$

Conversely, given a $z \in \{-1,1\}^n$, then for $S = \{i \in \mathcal{V} | z_i = 1\}$ the same reasoning implies (22.6). Taking maximum on both sides of (22.6), we obtain (22.5) in view of (22.3).

Hence, given a positive integer L, we have

$$c(S) \geq L \tag{22.7}$$

for some $S \subseteq \mathcal{V}$ if and only if

$$z^T A z \geq 4L - 2\mathrm{Card}(\mathcal{E}) + n^2$$

for some $z \in \{-1,1\}^n$. Since the decision problem (22.7) is NP-complete ("simple max-cut problem", Garey, Johnson and Stockmeyer [122]), we obtain that the decision problem

$$z^T A z \geq \ell \tag{22.8}$$

(ℓ positive integer) is NP-hard. It is NP-complete since for a guessed solution $z \in \{-1,1\}^n$ the validity of (22.8) can be checked in polynomial time. The theorem is proven.

In this way, we have also proven that computing $||A||_{\infty,1}$ is NP-hard for MC-matrices, and thus, we have proven Theorem 21.23.

To facilitate formulations of some subsequent results, it is advantageous to remove the integer parameter ℓ from the formulation of Theorem 22.2. This can be done by using M-matrices instead of MC-matrices. Let us recall that $A = (a_{ij})$ is called an M-matrix if $a_{ij} \le 0$ for $i \neq j$ and $A^{-1} \ge 0$ (a number of equivalent formulations may be found in Berman and Plemmons [33]); hence each MC-matrix is an M-matrix (Theorem 22.1). Since a symmetric M-matrix is positive definite Berman and Plemmons [33], this property is not explicitly mentioned in the following theorem:

Theorem 22.3. *The following decision problem is NP-hard:*
Instance. *An $n \times n$ symmetric rational M-matrix A with $||A||_1 \le 2n - 1$.*
Question. *Is $||A||_{\infty,1} \ge 1$?*

Proof. Given an MC-matrix A and a positive integer ℓ, the assertion

$$z^T Az \ge \ell \text{ for some } z \in \{-1,1\}^n$$

is equivalent to $||A||_{\infty,1} \ge \ell$ and thereby also to

$$\left\|\frac{1}{\ell}A\right\|_{\infty,1} \ge 1,$$

where $\frac{1}{\ell}A$ is a symmetric rational M-matrix with $||\frac{1}{\ell}A||_1 \le ||A||_1 \le 2n - 1$. Hence the decision problem of Theorem 22.2 can be reduced in polynomial time to the current one, which is then NP-hard. The theorem is proven.

Thus, we have also proven the first statement of Theorem 21.25, that checking whether $||A||_{\infty,1} \ge 1$ for a given matrix A is NP-hard.

Finally we will show that even computing a sufficiently close approximation of $||A||_{\infty,1}$ is NP-hard:

Theorem 22.4. *Suppose there exists a polynomial-time algorithm which for each MC-matrix A computes a rational number $\nu(A)$ satisfying*

$$|\nu(A) - ||A||_{\infty,1}| < \frac{1}{2}.$$

Then P=NP.

Proof. If such an algorithm exists, then $||A||_{\infty,1} < \nu(A) + \frac{1}{2} < ||A||_{\infty,1} + 1$, therefore,

$$||A||_{\infty,1} = \left\lfloor \nu(A) + \frac{1}{2} \right\rfloor$$

(since $||A||_{\infty,1}$ is integer for an MC-matrix A, see (22.3)), hence the NP-hard problem of Theorem 22.3 can be solved in polynomial time, implying P=NP. The theorem is proven.

By considering matrices $2\delta \cdot A$ for MC-matrices A, we thus prove the first statement of Theorem 21.26, that for every $\delta > 0$, computing a rational number that is δ-close to $||A||_{\infty,1}$ is NP-hard.

22.3. Regularity

22.3.1. *Checking regularity is NP-hard*

The basic relationship of regularity to the previous section is provided by the following equivalence:

Theorem 22.5. *For a symmetric positive definite matrix A, the following statements are equivalent to each other:*

(i) $||A||_{\infty,1} \geq 1$,

(ii) *the interval matrix*

$$[A^{-1} - E, A^{-1} + E] \tag{22.9}$$

is not regular,

(iii) *the interval matrix* (22.9) *contains a symmetric singular matrix A' of the form*

$$A' = A^{-1} - \frac{zz^T}{z^T A z} \tag{22.10}$$

for some $z \in \{-1, 1\}^n$.

Proof. (i)$\Rightarrow$(iii): Due to Theorem 21.27, if (i) holds, then

$$||A||_{\infty,1} = \max_{z \in \{-1,1\}^n} z^T A z \geq 1,$$

hence $z^T Az \geq 1$ for some $z \in \{-1,1\}^n$. Since

$$\left|\frac{zz^T}{z^T Az}\right| \leq E,$$

the matrix A' defined by (22.10) belongs to $[A^{-1} - E, A^{-1} + E]$ and satisfies

$$A'Az = z - \frac{z(z^T Az)}{z^T Az} = 0,$$

where $Az \neq 0$ (A is nonsingular since it is positive definite), hence A' is singular, and obviously also symmetric.

(iii)⇒(ii) is obvious.

(ii)⇒(i): Let $A''x = 0$ for some $A'' \in [A^{-1} - E, A^{-1} + E]$ and $x \neq 0$. Define $z \in \{-1,1\}^n$ by $z_j = 1$ if $x_j \geq 0$ and $z_j = -1$ otherwise ($j = 1, \ldots, n$). Then we have

$$e^T|x| = z^T x = z^T A(A^{-1} - A'')x \leq |z^T A(A^{-1} - A'')x| \leq |z^T A|ee^T|x|,$$

hence

$$1 \leq |z^T A|e = ||Az||_1 \leq ||A||_{\infty,1},$$

which is (i). The theorem is proven.

Theorem 22.6. *The following problem is NP-complete:*
Instance. *A nonnegative symmetric positive definite rational matrix A.*
Question. *Is $[A - E, A + E]$ non-regular?*

Proof. For a symmetric rational M-matrix A (which is positive definite [33]),

$$||A||_{\infty,1} \geq 1 \tag{22.11}$$

is according to Theorem 22.5 equivalent to non-regularity of

$$[A^{-1} - E, A^{-1} + E],$$

where A^{-1} is rational, nonnegative and symmetric positive definite. Since computing A^{-1} can be done (by a modified Gaussian elimination) in polynomial time (Edmonds [98]), we have a polynomial-time reduction of the NP-hard problem (22.11) (Theorem 22.3) to the current problem, which is thus also NP-hard.

If $[A-E, A+E]$ is not regular, then it contains a rational singular matrix of the form

$$A - \frac{zz^T}{z^T A^{-1} z}$$

for some $z \in \{-1, 1\}^n$ (Theorem 22.5, (ii)⇔(iii)) which can be guessed (generated by a nondeterministic polynomial-time algorithm) and then checked for singularity by modified Gaussian elimination in polynomial time [98]. Thus the problem is in the class NP, hence it it NP-complete. The theorem is proven.

This result immediately implies NP-hardness of checking regularity (Theorem 21.2); to be more precise, we have the following result:

Theorem 22.7. *The following problem is NP-hard:*
Instance. *A nonnegative symmetric positive definite rational matrix A.*
Question. *Is $[A-E, A+E]$ regular?*

As a by-product of the equivalence (ii)⇔(iii) of Theorem 22.5 we obtain that the problem of checking regularity of all *symmetric* matrices contained in $[A-E, A+E]$ is also NP-hard (Theorem 21.3).

22.3.2. *Necessary and/or sufficient conditions*

Let us prove the results about checking regularity for small n (that were formulated in Chapter 21). These results are based on the following corollary of the Oettli-Prager theorem [317]:

Theorem 22.8. *An interval matrix $\mathbf{A} = [\widetilde{A} - \Delta, \widetilde{A} + \Delta]$ is not regular if and only if the inequality*

$$|\widetilde{A}x| \leq \Delta|x| \qquad (22.12)$$

has a nontrivial solution.

Proof. If $\mathbf{A}$ contains a singular matrix A, then $Ax = 0$ for some $x \neq 0$, which implies

$$|\widetilde{A}x| = |(\widetilde{A} - A)x| \leq \Delta|x|.$$

Conversely, let (22.12) hold for some $x \neq 0$. Define $y \in R^n$ and $z \in \{-1, 1\}^n$ by

$$y_i = \begin{cases} (\widetilde{A}x)_i/(\Delta|x|)_i & \text{if } (\Delta|x|)_i > 0, \\ 1 & \text{if } (\Delta|x|)_i = 0 \end{cases}$$

and

$$z_j = \begin{cases} 1 & \text{if } x_j \geq 0, \\ -1 & \text{if } x_j < 0 \end{cases}$$

$(i, j = 1, \ldots, n)$. In Chapter 21, we have used the formula (21.11) to define, for each interval matrix $\mathbf{A}$, and for each pair of vectors $y, z \in \{-1, 1\}^n$, a numerical matrix $A^{(y,z)}$. We can use the same formula (21.11) to define a matrix $A^{(y,z)}$ for *arbitrary* vectors y and z. For thus defined matrix, we have

$$(A^{(y,z)}x)_i = (\widetilde{A}x)_i - y_i(\Delta|x|)_i = 0$$

for each i, hence $A^{(y,z)}$ is singular, and since $|y_i| \leq 1$ for each i due to (22.12), from (21.11) it follows that $A^{(y,z)} \in \mathbf{A}$, hence $\mathbf{A}$ is not regular. The theorem is proven.

Proof of Theorem 21.4. Let us prove that an interval matrix $\mathbf{A}$ is regular if and only if determinants of all the matrices $A^{(y,z)}$, $y, z \in \{-1, 1\}^n$ are nonzero and of the same sign.

Let $\mathbf{A}$ be regular and assume that

$$(\det A^{(y,z)})(\det A^{(y',z')}) \leq 0$$

holds for some $y, z, y', z' \in \{-1, 1\}^n$. Define a real function φ of one real variable by

$$\varphi(t) = \det(A^{(y,z)} + t(A^{(y',z')} - A^{(y,z)})), \quad t \in [0, 1].$$

Then $\varphi(0)\varphi(1) \leq 0$, hence there exists a $\tau \in [0, 1]$ with $\varphi(\tau) = 0$. Thus the matrix $A^{(y,z)} + \tau(A^{(y',z')} - A^{(y,z)})$ is singular and belongs to $\mathbf{A}$ (due to its convexity), which is a contradiction. Hence

$$(\det A^{(y,z)})(\det A^{(y',z')}) > 0$$

holds for each $y, z, y', z' \in \{-1, 1\}^n$.

Conversely, let $\mathbf{A}$ be not regular. From the proof of Theorem 22.8 we know that there exists a singular matrix of the form $A^{(y,z)}$ for some $|y| \leq e$, $z \in \{-1, 1\}^n$. Let us introduce the function

$$f(s) = \det A^{(s,z)}$$

for $s \in R^n$, and define a vector $\overline{y} = (\overline{y}_j) \in \{-1, 1\}^n$ componentwise by induction on $j = 1, \ldots, n$ as follows: if the function of one real variable

$$f(\overline{y}_1, \ldots, \overline{y}_{j-1}, t, y_{j+1}, \ldots, y_n) \tag{22.13}$$

is increasing in t, set $\overline{y}_j = 1$, otherwise set $\overline{y}_j = -1$. Since the function (22.13) is linear in t due to (21.11), we have

$$f(\overline{y}_1, \ldots, \overline{y}_{j-1}, y_j, y_{j+1}, \ldots, y_n) \le f(\overline{y}_1, \ldots, \overline{y}_{j-1}, \overline{y}_j, y_{j+1}, \ldots, y_n)$$

for each j, and by induction

$$0 = \det A^{(y,z)} = f(y_1, \ldots, y_n) \le f(\overline{y}_1, \ldots, \overline{y}_n) = \det A^{(\overline{y},z)},$$

hence $0 \le \det A^{(\overline{y},z)}$, $\overline{y}, z \in \{-1,1\}^n$. In an analogous way we may construct a $\underline{y} \in \{-1,1\}^n$ satisfying $\det A^{(\underline{y},z)} \le 0$. Hence

$$(\det A^{(\underline{y},z)})(\det A^{(\overline{y},z)}) \le 0$$

for some $\underline{y}, \overline{y}, z \in \{-1,1\}^n$, which concludes the proof of the second implication. The theorem is proven.

Proof of Theorem 21.5. Let us show that the conditions described in Theorem 21.5 indeed imply that the matrix $\mathbf{A}$ is, correspondingly, regular or not regular.

(i) Let condition (i) hold, i.e., let $\varrho(|\widetilde{A}^{-1}|\Delta) < 1$. Assume to the contrary that $\mathbf{A}$ is not regular, then

$$|\widetilde{A}x| \le \Delta|x| \tag{22.14}$$

for some $x \ne 0$ (Theorem 22.8), hence

$$|x'| \le \Delta|\widetilde{A}^{-1}x'| \le \Delta|\widetilde{A}^{-1}||x'|$$

holds for $x' = \widetilde{A}x \ne 0$, which implies

$$1 \le \varrho(\Delta|\widetilde{A}^{-1}|) = \varrho(|\widetilde{A}^{-1}|\Delta)$$

(Neumaier [303]), a contradiction.

(ii) Let now condition (ii) hold, i.e., $\sigma_{\max}(\Delta) < \sigma_{\min}(\widetilde{A})$. Again assuming to the contrary that $\mathbf{A}$ is not regular, we have that (22.14) holds for some $x \ne 0$ which may be normalized so that $\|x\|_2 = 1$, hence also

$$|\widetilde{A}x|^T|\widetilde{A}x| \le (\Delta|x|)^T(\Delta|x|),$$

which implies

$$\begin{aligned}
\sigma^2_{\min}(\widetilde{A}) &= \lambda_{\min}(\widetilde{A}^T\widetilde{A}) = \min_{\|x\|_2=1} x^T\widetilde{A}^T\widetilde{A}x \le (\widetilde{A}x)^T(\widetilde{A}x) \\
&\le |\widetilde{A}x|^T|\widetilde{A}x| \le (\Delta|x|)^T(\Delta|x|) = |x|^T\Delta^T\Delta|x| \\
&\le \max_{\|x\|_2=1} x^T\Delta^T\Delta x = \lambda_{\max}(\Delta^T\Delta) = \sigma^2_{\max}(\Delta),
\end{aligned}$$

hence

$$\sigma_{\min}(\widetilde{A}) \leq \sigma_{\max}(\Delta),$$

which is a contradiction.

(iii) Let condition (iii) hold, i.e., $\max_j(\Delta|\widetilde{A}^{-1}|)_{jj} \geq 1$. This means that $(\Delta|\widetilde{A}^{-1}|)_{jj} \geq 1$ for some j and let e_j denote the jth column of the unit matrix I. Then

$$e_j \leq \Delta|\widetilde{A}^{-1}|e_j = \Delta|\widetilde{A}^{-1}e_j|$$

holds, hence for $x = \widetilde{A}^{-1}e_j \neq 0$ we have

$$|\widetilde{A}x| \leq \Delta|x|$$

and **A** is not regular due to Theorem 22.8.

(iv) Let condition (iv) hold, i.e., $(\Delta - |\widetilde{A}|)^{-1} \geq 0$. Then for $x = (\Delta - |\widetilde{A}|)^{-1}e$ we have $x > 0$ and $(\Delta - |\widetilde{A}|)x = e > 0$, hence

$$|\widetilde{A}x| \leq |\widetilde{A}|x < \Delta x = \Delta|x|$$

and Theorem 22.8 implies that the matrix **A** is not regular. The theorem is proven.

22.3.3. *Radius of non-singularity*

Let us now prove results about the radius of non-singularity, the first being Theorem 21.22 about the relation between this radius and the matrix norm.

Proof of Theorem 21.22. Let us prove that $d(A) = 1/||A^{-1}||_{\infty,1}$.

Since $||A||_{1,\infty} = \max_{ij}|a_{ij}|$ (see (21.18)), Kahan's theorem [166] gives

$$\begin{aligned} d(A) &= \min\{\varepsilon \geq 0 | [A - \varepsilon E, A + \varepsilon E] \text{ is not regular}\} \\ &= \min\{||A - A'||_{1,\infty} | A' \text{ is singular}\} \\ &= \frac{1}{||A^{-1}||_{\infty,1}}. \end{aligned}$$

The theorem is proven.

This theorem implies the following complexity result:

Theorem 22.9. *The following problem is NP-hard:*
Instance. *A nonnegative symmetric positive definite rational matrix A.*
Question. *Is $d(A) \leq 1$?*

Proof. For a symmetric M-matrix A,

$$\|A\|_{\infty,1} \geq 1$$

is according to Theorem 21.22 equivalent to

$$d(A^{-1}) \leq 1,$$

where A^{-1} is rational, nonnegative symmetric positive definite, hence the NP-hard problem of Theorem 22.3 can be reduced in polynomial time to the current one, which is thus NP-hard as well. The theorem is proven.

As an immediate consequence we obtain Theorem 21.24 (that computing the radius of non-singularity $d(A, \Delta)$ is NP-hard even in the special case $\Delta = E$) and the second statement of Theorem 21.25 (that checking whether $d(A) \leq 1$ for a given matrix A is NP-hard).

Proof of the second statement of Theorem 21.26. Let us prove that the existence of a polynomial-time algorithm which for each nonnegative symmetric positive definite rational matrix A computes a rational approximation $\tilde{d}(A)$ to $d(A)$ satisfying

$$\left| \frac{\tilde{d}(A) - d(A)}{d(A)} \right| \leq \frac{1}{4n^2},$$

implies P=NP.

Let A be an $n \times n$ MC-matrix, then A^{-1} is rational nonnegative symmetric positive definite, hence we have

$$\left| \frac{\tilde{d}(A^{-1}) - d(A^{-1})}{d(A^{-1})} \right| \leq \frac{1}{4n^2}.$$

Since $\|A\|_{\infty,1} \leq n(2n-1)$ by Theorem 22.1, there holds $2\|A\|_{\infty,1} + 1 \leq 4n^2 - 2n + 1 < 4n^2$, hence

$$\left| \frac{\tilde{d}(A^{-1})}{d(A^{-1})} - 1 \right| \leq \frac{1}{4n^2} < \frac{1}{2\|A\|_{\infty,1} + 1} < \frac{1}{2\|A\|_{\infty,1} - 1},$$

which implies

$$\frac{2\|A\|_{\infty,1}}{2\|A\|_{\infty,1}+1} = 1 - \frac{1}{2\|A\|_{\infty,1}+1} < \frac{\tilde{d}(A^{-1})}{d(A^{-1})} <$$

$$1 + \frac{1}{2\|A\|_{\infty,1}-1} = \frac{2\|A\|_{\infty,1}}{2\|A\|_{\infty,1}-1}$$

and by Theorem 21.22,

$$\frac{2}{2\|A\|_{\infty,1}+1} < \tilde{d}(A^{-1}) < \frac{2}{2\|A\|_{\infty,1}-1}$$

and

$$\left|\frac{1}{\tilde{d}(A^{-1})} - \|A\|_{\infty,1}\right| < \frac{1}{2}.$$

Hence we have a polynomial-time algorithm for computing $\|A\|_{\infty,1}$ with accuracy better than $\frac{1}{2}$, which according to Theorem 22.4 implies that P=NP. The theorem is proven.

22.4. Positive definiteness

22.4.1. *Positive definiteness and regularity*

Proof of Theorem 21.9. Let us prove that an interval matrix $\mathbf{A}$ is positive definite if and only if its symmetrization $\mathbf{A}^{\text{sym}}$ is regular and contains at least one positive definite matrix.

Let $\mathbf{A} = [\widetilde{A} - \Delta, \widetilde{A} + \Delta]$, so that

$$\mathbf{A}^{\text{sym}} = [\widetilde{A}^{\text{sym}} - \Delta^{\text{sym}}, \widetilde{A}^{\text{sym}} + \Delta^{\text{sym}}].$$

We will first prove that if $\mathbf{A}$ is positive definite, then $\mathbf{A}^{\text{sym}}$ is also positive definite. Assume to the contrary that $\mathbf{A}^{\text{sym}}$ is not positive definite, so that $x^T A' x \le 0$ for some $A' \in \mathbf{A}^{\text{sym}}$ and $x \ne 0$. Since $|x^T(A' - \widetilde{A}^{\text{sym}})x| \le |x|^T \Delta^{\text{sym}} |x|$, we have

$$x^T \widetilde{A} x - |x|^T \Delta |x| = x^T \widetilde{A}^{\text{sym}} x - |x|^T \Delta^{\text{sym}} |x| \le$$

$$x^T \widetilde{A}^{\text{sym}} x + x^T (A' - \widetilde{A}^{\text{sym}}) x = x^T A' x \le 0. \tag{22.15}$$

Define a diagonal matrix T by $t_{jj} = 1$ if $x_j \geq 0$ and $t_{jj} = -1$ otherwise. Then $|x| = Tx$, and from (22.15) we have

$$x^T(\widetilde{A} - T\Delta T)x \leq 0,$$

where $|T\Delta T| = \Delta$, hence the matrix $\widetilde{A} - T\Delta T$ belongs to $\mathbf{A}$ and is not positive definite. This contradiction shows that positive definiteness of $\mathbf{A}$ implies positive definiteness of $\mathbf{A}^{\mathrm{sym}}$, and thereby also regularity of $\mathbf{A}^{\mathrm{sym}}$.

Conversely, let $\mathbf{A}^{\mathrm{sym}}$ be regular and contain a positive definite matrix A_0. Assume to the contrary that some $A_1 \in \mathbf{A}$ is not positive definite. Let $\widetilde{A}_0 = \frac{1}{2}(A_0 + A_0^T)$, $\widetilde{A}_1 = \frac{1}{2}(A_1 + A_1^T)$, hence both $\widetilde{A}_0$ and $\widetilde{A}_1$ are symmetric and belong to $\mathbf{A}^{\mathrm{sym}}$, $\widetilde{A}_0$ is positive definite whereas $\widetilde{A}_1$ is not. Put

$$\tau = \sup\{t \in [0,1] | \widetilde{A}_0 + t(\widetilde{A}_1 - \widetilde{A}_0) \text{ is positive definite}\}.$$

Then $\tau \in (0, 1]$, hence the matrix

$$A^* = \widetilde{A}_0 + \tau(\widetilde{A}_1 - \widetilde{A}_0)$$

belongs to $\mathbf{A}^{\mathrm{sym}}$ (due to its convexity) and is symmetric positive semidefinite, but not positive definite, hence $\lambda_{\min}(A^*) = 0$, which shows that A^* is singular contrary to the assumed regularity of $\mathbf{A}^{\mathrm{sym}}$. Hence $\mathbf{A}$ is positive definite, which completes the proof. The theorem is proven.

Proof of Theorem 21.10. Let us now prove that a matrix $\mathbf{A}$ is positive definite if and only if $\mathbf{A}^{\mathrm{sym}}$ is positive definite.

Indeed, according to Theorem 21.9, $\mathbf{A}$ is positive definite if and only if $\mathbf{A}^{\mathrm{sym}}$ is regular and contains a positive definite matrix. If we apply the same theorem to $\mathbf{A}^{\mathrm{sym}}$ instead of $\mathbf{A}$, in view of the obvious fact that $(\mathbf{A}^{\mathrm{sym}})^{\mathrm{sym}} = \mathbf{A}^{\mathrm{sym}}$ we obtain that $\mathbf{A}^{\mathrm{sym}}$ is positive definite if and only if $\mathbf{A}^{\mathrm{sym}}$ is regular and contains a positive definite matrix. These two equivalences show that $\mathbf{A}$ is positive definite if and only if $\mathbf{A}^{\mathrm{sym}}$ is positive definite. The theorem is proven.

22.4.2. *Checking positive definiteness is NP-hard*

To prove NP-hardness of checking positive definiteness (Theorems 21.2 and 21.3), we will use the following auxiliary result:

Theorem 22.10. *Let A be a symmetric positive definite matrix. Then the interval matrix $[A - E, A + E]$ is positive definite if and only if it is regular.*

Proof. Under the assumption, the interval matrix $\mathbf{A} = [A - E, A + E]$ satisfies $\mathbf{A}^{\text{sym}} = \mathbf{A}$ and contains a symmetric positive definite matrix A. Hence according to Theorem 21.9, $\mathbf{A}$ is positive definite if and only if it is regular. The theorem is proven.

As a direct consequence we prove NP-hardness of checking positive definiteness:

Theorem 22.11. *The following problem is NP-hard:*

Instance. *A nonnegative symmetric positive definite rational matrix A.*
Question. *Is $[A - E, A + E]$ positive definite?*

Proof. In view of Theorem 22.10, such an interval matrix is positive definite if and only if it is regular. Checking regularity was proved to be NP-hard for this class of interval matrices in Theorem 22.7. Hence the same is true for checking positive definiteness. The theorem is proven.

22.4.3. *Necessary and/or sufficient conditions*

Proof of Theorem 21.11. Let us show that $\mathbf{A}$ is positive definite if and only if each $A^{(z,z)}$, $z \in \{-1, 1\}^n$ is positive definite.

The "only if" part is obvious since $A^{(z,z)} \in \mathbf{A}$ for each $z \in \{-1,1\}^n$. The "if" part was proved in the first part of the proof of Theorem 21.9 (a matrix $\widetilde{A} - T\Delta T$ is of the form $A^{(z,z)}$ where z is the diagonal vector of T). The theorem is proven.

Proof of Theorem 21.12. Let us prove that an interval matrix $\mathbf{A} = [\widetilde{A} - \Delta, \widetilde{A} + \Delta]$ is positive definite if $\varrho(\Delta^{\text{sym}}) < \lambda_{\min}(\widetilde{A}^{\text{sym}})$.

Indeed, for each $A \in \mathbf{A}$ and x with $\|x\|_2 = 1$ we have

$$\begin{aligned} x^T A x &= x^T \widetilde{A} x + x^T (A - \widetilde{A}) x \geq x^T \widetilde{A} x - |x|^T \Delta |x| \\ &= x^T \widetilde{A}^{\text{sym}} x - |x|^T \Delta^{\text{sym}} |x| \\ &\geq \lambda_{\min}(\widetilde{A}^{\text{sym}}) - \lambda_{\max}(\Delta^{\text{sym}}) = \lambda_{\min}(\widetilde{A}^{\text{sym}}) - \varrho(\Delta^{\text{sym}}) > 0, \end{aligned}$$

hence $\mathbf{A}$ is positive definite. The theorem is proven.

22.5. Positive semi-definiteness

Let us prove that checking positive semi-definiteness of an interval matrix is NP-hard (Theorems 21.2 and 21.3). By definition, all elements of an MC-matrix A are integers. Therefore, due to the the second part of Theorem 21.27, its norm $||A||_{\infty,1}$ is also an integer. Due to Theorem 22.2, it is NP-hard to decide, for any given ℓ, whether this norm is $\geq \ell$ or $< \ell$. If it is $< \ell$, then, since this norm is an *integer*, we conclude that it is $\leq \ell - 1$. According to the formula (22.5), this norm cannot exceed twice the total number of edges plus n^2. Since the number of edges, in its turn, cannot exceed the total number of pairs n^2, we conclude that the norm cannot exceed $3n^2$. Thus, it is sufficient to consider only the values $\ell \leq 3n^2$.

For a given MC-matrix A, and for a given ℓ, either $||A||_{\infty,1} \geq \ell$ or $||A||_{\infty,1} \leq \ell - 1$, and it is NP-hard to tell which is the case. Thus, for an MC-related matrix $B = \ell^{-1}A$, either $||B||_{\infty,1} \geq 1$ or $||B||_{\infty,1} \leq 1 - 1/\ell$, and it is hard to tell which is the case. Since $\ell \leq 3n^2$, in the second case, we have $||B||_{\infty,1} \leq 1 - 1/(3n^2)$.

According to Theorem 22.5, $||A||_{\infty,1} \geq 1$ if and only if the interval matrix $[A^{-1} - E, A^{-1} + E]$ is not regular. Thus, in the first case, the interval matrix $\mathbf{M} = [M - E, M + E]$, where $M = B^{-1}$, is not regular (and hence, the smallest eigenvalue $\lambda_{\min}$ of symmetric matrices from this interval matrix is non-positive), while in the second case, this interval matrix regular (hence, $\lambda_{\min} > 0$).

To prove our result, we will consider an auxiliary matrix

$$B' = \frac{B}{1 - 1/(6n^2)};$$

its inverse is

$$M' = (B')^{-1} = \left(1 - \frac{1}{6n^2}\right) \cdot B^{-1} = \left(1 - \frac{1}{6n^2}\right) \cdot M.$$

Let us show that in the first case (when $||B||_{\infty,1} \geq 1$), the interval matrix $\mathbf{M}' = [M' - E, M' + E]$ is *not* positive semi-definite, while in the second case, it is. Then, from the NP-hardness of distinguishing between these two cases, we would conclude that checking positive semi-definiteness is also NP-hard. Indeed:

- In the *first* case, when $||B||_{\infty,1} \geq 1$, then $[M - E, M + E]$ is not a regular interval matrix, which means that it contains a singular symmetric numerical matrix S, i.e., a symmetric matrix for which 0 is an eigenvalue. If we multiply a singular matrix $S \in [M - E, M + E]$ by a positive constant $1 - 1/(6n^2)$, we get a new singular symmetric matrix

$$S_1 = \left(1 - \frac{1}{6n^2}\right) \cdot S \in \left[M' - \left(1 - \frac{1}{6n^2}\right) \cdot E, M' + \left(1 - \frac{1}{6n^2}\right) \cdot E\right],$$

for which $\lambda = 0$ is also an eigenvalue. Hence, for a matrix

$$S_2 = S_1 - \frac{1}{6n^2} \cdot I$$

which still belongs to the desired interval matrix $\mathbf{M}' = [M' - E, M' + E]$, one of the eigenvalues is negative (equal to $-1/(6n^2)$), and therefore, this interval matrix is *not* positive semi-definite.

- In the *second* case, when $\|B\|_{\infty,1} \leq 1 - 1/(3n^2)$, then

$$\|B'\|_{\infty,1} = \frac{\|B\|_{\infty,1}}{1 - 1/(6n^2)} \leq \frac{1 - 1/(3n^2)}{1 - 1/(6n^2)} < 1;$$

therefore, due to Theorem 22.5, the interval matrix $\mathbf{M}' = [M' - E, M' + E]$ is regular and hence, positive definite.

The NP-hardness is proven.

22.6. P-property (for interval matrices)

22.6.1. *Necessary and sufficient conditions*

Proof of Theorem 21.13. Let us prove that $\mathbf{A}$ is a P-matrix if and only if each $A^{(z,z)}, z \in \{-1, 1\}^n$ is a P-matrix.

Indeed, if $\mathbf{A}$ is a P-matrix, then each $A^{(z,z)}$ is a P-matrix since $A^{(z,z)} \in \mathbf{A}, z \in \{-1, 1\}^n$. Conversely, let each $A^{(z,z)}, z \in \{-1, 1\}^n$ be a P-matrix. Fiedler and Pták proved, in [107], that A is a P-matrix if and only if for each $x \neq 0$ there exists an $i \in \{1, \ldots, n\}$ such that $x_i(Ax)_i > 0$. Take $A \in \mathbf{A}, x \neq 0$, and let $z \in \{-1, 1\}^n$ be defined by $z_j = 1$ if $x_j \geq 0$ and $z_j = -1$ otherwise $(j = 1, \ldots, n)$. Since $A^{(z,z)}$ is a P-matrix, according to the Fiedler-Pták theorem there exists an $i \in \{1, \ldots, n\}$ such that $x_i(A^{(z,z)}x)_i > 0$. Then we have

$$x_i(Ax)_i \;=\; \sum_j \widetilde{a}_{ij} x_i x_j + \sum_j (a_{ij} - \widetilde{a}_{ij}) x_i x_j$$

$$\geq \sum_j \widetilde{a}_{ij} x_i x_j - \sum_j \Delta_{ij} |x_i||x_j|$$
$$= \sum_j (\widetilde{a}_{ij} - \Delta_{ij} z_i z_j) x_i x_j = x_i (A^{(z,z)} x)_i > 0,$$

hence A is a P-matrix by the Fiedler-Pták theorem. This proves that **A** is a P-matrix. The theorem is proven.

22.6.2. *P-property and positive definiteness*

Proof of Theorem 21.14. Let us now prove that a symmetric interval matrix **A** is a P-matrix if and only if it is positive definite.

Indeed, all the matrices $A^{(z,z)}$, $z \in \{-1,1\}^n$ defined by (21.15) are symmetric for a symmetric interval matrix **A**. Hence, **A** is a P-matrix if and only if each $A^{(z,z)}$, $z \in \{-1,1\}^n$ is a P-matrix, which is the case if and only if each $A^{(z,z)}$, $z \in \{-1,1\}^n$ is positive definite, and this is equivalent to positive definiteness of **A** (Theorem 21.11). The theorem is proven.

22.6.3. *Checking P-property is NP-hard*

As a result, we get NP-hardness of checking P-property of interval matrices (Theorems 21.2 and 21.3):

Theorem 22.12. *The following problem is NP-hard:*
Instance. *A nonnegative symmetric rational P-matrix* A.
Question. *Is* $[A-E, A+E]$ *a P-matrix?*

Proof. Since A is symmetric positive definite, $[A-E, A+E]$ is a P-matrix if and only if it is positive definite (Theorem 21.14). Checking positive definiteness of this class of interval matrices was proved to be NP-hard in Theorem 22.11. The theorem is proven.

22.7. Stability

22.7.1. *Checking stability is NP-hard*

Proof of Theorem 21.6. Let us show that a symmetric interval matrix $\mathbf{A}$ is stable if and only if $-\mathbf{A}$ is positive definite.

First notice that $A \in \mathbf{A}$ if and only if $-A \in -\mathbf{A}$. Let $\mathbf{A}$ be stable, and consider a symmetric matrix $A \in -\mathbf{A}$. Then $-A \in \mathbf{A}$ is symmetric and stable, hence $\lambda_{\max}(-A) = -\lambda_{\min}(A) < 0$, so that $\lambda_{\min}(A) > 0$, which means that A is positive definite. Hence each symmetric $A \in -\mathbf{A}$ is positive definite, which in view of Theorem 21.11 implies that $-\mathbf{A}$ is positive definite.

Conversely, let $-\mathbf{A}$ be positive definite. Then a similar argument shows that each symmetric matrix in $\mathbf{A}$ is stable, and from Bendixson's theorem (see Stoer and Bulirsch [404]) we have that each eigenvalue λ of each $A \in \mathbf{A}$ satisfies

$$\mathrm{Re}\lambda \le \lambda_{\max}\left(\frac{1}{2}(A + A^T)\right) < 0$$

(since $\frac{1}{2}(A + A^T) \in \mathbf{A}$), hence $\mathbf{A}$ is stable. The theorem is proven.

NP-hardness of checking stability (Theorems 21.2 and 21.3) now follows easily:

Theorem 22.13. *The following problem is NP-hard:*
Instance. *A non-positive symmetric stable rational matrix A.*
Question. *Is $[A - E, A + E]$ stable?*

Proof. By Theorem 21.6, $[A - E, A + E]$ is stable if and only if

$$[-A - E, -A + E]$$

is positive definite, where $-A$ is a nonnegative symmetric positive definite rational matrix. Hence the result follows from Theorem 22.11. The theorem is proven.

22.7.2. *Necessary and/or sufficient conditions*

Proof of Theorem 21.7. Let us prove that a symmetric $\mathbf{A}$ is stable if and only if each $A^{(-z,z)}$, $z \in \{-1, 1\}^n$ is stable.

Indeed, $\mathbf{A}$ is stable if and only if $-\mathbf{A}$ is positive definite which, in view of (already proven) Theorem 21.11, is the case if and only if each $-A^{(-z,z)}$, $z \in$

$\{-1,1\}^n$ is positive definite, and this is equivalent to stability of all $A^{(-z,z)}$, $z \in \{-1,1\}^n$. The theorem is proven.

Proof of Theorem 21.8. Let us prove that an interval matrix $[\widetilde{A}-\Delta, \widetilde{A}+\Delta]$ is stable if $\lambda_{\max}(\widetilde{A}^{\text{sym}}) + \varrho(\Delta^{\text{sym}}) < 0$.

Indeed, if the inequality holds, then $\varrho(\Delta^{\text{sym}}) < \lambda_{\min}(-\widetilde{A}^{\text{sym}})$, hence $[-\widetilde{A}^{\text{sym}} - \Delta^{\text{sym}}, -\widetilde{A}^{\text{sym}} + \Delta^{\text{sym}}]$ is positive definite by (already proven) Theorem 21.12, and $[\widetilde{A}^{\text{sym}} - \Delta^{\text{sym}}, \widetilde{A}^{\text{sym}} + \Delta^{\text{sym}}]$ is stable by Theorem 21.6. Stability of $[\widetilde{A} - \Delta, \widetilde{A}+\Delta]$ then follows by using Bendixson's theorem as in the proof of Theorem 21.6. The theorem is proven.

22.7.3. *Radius of stability*

Proof of Theorem 21.28. Let us prove that if A is symmetric stable and Δ is symmetric nonnegative, then the radius of stability $s(A, \Delta)$ is equal to the radius of non-singularity $d(A, \Delta)$.

Indeed, an interval matrix $[A - \varepsilon\Delta, A + \varepsilon\Delta]$ is stable
if and only if $[-A - \varepsilon\Delta, -A + \varepsilon\Delta]$ is positive definite (Theorem 21.6)
if and only if $[-A - \varepsilon\Delta, -A + \varepsilon\Delta]$ is regular (Theorem 21.9)
if and only if $[A - \varepsilon\Delta, A + \varepsilon\Delta]$ is regular. Therefore the values of $s(A, \Delta)$ and $d(A, \Delta)$ are equal. The theorem is proven.

22.8. Semi-stability

NP-hardness of semi-stability (Theorems 21.2 and 21.3) is proven by a reduction to positive semi-definiteness that is similar to the above deduction of stability to positive definiteness.

22.9. Schur stability

NP-hardness of s-Schur stability (Theorem 21.3) is proven by using the following reduction to stability:

Theorem 22.14. *A symmetric interval matrix* $[\underline{A}, \overline{A}]$ *is stable if and only if the symmetric interval matrix*

$$[I + \alpha\underline{A}, I + \alpha\overline{A}]$$

is s-Schur stable, where

$$\alpha = \frac{2}{\|\underline{A}\|_1 + \|\overline{A} - \underline{A}\|_1 + 2}. \tag{22.16}$$

Proof. Let $[\underline{A}, \overline{A}]$ be s-stable. Then for each symmetric $A' \in [I + \alpha\underline{A}, I + \alpha\overline{A}]$ we have $A' = I + \alpha A$ for some symmetric $A \in [\underline{A}, \overline{A}]$, hence $\lambda_{\max}(A') = 1 + \alpha\lambda_{\max}(A) < 1$. Furthermore, from

$$|\lambda_{\min}(A)| \leq \varrho(A) \leq \|A\|_1 \leq \|\underline{A}\|_1 + \|\overline{A} - \underline{A}\|_1 < \frac{2}{\alpha}$$

we have

$$\lambda_{\min}(A') = 1 + \alpha\lambda_{\min}(A) > -1,$$

hence A' is Schur stable and thereby $[I + \alpha\underline{A}, I + \alpha\overline{A}]$ is s-Schur stable.

Conversely, if $[I+\alpha\underline{A}, I+\alpha\overline{A}]$ is s-Schur stable, then each symmetric $A \in [\underline{A}, \overline{A}]$ is of the form $A = \frac{1}{\alpha}(A' - I)$ for some symmetric $A' \in [I + \alpha\underline{A}, I + \alpha\overline{A}]$, hence $\lambda_{\max}(A) = \frac{1}{\alpha}(\lambda_{\max}(A') - 1) < 0$, and A is stable. Stability of all symmetric matrices in $[\underline{A}, \overline{A}]$ implies stability of $[\underline{A}, \overline{A}]$ due to Theorem 21.7. The theorem is proven.

As a consequence of Theorem 22.14 we obtain the desired NP-hardness result:

Theorem 22.15. *The following problem is NP-hard:*

Instance. *A symmetric Schur stable rational matrix A with $A \leq I$, and a rational number $\alpha \in [0, 1]$.*

Question. *Is $[A - \alpha E, A + \alpha E]$ s-Schur stable?*

Proof. For a non-positive symmetric stable rational matrix A, the symmetric interval matrix $[A-E, A+E]$ is stable if and only if $[(I+\alpha A)-\alpha E, (I+\alpha A)+\alpha E]$ is s-Schur stable, where α is given by (22.16). Here $I + \alpha A$ is a symmetric Schur stable rational matrix with $I + \alpha A \leq I$, and $\alpha \in [0, 1]$. Hence we have a polynomial-time reduction of the NP-hard problem of Theorem 22.13 to the current problem, which shows that it is NP-hard as well. The theorem is proven.

Comment. This result differs from the previous results where NP-hardness was established for the class of interval matrices of the form $[A - E, A + E]$. This is explained by the fact that regularity, positive definiteness and stability are *invariant* under multiplication by a positive parameter whereas Schur stability is not. This same invariance explains why in Theorem 21.2 we can take interval matrices with *narrow* intervals and still keep NP-hardness: because a matrix **A** has the desired property if and only if a "narrow" matrix $\delta \cdot \mathbf{A}$ has it.

22.10. Schur semi-stability

NP-hardness of checking s-Schur semi-stability (Theorem 21.3) can be proven by reducing this problem to (already proven) semi-stability. This reduction is similar to the one used in Theorem 22.14.

22.11. Checking eigenvalues

22.11.1. *Checking eigenvalues is NP-hard*

Theorem 21.17 follows from the following result:

Theorem 22.16. *The following problem is NP-hard:*

Instance. *A nonnegative symmetric positive definite rational matrix A and a rational number λ.*

Question. *Is λ an eigenvalue of some symmetric matrix in $[A-E, A+E]$?*

Proof. $[A - E, A + E]$ is not regular if and only if 0 is an eigenvalue of some symmetric matrix in $[A - E, A + E]$ (Theorem 22.5). Hence the NP-hard problem of Theorem 22.7 can be reduced in polynomial time to the current problem, which is thereby NP-hard. The theorem is proven.

Proof of Theorem 21.18. Suppose there exists a polynomial-time algorithm which for each symmetric interval matrix **A** computes a rational number $\widetilde{\lambda}(\mathbf{A})$ for which

$$\left| \frac{\widetilde{\lambda}(\mathbf{A}) - \overline{\lambda}(\mathbf{A})}{\overline{\lambda}(\mathbf{A})} \right| < 1$$

if $\overline{\lambda}(\mathbf{A}) \neq 0$ and $\widetilde{\lambda}(\mathbf{A}) \geq 0$ otherwise. Let us prove that P=NP.

Indeed, under the assumptions, $\widetilde{\lambda}(\mathbf{A}) < 0$ if and only if $\overline{\lambda}(\mathbf{A}) < 0$, and this is equivalent to stability of $\mathbf{A}$. Hence we have a polynomial-time algorithm for solving the NP-hard problem of Theorem 22.13, which implies P=NP. The theorem is proven.

22.11.2. *Computing the maximal eigenvalue is NP-hard*

Proof of Theorem 21.19. Let us prove that for a symmetric interval matrix $\mathbf{A}$, the set $\lambda_{\max}(\mathbf{A})$ of all values of $\lambda_{\max}(A)$ for all symmetric matrices $A \in \mathbf{A}$ is an interval (i.e., a finite and closed interval).

Indeed, let

$$\begin{aligned}\underline{\lambda}(\mathbf{A}) &= \min\{\lambda_{\max}(A) | A \text{ symmetric}, A \in \mathbf{A}\},\\ \overline{\lambda}(\mathbf{A}) &= \max\{\lambda_{\max}(A) | A \text{ symmetric}, A \in \mathbf{A}\}.\end{aligned}$$

By continuity argument, both bounds are attained, hence $\underline{\lambda}(\mathbf{A}) = \lambda_{\max}(A_1)$ and $\overline{\lambda}(\mathbf{A}) = \lambda_{\max}(A_2)$ for some symmetric $A_1, A_2 \in \mathbf{A}$. Define a real function φ of one real variable by $\varphi(t) = f(A_1 + t(A_2 - A_1))$, $\quad t \in [0,1]$, where

$$f(A) = \max_{\|x\|_2=1} x^T A x.$$

This function φ is continuous since $f(A)$ is continuous (Rohn [350]), and $\varphi(0) = f(A_1) = \lambda_{\max}(A_1) = \underline{\lambda}(\mathbf{A}), \varphi(1) = f(A_2) = \lambda_{\max}(A_2) = \overline{\lambda}(\mathbf{A})$, hence for each $\lambda \in [\underline{\lambda}(\mathbf{A}), \overline{\lambda}(\mathbf{A})]$ there exists a $t_\lambda \in [0,1]$ such that

$$\lambda = \varphi(t_\lambda) = f(A_1 + t_\lambda(A_2 - A_1)) = \lambda_{\max}(A_1 + t_\lambda(A_2 - A_1)).$$

Hence each $\lambda \in [\underline{\lambda}(\mathbf{A}), \overline{\lambda}(\mathbf{A})]$ is the maximal eigenvalue of some symmetric matrix in $\mathbf{A}$, and we have $\lambda_{\max}(\mathbf{A}) = [\underline{\lambda}(\mathbf{A}), \overline{\lambda}(\mathbf{A})]$. The theorem is proven.

22.11.3. *Checking enclosures is NP-hard*

Finally, Theorem 21.20 is a corollary of the following result:

Theorem 22.17. *The following problem is NP-hard:*

Instance. *A non-positive symmetric stable rational matrix A, and rational numbers $a, b, a < b$.*

Question. *Is $\lambda_{\max}([A - E, A + E]) \subset (a, b)$?*

Proof. For each symmetric $A' \in [A - E, A + E]$ we have

$$|\lambda_{\max}(A')| \leq \varrho(A') \leq \|A'\|_1 \leq \|A\|_1 + \|E\|_1 = \|A\|_1 + n < \alpha,$$

where we denoted $\alpha = ||A||_1 + n + 1$. Hence due to Theorem 21.7, $[A - E, A + E]$ is stable if and only if

$$\lambda_{\max}([A - E, A + E]) \subset (-\alpha, 0)$$

holds. This shows that the NP-hard problem of checking stability of $[A - E, A + E]$ (Theorem 22.13) can be reduced in polynomial time to the current problem, which is thus NP-hard. The theorem is proven.

22.12. Determinants

22.12.1. *Computing extremal values of determinant is NP-hard*

Proof of Theorem 21.15. Let us show that computing the exact range of the determinant of an interval matrix is an NP-hard problem.

Indeed, an interval matrix of the form $\mathbf{A} = [A - E, A + E]$, where A is a nonnegative symmetric positive definite rational matrix, is not regular if and only if

$$\overline{\det}(\mathbf{A}_0) \geq 0, \tag{22.16}$$

where $\mathbf{A}_0 = \mathbf{A}$ if $\det A \leq 0$ and $\mathbf{A}_0$ is constructed by swapping the first two rows of $\mathbf{A}$ otherwise (which changes the sign of the determinant). Here $\mathbf{A}_0 = [A_0 - E, A_0 + E]$, where A_0 is a nonnegative rational matrix. Hence the NP-hard problem of checking regularity (Theorem 22.7) can be reduced in polynomial time to the decision problem (22.16) which shows that computing $\overline{\det}(\mathbf{A})$ is NP-hard in this class of interval matrices. The proof for $\underline{\det}(\mathbf{A})$ is analogous. The theorem is proven.

22.12.2. *Edge theorem*

To prove other results about $\det(\mathbf{A})$, we will need the following auxiliary "edge theorem":

Theorem 22.18. (Rohn [344]) *Let* $\mathbf{A} = [\underline{A}, \overline{A}]$ *be an interval matrix. Then for each* $A \in \mathbf{A}$ *there exists an edge matrix* $A' \in \mathbf{A}$ *with the same value of the determinant* ($\det A = \det A'$).

Proof. For each $\widetilde{A} \in \mathbf{A}$ denote by $h(\widetilde{A})$ the number of matrix entries with $\widetilde{a}_{ij} \notin \{\underline{a}_{ij}, \overline{a}_{ij}\}$, $i, j = 1, \ldots, n$. Given an $A \in \mathbf{A}$, let A' be a matrix satisfying $A' \in \mathbf{A}$, $\det A' = \det A$ and

$$h(A') = \min\{h(\widetilde{A}) | \widetilde{A} \in \mathbf{A}, \det \widetilde{A} = \det A\}. \tag{22.17}$$

If $h(A') \geq 2$, then there exist indices $(p,q), (r,s), (p,q) \neq (r,s)$ such that $a'_{pq} \in (\underline{a}_{pq}, \overline{a}_{pq})$, $a'_{rs} \in (\underline{a}_{rs}, \overline{a}_{rs})$. Then we can move these two entries within their intervals in such a way that at least one attains its bound, and the determinant is kept unchanged. Then the resulting matrix A'' satisfies $h(A'') < h(A')$, which is a contradiction. Hence A' defined by (22.17) satisfies $h(A') \leq 1$ (i.e., A' is an edge matrix), and $\det A = \det A'$. The theorem is proven.

As a corollary, for $\det A = 0$, we have the following results:

Theorem 22.19. (Rohn [347]) *If an interval matrix* $\mathbf{A}$ *is not regular, then it contains a singular edge matrix.*

Proof of Theorem 21.21. Let us prove that if a real number λ is an eigenvalue of some $A \in \mathbf{A}$, then it is also an eigenvalue of some edge matrix.

Indeed, if λ is a real eigenvalue of some $A \in \mathbf{A} = [\underline{A}, \overline{A}]$, then $A - \lambda I$ is a singular matrix belonging to the interval matrix $[\underline{A} - \lambda I, \overline{A} - \lambda I]$, which is thus not regular; hence by Theorem 22.19 this new interval matrix contains a singular edge matrix $A' - \lambda I$. Hence, A' is also an edge matrix of the matrix $\mathbf{A}$, and λ is an eigenvalue of A'. The theorem is proven.

Proof of Theorem 21.16. Let us prove that for every interval matrix $\mathbf{A} = [\underline{A}, \overline{A}]$, each of the extremal values of $\det A$, $A \in \mathbf{A}$, is attained at one of the vertex matrices (as defined in Chapter 21; i.e., matrix for which for all i and j, the corresponding value a_{ij} is either equal to $\underline{a}_{ij}$, or equal to $\overline{a}_{ij}$).

Indeed, since the determinant is linear in each entry, Theorem 22.18 implies that the extremal values of the determinant are attained at some of the 2^{n^2} vertex matrices. The theorem is proven.

22.13. Nonnegative invertibility and M-matrices

Let us describe results that lead to feasible algorithms for checking whether a given interval matrix is nonnegative invertible or an M-matrix. This result was originally proven by Kuttler [242]; we will use an elementary proof from Rohn [346]. This algorithm is based on the following result:

Theorem 22.20. *An interval matrix* $\mathbf{A} = [\underline{A}, \overline{A}]$ *is nonnegative invertible if and only if* $\underline{A}^{-1} \geq 0$ *and* $\overline{A}^{-1} \geq 0$.

Proof. The "only if" part is obvious. To prove the "if" part, denote $D_0 = \overline{A}^{-1}(\overline{A} - \underline{A})$, then $D_0 \geq 0$ and

$$(I - D_0)^{-1} = (\overline{A}^{-1}\underline{A})^{-1} = \underline{A}^{-1}\overline{A} = I + \underline{A}^{-1}(\overline{A} - \underline{A}) \geq 0,$$

hence $\varrho(D_0) < 1$. Then for each $A \in \mathbf{A}$ we have $\varrho(\overline{A}^{-1}(\overline{A} - A)) \leq \varrho(D_0) < 1$, and from the identity

$$A = \overline{A}(I - \overline{A}^{-1}(\overline{A} - A))$$

we obtain

$$A^{-1} = \sum_{j=0}^{\infty}(\overline{A}^{-1}(\overline{A} - A))^j\overline{A}^{-1} \geq 0.$$

The theorem is proven.

Hence, checking nonnegative invertibility of an interval matrix $\mathbf{A}$ with rational bounds can be performed in polynomial time (Edmonds [98]). For M-matrices, we have a similar result:

Theorem 22.21. *An interval matrix* $\mathbf{A} = [\underline{A}, \overline{A}]$ *is an M-matrix if and only if* $\underline{A}$ *and* $\overline{A}$ *are M-matrices.*

Proof. The "only if" part is obvious. Conversely, if both $\underline{A}$ and $\overline{A}$ are M-matrices, then $\underline{A}^{-1} \geq 0$ and $\overline{A}^{-1} \geq 0$, hence each $A \in \mathbf{A}$ satisfies $A^{-1} \geq 0$ (Theorem 22.20) and $A_{ij} \leq \overline{A}_{ij} \leq 0$ for $i \neq j$, i.e. A is an M-matrix. The theorem is proven.

22.14. P-property (for numerical matrices)

Let us prove the last statement from Theorem 21.1, that checking P-property for numerical matrices is NP-hard. This proof is based on an interesting equivalence of regularity of *interval* matrices and P-property of associated *numerical* matrices.

22.14.1. *Regularity and P-property*

Consider an $n \times n$ interval matrix $\mathbf{A} = [\underline{A}, \overline{A}] = [\underline{A}, \underline{A} + 2\Delta]$. Assuming nonsingularity of $\underline{A}$, for each $i, j \in \{1, \ldots, n\}$ define the vector

$$c_{ij} = 2(\Delta_{i1}m_{1j}, \Delta_{i2}m_{2j}, \ldots, \Delta_{in}m_{nj})^T$$

(where we denoted $M = \underline{A}^{-1}$), and the matrix $C_{ij} = c_{ij}e^T$ (where e is the n-vector of all ones). Hence, C_{ij} is an $n \times n$ matrix whose all columns are identical and equal to the vector c_{ij}. Finally, define the numerical matrix

$$C(\mathbf{A}) = \begin{pmatrix} I & 0 & \ldots & 0 \\ 0 & I & \ldots & 0 \\ \vdots & \vdots & \ddots & \vdots \\ 0 & 0 & \ldots & I \end{pmatrix} + \begin{pmatrix} C_{11} & C_{12} & \ldots & C_{1n} \\ C_{21} & C_{22} & \ldots & C_{2n} \\ \vdots & \vdots & \ddots & \vdots \\ C_{n1} & C_{n2} & \ldots & C_{nn} \end{pmatrix}$$

This matrix consists of $n \times n$ blocks, each of which is a $n \times n$ matrix; therefore, $C(\mathbf{A})$ is of size $n^2 \times n^2$. For each $y, z \in \{-1, 1\}^n$, let us define the *yz-minor* of $C(\mathbf{A})$ as the determinant of the principal submatrix of $C(\mathbf{A})$ consisting of rows and columns with indices $(i-1)n + j$, where $y_i z_j = -1$.

Theorem 22.22. *For an interval matrix $\mathbf{A} = [\underline{A}, \overline{A}]$ with $\underline{A} < \overline{A}$, the following conditions are equivalent:*

(i) *$\mathbf{A}$ is regular,*

(ii) *$\underline{A}$ is nonsingular and $C(\mathbf{A})$ is a P-matrix,*

(iii) *$\underline{A}$ is nonsingular and each yz-minor of $C(\mathbf{A})$ is positive, $y, z \in \{-1, 1\}^n$.*

Historical comment. The equivalence (i)⇔(ii) of the above theorem is due to Coxson [77], equivalence (i)⇔(iii) is added here as a consequence of the Baumann's Theorem 21.4 to show that the number of determinants to be checked for positivity can be decreased from $2^{n^2} - 1$ to $2^{2n-1} - 1$. The specific feature of this result consists in the fact that regularity of an $n \times n$ interval matrix $\mathbf{A}$ is characterized in terms of an $n^2 \times n^2$ numerical matrix $C(\mathbf{A})$. Nevertheless, the number of operations involved still remains exponential in n.

Proof. (i)⇔(ii): Let

$$F = \begin{pmatrix} e^T & 0^T & \dots & 0^T \\ 0^T & e^T & \dots & 0^T \\ \vdots & \vdots & \ddots & \vdots \\ 0^T & 0^T & \dots & e^T \end{pmatrix},$$

where all the blocks are n-dimensional vectors, hence F is of size $n \times n^2$, and

$$G = \begin{pmatrix} \Delta_{11}e_1 & \Delta_{12}e_2 & \dots & \Delta_{1n}e_n \\ \Delta_{21}e_1 & \Delta_{22}e_2 & \dots & \Delta_{2n}e_n \\ \vdots & \vdots & \ddots & \vdots \\ \Delta_{n1}e_1 & \Delta_{n2}e_2 & \dots & \Delta_{nn}e_n \end{pmatrix},$$

where e_j denotes the jth column of the $n \times n$ unit matrix I, hence G is of size $n^2 \times n$. Consider any vertex matrix A of $\mathbf{A}$, i.e. a matrix for which $a_{ij} \in \{\underline{a}_{ij}, \overline{a}_{ij}\}$ for all $i, j = 1, \dots, n$. A straightforward computation shows that A can be written in the form $A = \underline{A} + 2FDG$, where D is the $n^2 \times n^2$ diagonal matrix satisfying

$$D_{(i-1)n+j,(i-1)n+j} = \begin{cases} 1 & \text{if } a_{ij} = \overline{a}_{ij}, \\ 0 & \text{if } \dot{a}_{ij} = \underline{a}_{ij} \end{cases}$$

$(i, j = 1, \dots, n)$. Then we have

$$\det A = (\det \underline{A})(\det(I + 2\underline{A}^{-1}FDG)). \tag{22.18}$$

Since

$$\det(I + 2\underline{A}^{-1}FDG) = \det(I_{n^2} + 2DG\underline{A}^{-1}F) \tag{22.19}$$

(see Gantmacher [119]; I_{n^2} is the $n^2 \times n^2$ unit matrix), and since

$$2G\underline{A}^{-1}F = C(\mathbf{A}) - I_{n^2} \tag{22.20}$$

(as it can be easily verified), from (22.18)–(22.20) we obtain

$$\det A = (\det \underline{A})(\det(I_{n^2} + D(C(\mathbf{A}) - I_{n^2}))), \tag{22.21}$$

where

$$\det(I_{n^2} + D(C(\mathbf{A}) - I_{n^2})) \tag{22.22}$$

is obviously the determinant of the principal submatrix formed from rows and columns of $C(\mathbf{A})$ with indices $(i-1)n+j$ for which $a_{ij} = \overline{a}_{ij}$.

Now, if $\mathbf{A}$ is regular, then each principal minor of $C(\mathbf{A})$ can be written in the form (22.22) for an appropriately chosen vertex matrix A. Since $(\det A)(\det \underline{A}) > 0$ due to regularity, (22.21) implies that (22.22) is positive. Conversely, if each principal minor of $C(\mathbf{A})$ is positive, then $(\det A)(\det \underline{A}) > 0$ for each vertex matrix A of $\mathbf{A}$ due to (22.21), which implies that $\mathbf{A}$ is regular (Theorem 21.4). Hence (i) and (ii) are equivalent.

To prove (i)$\Leftrightarrow$(iii), notice that each matrix $A^{(y,z)} \in \mathbf{A}$, $y, z \in \{-1,1\}^n$ defined by (21.11) satisfies

$$a_{ij}^{(y,z)} = \underline{A}_{ij} + (1 - y_i z_j)\Delta_{ij}, \quad i,j = 1,\ldots,n,$$

hence it can be written as

$$A^{(y,z)} = \underline{A} + FD^{(y,z)}G,$$

where F and G are as above and $D^{(y,z)}$ is the $n^2 \times n^2$ diagonal matrix for which

$$d^{(y,z)}_{(i-1)n+j,(i-1)n+j} = 1 - y_i z_j, \quad i,j = 1,\ldots,n.$$

Then we obtain as before that

$$\det A^{(y,z)} = (\det \underline{A}) \left(\det \left(I_{n^2} + \frac{1}{2} D^{(y,z)} (C(\mathbf{A}) - I_{n^2}) \right) \right),$$

where

$$\det \left(I_{n^2} + \frac{1}{2} D^{(y,z)} (C(\mathbf{A}) - I_{n^2}) \right)$$

is exactly the yz-minor of $C(\mathbf{A})$ defined earlier in this section. Hence an obvious reasoning based on Baumann's Theorem 21.4 leads to the conclusion that $\mathbf{A}$ is regular if and only if all the yz-minors of $C(\mathbf{A})$ are positive, $y, z \in \{-1,1\}^n$. The theorem is proven.

22.14.2. *Checking P-property is NP-hard for numerical matrices*

Proof of Theorem 21.1. We have already shown, in the main text, that all the properties of numerical matrices except for the P-property can be checked in polynomial time. To complete the proof of Theorem 21.1, we must thus prove that checking P-property is NP-hard.

This is an immediate consequence of the previous characterization of P-property. Indeed, according to the equivalence (i)⇔(ii) of Theorem 22.22, the problem of checking regularity of an interval matrix $\mathbf{A}$ with rational bounds can be reduced in polynomial time to the problem of checking P-property of a rational matrix $C(\mathbf{A})$. Since the former problem is NP-hard (Theorem 22.7), the same is true for the latter one as well. The theorem is proven.

22.14.3. *Regularity and P-property: additional connection*

It should be noted that there also exists another relationship between regularity and the P-property, which proved to be a very useful tool for deriving some nontrivial properties of inverse interval matrices and of systems of linear interval equations. The following theorem was published in a report form [342] in 1984 and in a journal form [348] in 1989.

Theorem 22.23. *If* $\mathbf{A}$ *is a regular interval matrix and* $A_1, A_2 \in \mathbf{A}$*, then* $A_1^{-1}A_2$ *is a P-matrix.*

Proof. Assume to the contrary that $A_1^{-1}A_2$ is not a P-matrix for some $A_1, A_2 \in \mathbf{A} = [\widetilde{A} - \Delta, \widetilde{A} + \Delta]$. Then according to the Fiedler-Pták characterization of P-matrices [107] (quoted in the proof of Theorem 21.13) there exists an $x \neq 0$ such that $x_i(A_1^{-1}A_2x)_i \leq 0$ for each i. Put $x' = A_1^{-1}A_2x$, then

$$x_i x_i' \leq 0 \quad (i = 1, \ldots, n) \tag{22.23}$$

and

$$x \neq x' \tag{22.24}$$

holds. In fact, since $x \neq 0$, there exists a j with $x_j \neq 0$; then $x_j^2 > 0$ whereas (22.23) implies $x_j x_j' \leq 0$, hence $x_j \neq x_j'$. Now we have

$$|\widetilde{A}(x' - x)| = |(\widetilde{A} - A_1)x' + (A_2 - \widetilde{A})x| \leq \Delta|x'| + \Delta|x| = \Delta|x' - x| \tag{22.25}$$

since $|x'| + |x| = |x' - x|$ due to (22.23). Hence Theorem 22.8 in the light of (22.25) and (22.24) implies that $\mathbf{A}$ is not regular, which is a contradiction. The theorem is proven.

For applications of this result, see Rohn [347].

23

NON-INTERVAL UNCERTAINTY I: ELLIPSOID UNCERTAINTY AND ITS GENERALIZATIONS

In the previous chapters, we considered the problem of estimating the range of a function $f(x_1, \ldots, x_n)$ under the assumption that each variable x_i is known to belong to a given interval $\mathbf{x}_i = [\tilde{x}_i - \Delta_i, \tilde{x}_i + \Delta_i]$. So far, we have analyzed this problem under the assumption that we do not know of any dependency between the variables x_i. Under this assumption, the set of all possible values of $\vec{x} = (x_1, \ldots, x_n)$ forms a multi-dimensional interval (box) $\mathbf{x}_1 \times \ldots \times \mathbf{x}_n$.

In many practical cases, however, there is a known dependency between the variables x_i. For the simplest (quadratic) type of this dependency, the set of all possible values of $\vec{x}$ forms an ellipsoid. In this chapter, we analyze the computational complexity and feasibility of data processing under such ellipsoid uncertainty.

23.1. Why non-interval uncertainty?

So far, we considered the problem of estimating the range of a function $f(x_1, \ldots, x_n)$ under the assumption that each variable x_i is known to belong to a given interval $\mathbf{x}_i = [\tilde{x}_i - \Delta_i, \tilde{x}_i + \Delta_i]$, and we have analyzed this problem under the assumption that we do not know of any dependency between the variables x_i. Under this assumption, the set of all possible values of $\vec{x} = (x_1, \ldots, x_n)$ forms a *multi-dimensional interval* $\mathbf{X} = \mathbf{x}_1 \times \ldots \times \mathbf{x}_n$, also called a *box*.

In many practical cases, however, there is a known dependency between the variables x_i. As a result, not all vectors $\vec{x}$ from the box $\mathbf{X}$ are possible, and the set of all possible vectors forms a *proper (non-interval) subset* of this box.

With this chapter, we will start to analyze the computational complexity and feasibility of data processing problems under such non-interval uncertainty.

23.2. Why ellipsoids?

23.2.1. *In the Simplest Case, Additional Information Leads to Ellipsoids*

Additional *information* about the values of x_i usually comes from additional *measurements*, i.e., from measuring an additional quantity z whose value is uniquely determined by the values of x_i's: $z = g(x_1, \ldots, x_n)$. As a result of each measurement of this type, we get the value $\tilde{z}$ with some measurement accuracy Δ. So, we can conclude that $g(x_1, \ldots, x_n) \in [\tilde{z} - \Delta, \tilde{z} + \Delta]$, i.e., that

$$\tilde{z} - \Delta \leq g(x_1, \ldots, x_n) \leq \tilde{z} + \Delta. \tag{23.1}$$

If error bounds Δ_i in directly measuring x_i are small enough, then the actual errors of direct measurements $\Delta x_i = \tilde{x}_i - \Delta x_i$ are small. Hence, we can expand the expression $g(x_1, \ldots, x_n) = g(\tilde{x}_1 - \Delta x_1, \ldots, \tilde{x}_n - \Delta x_n)$ into Taylor series and keep only quadratic terms in the resulting expansion. In other words, we replace the *original* function $g(x_1, \ldots, x_n)$ in the restriction (23.1) by a *quadratic* function $\tilde{g}(x_1, \ldots, x_n)$. Hence, restrictions of the type (23.1) become inequalities of the type $\tilde{g}(x_1, \ldots, x_n) \leq C$ or $\tilde{g}(x_1, \ldots, x_n) \geq C$ for a quadratic function $\tilde{g}(x_1, \ldots, x_n)$ and a real number C. Geometrically, depending on the quadratic function, each inequality describes an *ellipsoid*, or a *hyperboloid*, or a *plane*, etc.

The *simplest* case is when we have only *one* such inequality, i.e., when the set of possible values of $\vec{x} = (x_1, \ldots, x_n)$ is described by a single inequality of this type. Geometrically, when a single quadratic inequality describes a bounded set, this set is an *ellipsoid*

$$\sum a_{ij} \cdot x_i \cdot x_j + \sum a_i \cdot x_i + a_0 \leq C. \tag{23.2}$$

23.2.2. *In Addition to Error Bounds, We Often Know Probabilities of Different Errors*

In many cases, in addition to the ***intervals***, we know the ***probabilities*** of different values of errors. Measurement errors are usually assumed to be normally distributed. In other words, if we denote by x the (unknown) actual value of the physical quantity x, and if we denote by $\tilde{x}$ the result of the measurement, then for a given x, the probability density of $\tilde{x}$ is equal to

$$\rho(\tilde{x}) = \text{const} \cdot \exp\left(-\frac{(x-\tilde{x})^2}{2\sigma^2(x)}\right),$$

where $\sigma(x)$ is the standard deviation.

The errors of different measurements are usually assumed to be independent. As a result, if we make several measurements, then the probability density $\rho(\vec{\tilde{x}})$ on the set of possible measurement results $\vec{\tilde{x}} = (\tilde{x}_1, \tilde{x}_2, \ldots, \tilde{x}_n)$ is Gaussian (normal):

$$\rho(\vec{\tilde{x}}) = \rho(\tilde{x}_1) \cdot \rho(\tilde{x}_2) \cdot \ldots \cdot \rho(\tilde{x}_n) =$$

$$\left(\text{const}_1 \cdot \exp\left(-\frac{(x_1-\tilde{x}_1)^2}{2\cdot\sigma^2(x_1)}\right)\right) \cdot \left(\text{const}_2 \cdot \exp\left(-\frac{(x_2-\tilde{x}_2)^2}{2\cdot\sigma^2(x_2)}\right)\right) \cdot \ldots \cdot$$

$$\left(\text{const}_n \cdot \exp\left(-\frac{(x_n-\tilde{x}_n)^2}{2\cdot\sigma^2(x_n)}\right)\right) = \text{const} \cdot \exp\left(-\sum_{i=1}^{n} \frac{(x_i-\tilde{x}_i)^2}{2\cdot\sigma^2(x_i)}\right), \quad (23.3)$$

where x_i is the (unknown) actual value of i–th physical quantity, $\tilde{x}_i$ is the measured value of this quantity, and $\sigma(x_i)$ is its standard deviation.

For Gaussian distribution, the probability density $\rho(\vec{\tilde{x}})$ is everywhere positive; this means that, in principle, for any given set of measurement results $\vec{\tilde{x}}$, an arbitrary tuple $\vec{x}$ is possible. In practical statistics, however, tuples with very low probability density $\rho(\vec{\tilde{x}})$ are considered impossible.

For example, in 1-dimensional case, we have a "three sigma" rule: values for which $|x - \tilde{x}| > 3\sigma(x)$ are impossible. In multi-dimensional case, it is natural to choose some value $\alpha > 0$, and consider only tuples for which

$$\rho(\vec{\tilde{x}}) \geq \alpha \quad (23.4)$$

as possible ones. For Gaussian distribution, formula (23.4) can be simplified. Indeed, we can do the following:

- substitute formula (23.3) into the condition (23.4) that describes possible tuples;
- divide both sides of the resulting inequality by a positive constant const (from (23.4));
- take natural logarithms of both sides, and
- change the signs of both sides of the resulting inequality;

then, we will arrive at the inequality

$$\sum_{i=1}^{n} \frac{(x_i - \tilde{x}_i)^2}{2 \cdot \sigma^2(x_i)} \le C,$$

where by C, we denoted $-\ln(\alpha/\text{const})$. This inequality describes an *ellipsoid.* Therefore, for given measurement results $\tilde{x}_1, \ldots, \tilde{x}_n$, the possible set of values of $\vec{x}$ is an ellipsoid.

Comment. If the measurement errors are ***not independent***, then we ***also*** have an ***ellipsoid***, but with a general quadratic form in the left-hand side of the inequality.

23.2.3. *Ellipsoids as an Optimal Approximation*

Ellipsoids are also known to be the ***optimal*** approximation sets for different problems [401, 109].

23.2.4. *Ellipsoid Uncertainty is Successfully Used in Practical Problems*

Ellipsoid error estimates are actively used in different applications; see, e.g., Schweppe [386, 387], Fogel *et al.* [110], Belforte *et al.* [28], Norton [309], Chernousko [65, 66], Soltanov [402], Utyubaev [416], Filippov [108], and references therein.

23.3. Computational complexity and feasibility of data processing under ellipsoid uncertainty

23.3.1. *Why Ellipsoid Uncertainty May Be Computationally Easier to Process Than Interval Uncertainty*

For a smooth function $f(x)$ of one variable x, it is usually easy to find a maximum on a given *interval* $[a,b]$: this maximum is attained either inside the interval, in which case it is a zero of the derivative $f'(x) = 0$, or at one of the endpoints (a or b). A similar result is true if we look for a maximum of a function $f(x_1,\ldots,x_n)$ of n variables on a box $\mathbf{x}_1 \times \ldots \times \mathbf{x}_n$: this maximum is either inside the box, or on one of its faces. The only problem, as we have mentioned in Chapter 1, is that we have 2^n *possible faces*, so this approach leads to an exponentially long (thus, non-feasible) algorithm. An *ellipsoid* does not have many different faces, so for ellipsoids, the corresponding problem must be simpler.

23.3.2. *Definitions and Main Results*

Definition 23.1. *By the ellipsoid computation problem, we mean the following problem:*

GIVEN:

- *a positive integer n;*
- *a polynomial $f(x_1,\ldots,x_n)$; and*
- *an ellipsoid E defined by the formula (23.2) with rational coefficients a_{ij}, a_i, a_0, and C.*

COMPUTE the values $\overline{y} = \max f(x_1,\ldots,x_n)$ and $\underline{y} = \min f(x_1,\ldots,x_n)$, where the maximum and minimum are taken over all points $\vec{x} = (x_1,\ldots,x_n) \in E$.

Definition 23.2. *By the ε-approximate ellipsoid computation problem, we mean the following problem:*

GIVEN:

- *a positive integer n;*
- *a polynomial $f(x_1, \ldots, x_n)$;*
- *a rational number $\varepsilon > 0$; and*
- *an ellipsoid E defined by the formula (23.2) with rational coefficients a_{ij}, a_i, a_0, and C.*

COMPUTE rational numbers $\widetilde{\overline{y}}$ and $\widetilde{\underline{y}}$ that are ε-close to $\overline{y} = \max f(x_1, \ldots, x_n)$ and $\underline{y} = \min f(x_1, \ldots, x_n)$, and the maximum and minimum are taken over all points $\vec{x} = (x_1, \ldots, x_n) \in E$.

Our first comment is that this problem is algorithmically solvable:

Proposition 23.1. *There exists an algorithm that solves an arbitrary ellipsoid computation problem.*

For example, we can use Tarski's algorithm (mentioned in Chapter 4) or one of its modern faster versions to compute the desired minimum and maximum.

Theorem 23.1.

- *There exists a polynomial time algorithm that solves the ellipsoid computations problem for all quadratic polynomials.*
- *For quartic polynomials, and for every $\varepsilon > 0$, the $\varepsilon-$approximate ellipsoid computation problem is NP-hard.*

Historical comment. The first part of the theorem was proven in Vavasis [418]; the second part was (partly) announced in [204].

The resulting computations are clearly simpler than interval computations that are NP-hard already for quadratic polynomials:

	Interval computations	Ellipsoid computations
Linear	Linear time	Polynomial time
Quadratic	NP-hard	Polynomial time
Cubic	NP-hard	?
Quartic	NP-hard	NP-hard
5-th and higher degree	NP-hard	NP-hard

23.3.3. *Auxiliary Results*

In Theorem 23.1, we considered polynomials with *arbitrary coefficients*. It turns out that our computational complexity and feasibility results do not change if we impose *a priori bounds* on the values of these *coefficients*:

Theorem 23.2. *For quartic polynomials $f(x_1, \ldots, x_n)$ with coefficients from the set $\{0, 1, 2, 3\}$, and for every $\varepsilon > 0$, the $\varepsilon-$approximate ellipsoid computation problem is NP-hard.*

Theorem 23.1 shows what happens if we restrict the *degrees* of the polynomials. If, instead, we restrict the *number of variables* n, then we get the following results:

Theorem 23.3. *For every n, there exists a polynomial-time algorithm that solves the ellipsoid optimization problem for all polynomials of n variables.*

Comment. This algorithm is similar to the one presented in Chapter 4: it is polynomial time, but it is not yet practical.

23.4. Linear systems under ellipsoid uncertainty

In Chapter 11, we have considered systems of linear equations under *interval* uncertainty. Similarly, we can describe systems of linear equations under *ellipsoid* uncertainty. There are two possible definitions of such a system, and for each of these definitions, the resulting problem is NP-hard:

23.4.1. *Ellipsoid Uncertainty: First Definition*

Definition 23.3.

- *By a linear system under ellipsoid uncertainty, we mean a tuple* $\langle m, n, E\rangle$, *where* m *and* n *are positive integers, and* E *is a* $(m \times n + m)$*-dimensional ellipsoid with rational coefficients.*

- *We say that a vector* $(x_1, \ldots, x_n)$ *is a possible solution of a linear system* $\langle m, n, E\rangle$ *if for some values*

$$(a_{11}, a_{12}, \ldots, a_{1n}, \ldots, a_{m1}, \ldots, a_{mn}, b_1, \ldots, b_m) \in E,$$

and for all $i = 1, \ldots, m$, *we have*

$$\sum_{j=1}^{n} a_{ij} \cdot x_j = b_i.$$

- *We say that a linear system with ellipsoid uncertainty is consistent if it has a possible solution.*

Comment. If a system *has* a possible solution, there usually are *several* different possible solutions. When we talk about *solving* a linear system under uncertainty, we want to describe the *entire solution*, i.e., the entire set of possible solutions:

Definition 23.4. *By a problem of solving linear systems under ellipsoid uncertainty, we mean the following problem:*

GIVEN:

- *a linear system with ellipsoid uncertainty; and*
- *a positive integer* $i \leq n$;

FIND: the largest $\overline{x}_i$ *and the smallest* $\underline{x}_i$ *values of* x_i *for all possible solutions* $(x_1, \ldots, x_{i-1}, x_i, x_{i+1}, \ldots, x_n)$ *of the given linear system.*

Theorem 23.4. (Kreinovich *et al.* [222]) *Checking consistency of linear systems under ellipsoid uncertainty (as described in Definitions* 23.3–23.4*) is NP-hard.*

Theorem 23.5. (Kreinovich *et al.* [222]) *The problem of solving linear systems under ellipsoid uncertainty (as described in Definitions* 23.3-23.4*) is NP-hard.*

Comment. As we explained in Chapter 1, due to book size limitations, we had to omit some easily accessible proofs. In particular, we do not present the proofs of this and the following theorems. These proofs are described, in detail, in the papers Kreinovich *et al.* [222] and Lakeyev *et al.* [244] published in an easily accessible journals *Linear Algebra and its Applications* and Kluwer's *Reliable Computing.*

23.4.2. *Ellipsoid Uncertainty: Second Definition*

Definition 23.3′.

- *By a linear system under ellipsoid uncertainty,* we mean a tuple $\langle m, n, E_a, E_b\rangle$, *where m and n are positive integers, E_a is a $(m \times n)$-dimensional ellipsoid with rational coefficients, and E_b is an m-dimensional ellipsoid with rational coefficients.*
- *We say that a vector $(x_1, \ldots, x_n)$ is a possible solution of a linear system $\langle m, n, E_a, E_b\rangle$ if for some values $(a_{11}, a_{12}, \ldots, a_{1n}, \ldots, a_{m1}, \ldots, a_{mn}) \in E_a$ and $(b_1, \ldots, b_m) \in E_b$, and for all $i = 1, \ldots, m$, we have*

$$\sum_{j=1}^{n} a_{ij} \cdot x_j = b_i.$$

- *We say that a linear system with ellipsoid uncertainty is consistent if it has a possible solution.*

Theorem 23.4′. (Kreinovich *et al.* [222]) *Checking consistency of linear systems under ellipsoid uncertainty (as described in Definitions* 23.3′–23.4*) is NP-hard.*

Theorem 23.5′. (Kreinovich *et al.* [222]) *The problem of solving linear systems under ellipsoid uncertainty (as described in Definitions* 23.3′–23.4*) is NP-hard.*

Comment. Similar results hold when we consider the situations of combined interval and ellipsoid uncertainty:

23.4.3. *Mixed Uncertainty: Ellipsoid Coefficients, Interval Right-Hand Side*

Definition 23.5.

- *By a linear system with ellipsoid coefficients and interval right-hand side, we mean a tuple $\langle m, n, E_a, \mathbf{b}\rangle$, where m and n are positive integers, E_a is a $(m\times n)$-dimensional ellipsoid with rational coefficients, and $\mathbf{b} = \mathbf{b}_1, \ldots, \mathbf{b}_m$ is an m-dimensional interval vector.*

- *We say that a vector $(x_1, \ldots, x_n)$ is a possible solution of a linear system $\langle m, n, E_a, \mathbf{b}\rangle$ if for some values $(a_{11}, a_{12}, \ldots, a_{1n}, \ldots, a_{m1}, \ldots, a_{mn}) \in E_a$ and $b_1 \in \mathbf{b}_1, \ldots, b_m \in \mathbf{b}_m$, and for all $i = 1, \ldots, m$, we have*

$$\sum_{j=1}^{n} a_{ij} \cdot x_j = b_i.$$

- *We say that a linear system with ellipsoid coefficients and interval right-hand side is consistent if it has a possible solution.*

Definition 23.6. *By a problem of solving linear systems with ellipsoid coefficients and interval right-hand side, we mean the following problem:*

GIVEN:

- *a linear system with ellipsoid coefficients and interval right-hand side; and*
- *a positive integer $i \leq n$;*

FIND: the largest $\overline{x}_i$ and the smallest $\underline{x}_i$ values of x_i for all possible solutions $(x_1, \ldots, x_{i-1}, x_i, x_{i+1}, \ldots, x_n)$ of the given linear system.

Theorem 23.6. (Kreinovich *et al.* [222]) *Checking consistency of linear systems with ellipsoid coefficients and interval right-hand side is NP-hard.*

Theorem 23.7. (Kreinovich *et al.* [222]) *The problem of solving linear systems with ellipsoid coefficients and interval right-hand side is NP-hard.*

23.4.4. *Mixed Uncertainty: Interval Coefficients, Ellipsoid Right-Hand Side*

Definition 23.7.

- *By a linear system with interval coefficients and ellipsoid right-hand side,* we mean a tuple $\langle m, n, \mathbf{A}, E_b\rangle$, where m and n are positive integers, $\mathbf{A}$ is a $(m \times n)$-dimensional interval matrix with rational elements $\mathbf{a}_{ij}$, and E_b is an m-dimensional ellipsoid with rational coefficients.

- We say that a vector $(x_1, \ldots, x_n)$ is a *possible solution* of a linear system $\langle m, n, \mathbf{A}, E_b\rangle$ if for some values $a_{11} \in \mathbf{a}_{11}, \ldots, a_{mn} \in \mathbf{a}_{mn}$ and $(b_1, \ldots, b_m) \in E_b$, and for all $i = 1, \ldots, m$,, we have

$$\sum_{j=1}^{n} a_{ij} \cdot x_j = b_i.$$

- We say that a linear system with interval coefficients and ellipsoid right-hand side is *consistent* if it has a possible solution.

Definition 23.8. By a problem of *solving linear systems with interval coefficients and ellipsoid right-hand side,* we mean the following problem:

> GIVEN:
>
> - a linear system with ellipsoid coefficients and interval right-hand side; and
> - a positive integer $i \leq n$;
>
> FIND: the largest $\overline{x}_i$ and the smallest $\underline{x}_i$ values of x_i for all possible solutions $(x_1, \ldots, x_{i-1}, x_i, x_{i+1}, \ldots, x_n)$ of the given linear system.

Theorem 23.8. (Kreinovich *et al.* [222]) *Checking consistency of linear systems with interval coefficients and ellipsoid right-hand side is NP-hard.*

Theorem 23.9. (Kreinovich *et al.* [222]) *The problem of solving linear systems with interval coefficients and ellipsoid right-hand side is NP-hard.*

23.5. Sets more general than ellipsoids

23.5.1. *Non-Gaussian Probability Distributions and Resulting Sets*

Why Gaussian? One of the reasons why ellipsoids provide an adequate description of uncertainty is that ellipsoids naturally come from Gaussian distributions, the error probability distributions that are most widely used in practice (see, e.g., Rabinovich [333]). The main fundamental motivation to use *Gaussian* distributions is that according to the central limit theorem, under reasonable assumptions, the distribution of the sum of several (N) independent small random variables tends to the Gaussian distribution as $N \to \infty$. Therefore, if we eliminate major error components in the measurement error, the resulting error will be caused by the cumulative effect of many independent small components, and hence, its distribution will be close to Gaussian (see, e.g., [422]), Gnedenko *et al.* [129], and Arak *et al.* [14]).

Sometimes, errors are not Gaussian. In many cases, error distribution is Gaussian. However, in other cases, the distribution is *different* (see, e.g., Novitskii *et al.* [312], Orlov [320]. The reason why the above fundamental argument is not always applicable to measurements is that for some measurements, we know several major sources of error, but we cannot eliminate the corresponding error components.

> For *example*, in *indirect* geophysical measurements, without the very drilling that we are trying to avoid, we cannot measure the corresponding error-inducing characteristics.

As a result, the actual error distribution is often far from being Gaussian.

Weibull-type distributions: empirical fact. According to the experimental data analyzed in Novitskii *et al.* [312], for the majority of measuring instruments, the probability distribution of the measurement error Δx can be described by a *Weibull-type* distribution with the probability density $\text{const} \cdot \exp(-k \cdot |\Delta x|^p)$ for some $p > 0$. These distributions and the corresponding statistical methods are actively used in data processing, especially in geodesy and geophysics; see, e.g., Claerbout [68], Heindl *et al.* [145, 146, 147], Tarantola [407], Scales [374], Gomberg *et al.* [131], Gerstenberger [126], Doser *et al.* [95], and references therein.

Weibull-type distributions: theoretical justification. For a Weibull-type distribution, if we have several independent measurements $\tilde{x}^{(1)}, \ldots, \tilde{x}^{(n)}$ of the same quantity x, then for each x, the probability density is equal to the product

$$\rho = \left(\text{const} \cdot \exp(-k(|\tilde{x}^{(1)} - x|^p)\right) \cdot \ldots \cdot \left(\text{const} \cdot \exp(-k(|\tilde{x}^{(n)} - x|^p)\right) =$$

$$\text{const} \cdot \exp\left(-k(|\tilde{x}^{(1)} - x|^p + \ldots + |\tilde{x}^{(n)} - x|^p)\right),$$

and, as an estimate for x, it is natural to choose the *most probable* value, i.e., the value for which $\rho \to \max$. In statistics, the choice of the most probable value is called the *Maximum Likelihood Method.* For Weibull distribution, the maximum likelihood method is equivalent to

$$|\tilde{x}^{(1)} - x|^p + \ldots + |\tilde{x}^{(n)} - x|^p \to \min.$$

This condition is *scale-invariant* in the sense that it leads to the same x if we use a different unit for measuring x (i.e., if we replace x by $\lambda \cdot x$ and $\tilde{x}^{(j)}$ by $\lambda \cdot \tilde{x}^{(j)}$, where λ is the ratio of the old and the new units). It turns out that the above-described Weibull-type distributions are the only distributions for which the corresponding maximum likelihood method lead to a scale-invariant formula Bickel [47], Kirillova *et al.* [197, 178], and Shevlyakov *et al.* [394]. Thus, we get a *theoretical justification* for this class of distributions.

Sets resulting from Weibull-type distributions. If we make several independent measurements of *different* quantities, and the probability density of each measurement error $\Delta x_i = \tilde{x}_i - x_i$ is distributed according to the Weibull-like distribution (with the same p but with probably different const_i and k_i), then the resulting probability density $\rho(\vec{\tilde{x}})$ on the set of possible measurement results $\vec{\tilde{x}} = (\tilde{x}_1, \tilde{x}_2, \ldots, \tilde{x}_n)$ has the following form:

$$\rho(\vec{\tilde{x}}) = \text{const} \cdot \exp\left(-\sum_{i=1}^{n} k_i \cdot |x_i - \tilde{x}_i|^p\right), \tag{23.5}$$

Similarly to the Gaussian case, this probability density $\rho(\vec{\tilde{x}})$ is everywhere positive; this means that, *in principle*, for any given set of measurement results $\vec{\tilde{x}}$, an arbitrary tuple $\vec{x}$ is possible. In *practical* statistics, however, tuples with very low probability density $\rho(\vec{\tilde{x}})$ are considered impossible. It is natural to choose some value $\alpha > 0$, and consider only tuples for which $\rho(\vec{\tilde{x}}) \geq \alpha$ as possible ones. For Weibull-type distribution (23.5), the resulting formula (23.4) can be simplified, by taking logarithms of both sides, into a formula

$$\sum_{i=1}^{n} k_i \cdot |x_i - \tilde{x}_i|^p \leq C. \tag{23.6}$$

For $p = 2$, this formula turns into an ellipsoid; therefore, we will call the set of all vectors $(x_1, \ldots, x_n)$ that satisfy this formula a *p-generalized ellipsoid.*

Comment. If the measurement errors are ***not independent***, then we get an even more general class of sets. As we will show, the main computational problems of interval computations and data processing are NP-hard already for sets of type (23.5); therefore, they are NP-hard for more general sets as well.

23.5.2. *Definitions and the Main Results*

Definition 23.9. *Let $p \geq 1$ be a real number, and let n be a positive integer. By a p-generalized ellipsoid in an n-dimensional space R^n, we mean a tuple $\langle \tilde{\vec{x}}, \vec{k}, C \rangle$, where $\tilde{\vec{x}} \in R^n$, a vector $\vec{k} = (k_1, \ldots, k_n) \in R^n$ consists of positive values k_i, and $C > 0$ is a positive real number. We say that a vector $\vec{x} = (x_1, \ldots, x_n)$ belongs to the generalized ellipsoid if the formula (23.6) holds.*

Comment. For $p = 2$, a generalized ellipsoid becomes a normal ellipsoid.

Definition 23.10. *By the ε-approximate p-generalized ellipsoid computation problem, we mean the following problem:*

GIVEN:

- *a positive integer n;*
- *a polynomial $f(x_1, \ldots, x_n)$;*
- *a rational number $\varepsilon > 0$;*
- *a rational number $p \geq 1$; and*
- *a p-generalized ellipsoid E.*

COMPUTE rational numbers $\tilde{\overline{y}}$ and $\tilde{\underline{y}}$ that are ε-close to $\overline{y} = \max f(x_1, \ldots, x_n)$ and $\underline{y} = \min f(x_1, \ldots, x_n)$, and the maximum and minimum are taken over all points $\vec{x} = (x_1, \ldots, x_n) \in E$.

Theorem 23.10. *For quartic polynomials, for every $\varepsilon > 0$ and $p \geq 1$, the ε-approximate p-generalized ellipsoid computation problem is NP-hard.*

Definition 23.11.

- *By a linear system under (p_a, p_b)-generalized ellipsoid uncertainty,* we mean a tuple $\langle m, n, p_a, p_b, E_a, E_b\rangle$, where m and n are positive integers, E_a is a $(m \times n)$-dimensional p_a-generalized ellipsoid with rational coefficients, and E_b is an m-dimensional p_b-generalized ellipsoid with rational coefficients.

- We say that a vector $(x_1, \ldots, x_n)$ is a *possible solution* of a linear system $\langle m, n, p_a, p_b, E_a, E_b\rangle$ if for some values $(a_{11}, \ldots, a_{1n}, \ldots, a_{m1}, \ldots, a_{mn}) \in E_a$ and $(b_1, \ldots, b_m) \in E_b$, and for all $i = 1, \ldots, m$, we have

$$\sum_{j=1}^{n} a_{ij} \cdot x_j = b_i.$$

- We say that a linear system with ellipsoid uncertainty is *consistent* if it has a possible solution.

Definition 23.12. *By a problem of solving linear systems under (p_a, p_b)-generalized ellipsoid uncertainty,* we mean the following problem:

GIVEN:

- a linear system with (p_a, p_b)-generalized ellipsoid uncertainty; and
- a positive integer $i \le n$;

FIND: the largest $\overline{x}_i$ and the smallest $\underline{x}_i$ values of x_i for all possible solutions $(x_1, \ldots, x_{i-1}, x_i, x_{i+1}, \ldots, x_n)$ of the given linear system.

Theorem 23.11. (Lakeyev *et al.* [244]) *For every p_a and p_b, checking consistency of linear systems under (p_a, p_b)-generalized ellipsoid uncertainty is NP-hard.*

Theorem 23.12. (Lakeyev *et al.* [244]) *For every p_a and p_b, the problem of solving linear systems under (p_a, p_b)-generalized ellipsoid uncertainty is NP-hard.*

23.5.3. *Auxiliary Results: Linear Systems with Mixed Uncertainty*

Definition 23.23. *Let $p_a \geq 1$.*

- *By a linear system with p_a-generalized ellipsoid coefficients and interval right-hand side, we mean a tuple $\langle m, n, p_a, E_a, \mathbf{b}\rangle$, where m and n are positive integers, E_a is a $(m \times n)$-dimensional p_a-generalized ellipsoid with rational coefficients, and $\mathbf{b} = \mathbf{b}_1, \ldots, \mathbf{b}_m$ is an m-dimensional interval vector.*
- *We say that a vector $(x_1, \ldots, x_n)$ is a possible solution of a linear system $\langle m, n, p_a, E_a, \mathbf{b}\rangle$ if for some values $(a_{11}, \ldots, a_{1n}, \ldots, a_{m1}, \ldots, a_{mn}) \in E_a$ and $b_1 \in \mathbf{b}_1, \ldots, b_m \in \mathbf{b}_m$, and for all $i = 1, \ldots, m,$, we have*

$$\sum_{j=1}^{n} a_{ij} \cdot x_j = b_i.$$

- *We say that a linear system with p_a-generalized ellipsoid coefficients and interval right-hand side is consistent if it has a possible solution.*

Definition 23.14. *By a problem of solving linear systems with p_a-generalized ellipsoid coefficients and interval right-hand side, we mean the following problem:*

GIVEN:

- *a linear system with p_a-generalized ellipsoid coefficients and interval right-hand side; and*
- *a positive integer $i \leq n$;*

FIND: the largest $\overline{x}_i$ and the smallest $\underline{x}_i$ values of x_i for all possible solutions $(x_1, \ldots, x_{i-1}, x_i, x_{i+1}, \ldots, x_n)$ of the given linear system.

Theorem 23.23. (Lakeyev *et al.* [244]) *For every $p_a \geq 1$, checking consistency of linear systems with p_a-generalized ellipsoid coefficients and interval right-hand side is NP-hard.*

Theorem 23.14. (Lakeyev *et al.* [244]) *For every $p_a \geq 1$, the problem of solving linear systems with p_a-generalized ellipsoid coefficients and interval right-hand side is NP-hard.*

Definition 23.15. *Let $p_b \geq 1$.*

- *By a linear system with interval coefficients and p_b-generalized ellipsoid right-hand side,* we mean a tuple $\langle m, n, p_b, \mathbf{A}, E_b \rangle$, where m and n are positive integers, $\mathbf{A}$ is a $(m \times n)$-dimensional interval matrix with rational elements $\mathbf{a}_{ij}$, and E_b is an m-dimensional p_b-generalized ellipsoid with rational coefficients.

- We say that a vector $(x_1, \ldots, x_n)$ is a *possible solution* of a linear system $\langle m, n, p_b, \mathbf{A}, E_b \rangle$ if for some values $a_{11} \in \mathbf{a}_{11}, \ldots, a_{mn} \in \mathbf{a}_{mn}$ and $(b_1, \ldots, b_m) \in E_b$, and for all $i = 1, \ldots, m,$, we have

$$\sum_{j=1}^{n} a_{ij} \cdot x_j = b_i.$$

- We say that a linear system with interval coefficients and p_b-generalized ellipsoid right-hand side is *consistent* if it has a possible solution.

Definition 23.16. *By a problem of solving linear systems with interval coefficients and p_b-generalized ellipsoid right-hand side,* we mean the following problem:

GIVEN:

- *a linear system with p_b-generalized ellipsoid coefficients and interval right-hand side; and*
- *a positive integer $i \leq n$;*

FIND: the largest $\overline{x}_i$ and the smallest $\underline{x}_i$ values of x_i for all possible solutions $(x_1, \ldots, x_{i-1}, x_i, x_{i+1}, \ldots, x_n)$ of the given linear system.

Theorem 23.15. (Lakeyev *et al.* [244]) *For every $p_b \geq 1$, checking consistency of linear systems with interval coefficients and p_b-generalized ellipsoid right-hand side is NP-hard.*

Theorem 23.16. (Lakeyev *et al.* [244]) *For every $p_b \geq 1$, the problem of solving linear systems with interval coefficients and p_b-generalized ellipsoid right-hand side is NP-hard.*

Proofs

Proof of Theorem 23.1. Let us first show that the ellipsoid computation problem is *feasible* for *quadratic* functions $f(x_1, \ldots, x_n)$, i.e., that it is feasible to find the minimum and the maximum of a quadratic function $f(x_1, \ldots, x_n)$ under a quadratic constraint $g(x_1, \ldots, x_n) \le C$. It is sufficient to consider the minimum, because the maximum can be found in a similar fashion.

The minimum is attained either *inside* the ellipsoid, in which case all partial derivatives are equal to 0, and we have an easy-to-solve system of linear equations, or on its *border* $g(x_1, \ldots, x_n) = C$. To find the minimum of $f(x_1, \ldots, x_n)$ on the border $g(x_1, \ldots, x_n) = C$, we can use the Lagrange multiplier method, i.e., consider the unconstrained minimum $f(x_1, \ldots, x_n) + \lambda \cdot g(x_1, \ldots, x_n) \to \min$ and then find λ from the condition that $g(x_1, \ldots, x_n) = C$. The new objective function $f(x_1, \ldots, x_n) + \lambda \cdot g(x_1, \ldots, x_n)$ is also quadratic and therefore, to find a point $x = (x_1, \ldots, x_n)$ where its unconstrained minimum is attained, we can equate all its partial derivatives to 0 and get a system of linear equations. The coefficients of this system linearly depend on the (unknown) parameter λ; therefore, this system has the form $(F + \lambda \cdot G)x = f + \lambda \cdot g$ for some (known) matrices F, G, and vectors f and g. Hence, $x = (F + \lambda \cdot G)^{-1}(f + \lambda \cdot g)$. Substituting this x into the equation $g(x_1, \ldots, x_n) = C$, we get a polynomial equation with a single unknown λ; so, we can use a known polynomial-time algorithm for solving such equations (see Chapter 18). Thus, we can feasibly compute all points x in which the minimum can be attained. By comparing the values of the function $f(x_1, \ldots, x_n)$ in all these points, we get the desired minimum.

A simple algorithm that we have described requires polynomial time but may not be very practical. Vavasis [418] describes somewhat more complicated polynomial-time algorithms for solving this problem that are already very practical.

The second part of the theorem follows from the fact that, according to Theorem 17.1, the unconditional minimization problem is NP-hard for quartic polynomials. For polynomials $f(x_1, \ldots, x_n)$ considered in the proof of that theorem, we can easily find the bound B such that the minimum of the function $f(x_1, \ldots, x_n)$ is attained when each of the variables x_i is in the interval $[-B, B]$. The resulting box $[-B, B] \times \ldots \times [-B, B]$ is contained in an ellipsoid $x_1^2 + \ldots + x_n^2 \le n \cdot B^2$. Therefore, minimizing the objective function $f(x_1, \ldots, x_n)$ over this ellipsoid in equivalent to its unconditional minimization. Since we already know that unconditional minimization is NP-hard, we can thus conclude that ellipsoid computations are NP-hard. The theorem is proven.

Proof of Theorem 23.2. Theorem 23.2 follows from the first part of Theorem 17.3 in the same way as the first part of Theorem 23.1 follows from Theorem 17.1.

Proof of Theorem 23.10. This theorem is proven similarly to the second part of Theorem 23.1.

24

NON-INTERVAL UNCERTAINTY II: MULTI-INTERVALS AND THEIR GENERALIZATIONS

The basic problem of interval computations is to find all possible values of $y = f(x_1, \ldots, x_n)$ when we only have partial information about x_i.

*Traditional interval computation techniques deal with the situation in which the set of all possible values of each variable x_i is an **interval** $\mathbf{x}_i = [\underline{x}_i, \overline{x}_i]$ (where $\underline{x}_i$ is the smallest possible value of x_i and $\overline{x}_i$ is the largest possible value of x_i), and **all** possible combinations of these values are possible. In this case, the set of all possible values of $\vec{x} = (x_1, \ldots, x_n)$ is a **multi-dimensional interval (box)** $\mathbf{x}_1 \times \ldots \times \mathbf{x}_n$.*

*In the **previous chapter**, we considered the situations in which for **each** variable x_i, its set of possible values is (still) an interval, but **not all combinations** of x_i are possible. In this case, all possible values of $\vec{x}$ form a **subset** of the box, e.g., an ellipsoid.*

*In this chapter, we consider the case in which for some variables x_i, **not all** values from the interval $[\underline{x}_i, \overline{x}_i]$ are possible. In this case, the set of all possible values of each variable x_i becomes disconnected: it is, e.g., a union of two or more intervals. Such a union is called a **multi-interval**.*

We describe why such multi-intervals appear in practical problems, and analyze the computational complexity and feasibility of multi-interval (and more general) computations.

24.1. Not Only Intervals: Practical Motivation

Why non-interval sets: a theoretical possibility. The basic problem of interval computations is to find all possible values of $y = f(x_1, \ldots, x_n)$ when we only have partial information about x_i.

In the previous chapters, we considered the situations in which for each variable x_i, this uncertainty is of *interval* type, i.e., in which the set X_i of possible values of x_i coincides with the interval $\mathbf{x}_i = [\underline{x}_i, \overline{x}_i]$, where $\underline{x}_i$ is the smallest possible value of x_i and $\overline{x}_i$ is the largest possible value of x_i.

In practice, however, sometimes we know not only the bounds $\underline{x}_i$ and $\overline{x}_i$ on the possible values on x_i, but we also know that some values from the interval $[\underline{x}_i, \overline{x}_i]$ are impossible. In this case, the set X_i of all possible values of x_i is *not* an interval. Let us give a few examples of different possible origins of such situations.

Origin 1: non-interval noise. (Misane *et al.* [285], Nguyen *et al.* [306]) In *computer engineering*, especially in computer design and testing, we must perform measurements (of current, voltage, etc.) inside the computer. For each such measurement, we know the *upper bound* Δ on the measurement error, i.e., we know that the measurement error Δx is guaranteed to belong to the interval $[-\Delta, \Delta]$. In many measurements, this error is mainly caused by the influence of a nearby magnetic memory element, which can be in two possible states (corresponding to "0" and "1"). In this case, the error is either positive, or negative (depending on the state), but never 0; actually, the error can never be smaller than some value δ. Hence, the set X of possible values of the error is *not an interval*, but a *union* of two intervals: $X = [-\Delta, -\delta] \cup [\delta, \Delta]$. Such a union of *several* intervals is also called a *multi-interval.*

Origin 2: measuring x^2 (or a more complicated function) instead of directly measuring x. Another case of a non-interval uncertainty is when we do not exactly measure x_i directly. This may sound somewhat strange because the entire reason for measuring the quantities x_i is that we *can measure* these quantities *directly*, as opposed to the desired quantity y that we *cannot* easily *directly measure*. In reality, there is no inconsistency here: yes, we can measure x_i directly, but there are several *degrees* of directness. To measure an arbitrary quantity x_i, we must generate an electric signal that represents its value; such generation is done by a *sensor*.

For example, a *photo-sensor* generates the signal representing the light's intensity, an *accelerometer* generates a signal characterizing acceleration, etc.

In some (rare) cases, the signal coming from the sensor is exactly equal (or exactly proportional) to the measured value, but most often, the signal s is only a *function* of the measured value $s = g(x_i)$; so, to reconstruct the desired value x_i, we must *re-scale* the signal s (i.e., compute $x_i = g^{-1}(s)$). For intelligent sensors and measuring instruments, this re-scaling is performed by a built-in microprocessor and is, thus, hidden from us. All we get is the signal that is exactly equal to the measured value. From this viewpoint, this is truly a *direct measurement*. However, such a re-scaling is only possible when the function $g(x)$ is one-to-one. In some cases, this function is *not* one-to-one, and thus, one and the same value of the observed signal can correspond to *several* different values of the measured quantity.

For example, in *computer engineering*, it is very difficult to measure currents inside the chip, because the currents are very weak and traditional measurement techniques would change these very currents that we are trying to measure (and thus, disrupt the microprocessor). Therefore, it is desirable to measure each current x_i by measuring some quantity that is related to the current but that is, by itself, not electromagnetic (and thus, does not interfere with the workings of the chip). One such possibility is to measure the *heat* caused by the current. It is known that the power of this heat is proportional to the square of the current: $s = x^2$. Thus, e.g., if we have measured that the heat is $s \in [1, 4]$, then the only information that we have about the current x is that either x is between 1 and 2, or that x is between -2 and -1; in other words, the set of all possible values of the current is a *union* $[-2, -1] \cup [1, 2]$ of two (disjoint) intervals.

We can re-formulate this example in slightly different terms: we can assume that the current is the desired quantity y, and that the heat x_1 is the directly measured quantity. Similarly to the more traditional interval computation setting, there is a relationship between the quantity that we directly measure (i.e., x_1) and the quantity that we want to reconstruct (i.e., y), but in contrast to that setting, the value y is *not* uniquely determined by x_1: instead of a normal (one-valued) function $y = f(x_1)$, we have a two-valued "function" $y = \pm\sqrt{x_1}$.

Origin 3: automatic control (Buridan's ass). An important practical situation in which multi-intervals appear is *automated control* (see, e.g., Yen *et al.* [434, 435, 436] and Nguyen *et al.* [304]). If the analyzed system includes an optimally controlled part, then it is not necessary to actually measure all the parameters of this part (which may be fastly moving and thus difficult to measure): it is often sufficient to measure the same parameters as the system measures, and then use our knowledge of the system's objective function to accurately describe its behavior.

> For *example*, when a *spaceship* re-enters the Earth's atmosphere, its outside temperature gets so high that it is extremely difficult to measure any characteristics. However, if we know its initial position and velocity and we know exactly the objective function that its trajectory optimizes, then we can predict the trajectory of the spaceship with a reasonable accuracy, and meet it at the desired place.

Due to measurement uncertainty, we can only make *approximate* predictions. In most cases, this uncertainty can be well described by *intervals*, but in some cases, *multi-intervals* are a more adequate description. To avoid the complexity of a spaceship example, let us illustrate this "non-intervalness" on a much simpler example of a car.

> If a car, traveling on a lonely road, is approaching an obstacle, then we can either turn to the left, or turn to the right. If the situation is absolutely symmetric, then, from the viewpoint of any reasonable objective function, both turns are equally good. If we knew the precise values of all the parameters, then we would be able to recommend the optimal turn angle $\varphi > 0$, so that turning φ or $-\varphi$ degrees would lead to an optimal obstacle-avoiding trajectory. In this case, the set of possible angle values consists of two points $\{-\varphi, \varphi\}$. In reality, due to inevitable uncertainty, we can only describe the *interval* $[\underline{\varphi}, \overline{\varphi}]$ of possible values of this angle. Thus, the set of possible values of the actual turn angle is a two-component *multi-interval* $[-\overline{\varphi}, -\underline{\varphi}] \cup [\underline{\varphi}, \overline{\varphi}]$.

This situation formalizes the famous 14th century *Buridan's ass* example: when confronted between two equally accessible and equally attractive bales of hay, an ass must make an unpredictable choice between the two.

Comment. In addition to the above examples, there are numerous other practical applications of multi-intervals in data processing; see, e.g., the survey Dmitriev *et al.* [92] and references therein.

Multi-intervals are difficult to process. In spite of the practical importance of computations with multi-intervals, the existing techniques of processing multi-intervals require lots of computation time. It is not accidental that several methods of time-saving parallelization have been developed for computations with multi-intervals; see, e.g., Yakovlev [433], Shvetsov *et al.* [397], and [213].

It is important to analyze computational complexity and feasibility of computations with multi-intervals. Since the existing methods of processing multi-intervals require a lot of computation time, it is very important to analyze whether the excessive computation time is caused by the inefficiency of the existing algorithms or by the complexity of the problem itself. In other words, *it is necessary to analyze the computational complexity and feasibility of computations* involving *multi-intervals.*

24.2. Multi-Interval Computations: Definitions, Computational Complexity, and Feasibility

Definition 24.1. *By a multi-interval X, we mean a finite union of intervals. If all these intervals have rational endpoints, then this multi-interval is called rational. Unions of two intervals will be called two-component multi-intervals.*

Comment. The basic problem of *interval* computations is to compute the range of a given function $f(x_1, \ldots, x_n)$ when x_i run over given intervals $\mathbf{x}_i$. When the function $f(x_1, \ldots, x_n)$ is continuous, this range $\mathbf{y}$ is itself an interval, and therefore, to describe this range, it is sufficient to describe its lower and upper endpoints $\underline{y}$ and $\overline{y}$. If x_i belong to *multi-intervals*, then the set Y of possible values of y is *not* necessarily an *interval*: it is not an interval even for the simplest possible function $f(x_1) = x_1$. Therefore, it is *not sufficient* to describe the *endpoints* $\underline{y}$ and $\overline{y}$ of the range, we must also describe which numbers from the corresponding interval $[\underline{y}, \overline{y}]$ belong to the range and which numbers don't. In other words, we must be able to check, for each rational number y, whether $y \in Y$ or not. Thus, we arrive at the following definition:

Definition 24.2. *By the basic problem of multi-interval computations, we mean the following problem:*

GIVEN:

- *a positive integer n;*
- *a polynomial $f(x_1, \ldots, x_n)$ with rational coefficients;*
- *n rational multi-intervals $X_1, \ldots, X_n$;*
- *a rational number y.*

CHECK: whether there exist real numbers $x_1, \ldots, x_n$ for which:

- *$x_i \in X_i$ for all $i = 1, \ldots, n$; and*
- *$y = f(x_1, \ldots, x_n)$.*

Theorem 24.1. *For linear functions f, the basic problem of multi-interval computations is NP-hard.*

Comments.

- From the proof, we will see that this problem is NP-hard even for the simplest non-interval multi-intervals: namely, for two-component ones.
- Since this problem is NP-hard even for *linear* functions f, it is therefore NP-hard for any *more general* class of functions, e.g., for quadratic functions f.
- Since interval computations are feasible for linear functions $f(x_1, \ldots, x_n)$, this result shows that *multi-interval* computations are more complicated than *interval* computations. Thus, this result *explains* the above-mentioned *empirical fact*: that multi-intervals are more difficult to handle than intervals.

	Interval computations	Multi-interval computations
Linear f	Linear time	NP-hard
Quadratic f	NP-hard	NP-hard
Cubic f (and higher order)	NP-hard	NP-hard

Definition 24.3.

- *By a multi-interval vector, we mean a sequence* $(X_1, \ldots, X_n)$ *of multi-intervals.*
- *By a multi-interval matrix, we mean a collection of multi-intervals* X_{ij}, $1 \le i \le n$, $1 \le j \le m$.
- *By a multi-interval linear system, we mean a tuple* $\langle m, n, A, B\rangle$, *where* m *and* n *are positive integers,* A *is a* $(m \times n)$*-dimensional multi-interval matrix with elements* A_{ij}, *and* B *is an* m*-dimensional multi-interval vector with components* B_i.
- *We say that a vector* $(x_1, \ldots, x_n)$ *is a possible solution of the linear system* $\langle m, n, A, B\rangle$ *if for some values*

$$a_{11} \in A_{11}, a_{12} \in A_{12}, \ldots, a_{1n} \in A_{1n}, \ldots, a_{m1} \in A_{m1}, \ldots, a_{mn} \in A_{mn},$$

$$b_1 \in B_1, \ldots, b_m \in B_m,$$

and for all $i = 1, \ldots, m$, *we have*

$$\sum_{j=1}^{n} a_{ij} \cdot x_j = b_i.$$

- *We say that a multi-interval linear system is consistent if it has a possible solution.*

Definition 24.4. *By a problem of* $\varepsilon-$*approximately solving multi-interval linear systems, we mean the following problem:*

GIVEN:

- *a multi-interval linear system;*
- *a positive rational number* $\varepsilon > 0$; *and*
- *a positive integer* $i \le n$;

COMPUTE the rational numbers $\widetilde{\overline{x}}_i$ *and* $\widetilde{\underline{x}}_i$ *that are* ε*-close to, correspondingly, the largest* $\overline{x}_i$ *and the smallest* $\underline{x}_i$ *values of* x_i *for all possible solutions* $(x_1, \ldots, x_{i-1}, x_i, x_{i+1}, \ldots, x_n)$ *of the given linear system.*

Theorem 24.2. *Checking consistency of multi-interval linear systems is NP-hard.*

Theorem 24.3. *For every $\varepsilon > 0$, the problem of ε-approximately solving multi-interval linear systems is NP-hard.*

Comments.

- Both results remain true if we only allow two-component multi-intervals.
- We will see from the proof that both problems remain NP-hard even if we require that either A is a numerical matrix, or that B is a numerical vector. For *interval* computations, if the matrix A is numerical, the problem becomes feasible (namely, it becomes a particular case of linear programming). Thus, for solving linear systems, *multi-interval computations* are also *harder than interval computations.*

	Interval computations	Multi-interval computations
Numerical A_{ij}, numerical B_i	Polynomial time	Polynomial time
Numerical A_{ij}, non-numerical B_i	Polynomial time	NP-hard
Non-numerical A_{ij}, numerical B_i	NP-hard	NP-hard
Non-numerical A_{ij}, non-numerical B_i	NP-hard	NP-hard

24.3. The General Case of (Bounded) Non-Interval Uncertainty

In this section, we will show that if we add at least one non-interval multi-interval set to the family of all intervals, then some reasonable interval computation problems that were previously computationally feasible become intractable. Thus, we will provide a justification for using only intervals.

We know that the basic problem of interval computations is NP-hard for polynomials, i.e., for functions that can be obtained from variables x_i by addition, subtraction, and multiplication. We also know that if we only allow addition and multiplication, then we get *linear* functions $f(x_1, \ldots, x_n)$ for which the basic problem of interval computations is feasible.

A natural question is: What other operations can be added to addition and subtraction that still keep this computation feasible? If, in addition to $+$ and $-$, we allow *feasible monotonic functions* of one variable $g(x)$ (i.e., functions whose values can be computed in polynomial time), we get function $f(x_1, \ldots, x_n) = g_1(x_1) + \ldots + g_n(x_n)$. For such functions $f(x_1, \ldots, x_n)$ and for intervals $\mathbf{x}_1, \ldots, \mathbf{x}_n$, we can compute $Y = [\underline{y}, \overline{y}] = f(X_1, \ldots, X_n)$ in polynomial time; namely, $\underline{y} = \sum \underline{t}_i$ and $\overline{y} = \sum \overline{t}_i$, where:

- $\underline{t}_i = g_i(\underline{x}_i)$ if $g_i(x)$ is increasing, and $\underline{t}_i = g_i(\overline{x}_i)$ if $g_i(x)$ is decreasing;
- $\overline{t}_i = g_i(\overline{x}_i)$ if $g_i(x)$ is increasing, and $\overline{t}_i = g_i(\underline{x}_i)$ if $g_i(x)$ is decreasing.

Thus, for such functions, the basic problem of interval computation is also *feasible.*

We will prove that adding at least one non-interval (closed bounded) rational multi-interval S to the family of intervals makes this problem NP-hard, even for the functions $g_i(x)$ that are piecewise-linear functions with rational coefficients.

Definition 24.5. *Let S be a (closed bounded) rational multi-interval. We say that a set X is given if either we know that this set is a rational interval (and we are given its endpoints), or we know that $X = S$. By $\mathcal{P}_S$, we denote the following general problem:*

GIVEN:

- *n feasible piece-wise linear functions $g_1(x), \ldots, g_n(x)$ with rational coefficients;*
- *n given sets $X_1, \ldots, X_n$;*
- *a rational number y;*

CHECK whether $y \in Y$, where

$$Y = f(X_1, \ldots, X_n) = \{f(x_1, \ldots, x_n) | x_1 \in X_1, \ldots, x_n \in X_n\},$$

and $f(x_1, \ldots, x_n) = g_1(x_1) + \ldots + g_n(x_n)$.

Theorem 24.4. (Nogueira *et al.* [308])

- *If the set S is an interval, then the problem $\mathcal{P}_S$ is solvable in polynomial time.*
- *If the set S is not an interval, then the problem $\mathcal{P}_S$ is NP-hard.*

Comment. So, if we add even a *single non-interval* multi-interval to the family of all intervals, then a natural problem which was *originally feasible becomes NP-hard.* This result provides one more motivation for using the interval family: because if we increase this family, we lose computability. It should be mentioned that there are several alternative justifications of using intervals, such as:

- computational simplicity of interval computations Misane *et al.* [285, 286];
- invertibility of interval arithmetic as opposed to arithmetic operations with non-intervals Bouchon-Meunier *et al.* [52, 53], Kosheleva *et al.* [187];
- intervals naturally appear when the error is a result of several small causes [210], etc.

24.3. Infinite Multi-Intervals: Computational Complexity and Feasibility

In some cases, we get *infinite* intervals and multi-intervals: e.g., if we measure the quantity $1/x_i$ and it turns out that $1/x_i \in [-1, 1]$, then the possible values of x_i form an infinite multi-interval $(-\infty, -1] \cup [1, \infty)$.

When the set of possible values of each variable x_i is an infinite interval or multi-interval X_i, then the range $f(X_1, \ldots, X_n)$ of most functions $f(x_1, \ldots, x_n)$ is also infinite (and usually, simply coincides with the entire real line). Thus, after the corresponding indirect "measurement", arbitrary large and/or arbitrarily small values of the indirectly measured quantity $y = f(x_1, \ldots, x_n)$ are still possible. From the practical viewpoint, this is not what we would call a *measurement.*

More meaningful problems appear when we consider linear systems in which, in addition to interval uncertainty, we allow infinite multi-interval coefficients:

Definition 24.6.

- *By an infinite multi-interval X, we mean a set* $(-\infty, \underline{a}] \cup [\overline{a}, +\infty)$.
- *If both endpoints $\underline{a}$ and $\overline{a}$ are rational, then this infinite multi-interval is called rational.*

Definition 24.7.

- *By a mixed interval and infinite multi-interval vector (or simply a mixed vector, for short), we mean a sequence $(X_1, \ldots, X_n)$, in which each X_i is either a rational interval or a rational infinite multi-interval.*
- *By a mixed interval and infinite multi-interval matrix (or simply a mixed matrix, for short), we mean a collection X_{ij}, in which each X_{ij} is either a rational interval or a rational infinite multi-interval.*
- *By a mixed interval and infinite multi-interval linear system (or simply a mixed linear system, for short), we mean a tuple $\langle m, n, A, B\rangle$, where m and n are positive integers, A is a $(m \times n)$-dimensional mixed matrix with elements A_{ij}, and B is an m-dimensional mixed vector with components B_i.*
- *We say that a vector $(x_1, \ldots, x_n)$ is a possible solution of the linear system $\langle m, n, A, B\rangle$ if for some values*

$$a_{11} \in A_{11}, a_{12} \in A_{12}, \ldots, a_{1n} \in A_{1n}, \ldots, a_{m1} \in A_{m1}, \ldots, a_{mn} \in A_{mn},$$

$$b_1 \in B_1, \ldots, b_m \in B_m,$$

and for all $i = 1, \ldots, m$,, we have

$$\sum_{j=1}^{n} a_{ij} \cdot x_j = b_i.$$

- *We say that a mixed linear system is consistent if it has a possible solution.*

Definition 24.8. *By a problem of ε—approximately solving mixed linear systems, we mean the following problem:*

GIVEN:

- *a mixed linear system;*
- *a positive rational number $\varepsilon > 0$; and*
- *a positive integer $i \leq n$;*

COMPUTE the rational numbers $\widetilde{\overline{x}}_i$ and $\widetilde{\underline{x}}_i$ that are ε-close to, correspondingly, the largest $\overline{x}_i$ and the smallest $\underline{x}_i$ values of x_i for all possible solutions $(x_1, \ldots, x_{i-1}, x_i, x_{i+1}, \ldots, x_n)$ of the given linear system.

Theorem 24.5. *Checking consistency of mixed interval and infinite multi-interval linear systems is NP-hard.*

Theorem 24.6. *For every $\varepsilon > 0$, the problem of ε-approximately solving mixed interval and infinite multi-interval linear systems is NP-hard.*

Comments. We will see from the proof that both problems remain NP-hard even if we require that either A is a numerical matrix, or that B is a numerical vector. For *interval* computations, if the matrix A is numerical, the problem becomes feasible (namely, it becomes a particular case of linear programming). Thus, for solving linear systems, mixed computations are also harder than interval computations.

	Interval computations	Mixed interval and infinite multi-interval computations
Numerical A_{ij}, numerical B_i	Polynomial time	Polynomial time
Numerical A_{ij}, non-numerical B_i	Polynomial time	NP-hard
Non-numerical A_{ij}, numerical B_i	NP-hard	NP-hard
Non-numerical A_{ij}, non-numerical B_i	NP-hard	NP-hard

Proofs

Proof of Theorem 24.1. To prove this result, we will reduce the *PARTITION* problem (known to be NP-hard) to this problem. In the *PARTITION* problem, we are given a sequence of integers $s_1, \ldots, s_n$, and we must check whether there exist values $y_1, \ldots, y_n$ for which $y_i \in \{-1, 1\}$ and $s_1 \cdot y_1 + \ldots + s_n \cdot y_n = 0$. This is exactly a multi-interval computation problem for a linear function $f(x_1, \ldots, x_n) = s_1 \cdot x_1 + \ldots + s_n \cdot x_n$, (two-component) multi-intervals $X_i = [-1, -1] \cup [1, 1]$, and $y = 0$. Thus, the multi-interval computation problem is indeed NP-hard. The theorem is proven.

Proof of Theorem 24.2. For non-numerical A_{ij} and numerical B_i, NP-hardness was proven (in the corresponding chapter) even for the case of intervals. Thus, it is sufficient to prove NP-hardness for numerical A_{ij} and non-numerical B_i. This can be proven by a reduction from *PARTITION* similar to the reduction we used in the previous theorem. Namely, for every sequence $s_1, \ldots, s_n$, we consider the following equations: $x_1 = b_1$, with $B_1 = [-1, -1] \cup [1, 1]$, ..., $x_n = b_n$, with $B_n = [-1, -1] \cup [1, 1]$, and $s_1 \cdot x_1 + \ldots + s_n \cdot x_n = b_{n+1}$, where $B_{n+1} = [0, 0]$. This multi-interval system is consistent if and only if the original *PARTITION* problem has a solution. Thus, checking consistency of multi-linear systems is indeed NP-hard. The theorem is proven.

Proof of Theorem 24.3. For this result, we will also use reduction to *PARTITION*, but the system will be slightly different: for every instance $(s_1, \ldots, s_n)$, we will consider the following system of equations with $n + 1$ unknowns $x_1, \ldots, x_{n+1}$:

- $x_i = b_i$, $B_i = [-1, -1] \cup [1, 1]$, $1 \le i \le n + 1$;
- $s_1 \cdot x_1 + \ldots + s_n \cdot x_n + s_{n+1} \cdot x_{n+1} = s_{n+1}$, where we denoted $s_{n+1} = -0.5 \cdot (s_1 + \ldots + s_n)$.

This system has a possible solution $x_1 = \ldots = x_n = 1$, $x_{n+1} = -1$, and is, therefore, consistent. The variable x_{n+1} has two possible values: -1 and 1. The value -1 is always possible, but the value $x_{n+1} = 1$ is possible if and only if $s_1 \cdot x_1 + \ldots + s_n \cdot x_n = 0$ for some $x_1 \in \{-1, 1\}, \ldots, x_n \in \{-1, 1\}$, i.e., if and only this instance of the *PARTITION* problem is solvable. Thus, depending on whether this instance is solvable or not, we will get either $\overline{x}_{n+1} = 1$ or $\overline{x}_{n+1} = -1$. Therefore, if we are able to solve multi-interval linear systems with accuracy $\varepsilon < 1$, we can check whether the actual value of $\overline{x}_{n+1}$ is -1 or

1, and hence, solve the given instance of *PARTITION*. Since *PARTITION* is known to be NP-hard, our problem is thus also NP-hard.

For $\varepsilon > 1$, we can consider the system $Ax = 2\varepsilon \cdot B$. Solving this system with accuracy ε is equivalent to solving the system $Ax = B$ with accuracy 0.5, and thus, to solving the given instance of *PARTITION*. The theorem is proven.

Proof of Theorem 24.4. We have already shown that for a rational interval S, this problem is indeed solvable in polynomial time. Let us show that for all other rational multi-intervals S, this problem is NP-hard.

1°. For this proof, we will need the following fact: If S is a bounded closed set and not an interval, then there exist real numbers $\underline{a}$ and $\overline{a}$ that themselves belong to S, but no intermediate value $a \in (\underline{a}, \overline{a})$ belongs to S.

Indeed, every set S is contained in the interval (possibly infinite) $[\inf S, \sup S]$. Since S is bounded, the endpoints $\inf S$ and $\sup S$ of this interval are finite. Since S is closed, both endpoints belong to S. Since S is not an interval, it cannot coincide with this interval, so there must exist a point $a \in (\inf S, \sup S)$ that does not belong to S.

Now, we can take $\underline{a} = \sup\{s \in S | s < a\}$ and $\overline{a} = \inf\{s \in S | s > a\}$. By definition of inf and sup, no point in between $\underline{a}$ and $\overline{a}$ can belong to S. Since S is closed, both points $\underline{a}$ and $\overline{a}$ belong to S. The statement is proven.

For a rational multi-interval, these points $\underline{a}$ and $\overline{a}$ are rational numbers.

2°. Now, we are ready to prove NP-hardness of our problem by reducing *PARTITION* to it. Let $s_1, \ldots, s_n$ be an instance of *PARTITION*. Then, we can design the following particular case of our problem $\mathcal{P}_S$:

- $$g_i(x_i) = \begin{cases} -s_i & \text{if } x_i \le \underline{a} \\ -s_i + 2s_i \cdot (x_i - \underline{a})/(\overline{a} - \underline{a}) & \text{if } \underline{a} < x_i < \overline{a} \\ s_i & \text{if } x_i \ge \overline{a} \end{cases}$$
- all sets X_i are equal to S, and
- rational number $y = 0$.

We will show that for this case, $0 \in f(X_1, \ldots, X_n) \Leftrightarrow$ the given instance of *PARTITION* has a solution.

$\Rightarrow$ Assume that $0 \in f(X_1,\ldots,X_n)$. By definition of the set $f(X_1,\ldots,X_n)$, this means that there exists real numbers $x_1 \in X_1,\ldots,x_n \in X_n$ such that $f(x_1,\ldots,x_n) = 0$, i.e.,

$$g_1(x_1) + \ldots + g_n(x_n) = 0 \qquad (24.1).$$

The numbers x_i are within the set S, and therefore, these numbers cannot belong to the interval $(\underline{a},\overline{a})$ (due to our choice of $\underline{a}$ and $\overline{a}$). Hence, $x_i \le \underline{a}$ or $x_i \ge \overline{a}$. For every i, let us take $y_i = 1$ if $x_i \ge \overline{a}$, and $y_i = -1$ if $x_i \le \underline{a}$. By definition of the functions $g_i(x_i)$, we have $g_i(x_i) = s_i$ for $x_i \ge \overline{a}$ and $g_i(x_i) = -s_i$ for $x_i \le \underline{a}$. Thus, in both cases, $g_i(x_i) = s_i \cdot y_i$. Hence, from (24.1), we conclude that

$$s_1 \cdot y_1 + \ldots + s_n \cdot y_n = 0, \qquad (24.2)$$

i.e., that the given instance of *PARTITION* problem has a solution.

$\Leftarrow$ Vice versa, if (24.2) has a solution, we can take $x_i = \overline{a}$ when $y_i = 1$, and $x_i = \underline{a}$ when $y_i = -1$. Then, as before, $g_i(x_i) = s_i \cdot y_i$, and therefore, (24.2) implies (24.1), i.e., $f(x_1,\ldots,x_n) = 0$, and $0 \in f(X_1,\ldots,X_n)$.

Reduction is proven, so our problem $\mathcal{P}_S$ is indeed NP-hard. The theorem is proven.

Proof of Theorem 24.5. The proof is similar to the proof of Theorem 24.2, the only difference is that for each variable x_i, instead of a *single* equation $x_i = b_i$ with $B_i = [-1,-1] \cup [1,1]$, we have to consider *two* equations $x_i = b_i^{(1)}$ with $B_i^{(1)} = [-1,1]$ and $x_i = b_i^{(2)}$ with $B_i^{(2)} = (-\infty,-1] \cup [1,+\infty)$. These two equations hold if and only if $x_i \in B_i^{(1)} \cap B_i^{(2)} = \{-1,1\}$. Thus, they are indeed equivalent to the original equation. The theorem is proven.

Proof of Theorem 24.6. This theorem is proven by applying the same modification that we used for the previous theorem to the proof of Theorem 24.3.

25

WHAT IF QUANTITIES ARE DISCRETE?

25.1. Why Discrete Quantities?

The basic problem of interval computations that we analyze in this book is to compute the accuracy of the results of data processing. To be more precise, we know an algorithm $f(x_1, \ldots, x_n)$ that is used in data processing and the intervals $\mathbf{x}_i$ of possible values of the input quantities $x_1, \ldots, x_n$, and we want to describe the set of all possible values of $f(x_1, \ldots, x_n)$ when $x_i \in \mathbf{x}_i$.

In interval computations, it is customary to consider quantities $x_1, \ldots, x_n$ that can, in principle, take arbitrary real values. For any such quantity x_i, if the only information we have about its value is that this value belongs to the interval $[x_i^-, x_i^+]$, then each number from this interval can be a possible value of x_i. Therefore, the set of possible values of $f(x_1, \ldots, x_n)$ coincides with the *range*

$$Y = f(\mathbf{x}_1, \ldots, \mathbf{x}_n) = \{f(x_1, \ldots, x_n) | x_1 \in \mathbf{x}_1, \ldots, x_n \in \mathbf{x}_n\}$$

of the function f on intervals $\mathbf{x}_i$.

Our problem then is to *describe* this set Y. What does it mean to *describe this set*? It means, e.g., that for every number y, we must be able to tell whether y belongs to this set or not. For a continuous function f, the range Y is an interval; therefore, in order to describe this set Y, it is sufficient to describe its two endpoints y^-, y^+. If we know these endpoints, then in order to check whether a given number y belongs to the set $Y = [y^-, y^+]$, it is sufficient to check two inequalities $y \geq y^-$ and $y \leq y^+$.

There are, however, some real-life situations where some quantities x_i cannot take *all* possible real numbers as their values: they can only take values that are integer multiples of a certain value q_i (called a *quantum*). In other words, possible values of x_i are $\ldots, -nq_i, \ldots, -2q_i, -q_i, 0, q_i, 2q_i, \ldots, nq_i, \ldots$ Typical examples of such quantities are electric charge, spin (for which $q_i = 1/2$), and many other characteristics of elementary particles, ions, atoms, etc.

In such cases, if we know (from the measurements) the interval of possible values $[x_i^-, x_i^+]$ of such a quantity x_i, then only numbers that belong to this interval *and* that are integer multiples of q_i are possible values of x_i. When the set of possible values x_i is discrete, the set Y of possible values of $y = f(x_1, \ldots, x_n)$ is not necessarily an interval. Therefore, to describe this set Y, it is no longer sufficient to describe its greatest lower bound $\inf Y$ and its least upper bound $\sup Y$; we must also be able, for every rational number y, to check whether this number belongs to the set Y or not. (Similarly to the previous chapters, we restrict ourselves to rational numbers, because the existing computers usually represent only rational numbers. The following results will not change much if we use other computer-represented numbers.)

25.2. Precise Formulation of the Basic Problem of Discrete Interval Computations

Definition 25.1. *By the basic problem of discrete interval computations, we mean the following problem:*

GIVEN:

- *a positive integer n;*
- *a polynomial $f(x_1, \ldots, x_n)$;*
- *n positive rational numbers $q_1, \ldots, q_n$;*
- *n intervals $\mathbf{x}_i = [x_i^-, x_i^+]$, $1 \le i \le n$, with rational endpoints x_i^- and x_i^+.*
- *a rational number y.*

CHECK: whether there exist real numbers $x_1, \ldots, x_n$ for which:

- *$x_i \in \mathbf{x}_i$ for all $i = 1, \ldots, n$;*
- *the ratio x_i/q_i is an integer for all $i = 1, \ldots, n$;*
- *$y = f(x_1, \ldots . x_n)$.*

25.3. Computational Complexity of Discrete Interval Computations

The natural question is: what is the computational complexity of this problem?

Are discrete interval computations easier? At first glance, it may seem that this problem is easier than non-discrete interval computations. Intuitively, the complexity of a computational problem depends on the *size* of the area where a solution has to be found. This size describes the total number of objects that we need to analyze in order to find a solution; thus, the larger this size, the more complicated the problem.

This intuition is generally true; e.g., the more variables a problem has, the more difficult it is to solve it.

From this viewpoint, when we impose the additional condition that the ratio x_i/q_i must be an integer, then we drastically decrease the size of the area where the solution can be found: from continuum to a finite set. Therefore, it may seem at first glance that the discrete problem is computationally easier than the continuous problem of interval computations.

As we will soon see, this intuitive conclusion is false. The basic problem of discrete interval computations is *computationally harder* than its continuous analogue.

Computational complexity of the basic problem of discrete interval computations. For (continuous) interval computations, the basic problem is already NP-hard for quadratic functions

$$f(x_1, \ldots, x_n) = a_0 + a_1x_1 + \ldots + a_nx_n + a_{11}x_1^2 + a_{12}x_1x_2 + \ldots + a_{nn}x_n^2.$$

The major case when this problem can be solved in polynomial time is the case of linear functions $f(x_1, \ldots, x_n) = a_0 + a_1x_1 + \ldots + a_nx_n$. For this class, the basic problem of (continuous) interval computations is computable not only in *polynomial* time, but even in *linear* time. For *discrete* interval computations, the problem is much more difficult:

Theorem 25.1. *For linear functions f, thé basic problem of discrete interval computations is NP-hard.*

Comment. Since this problem is NP-hard even for linear functions f, it is therefore NP-hard for any more general class of functions, e.g., for quadratic functions f.

Computational complexity of an auxiliary problem of discrete interval computations. If we allow *infinite* intervals, then for *continuous* interval computations, the basic problem of interval computations turns into a problem of *unconstrained optimization*. This problem is known to be NP-hard, but *decidable* in the sense that there exists an algorithm that given a problem, returns its solution (for example, we can apply Tarski-Seidenberg algorithm [408, 388]).

It turns out that the discrete analogue of this problem is undecidable (i.e., does not allow any algorithm at all).

To formulate this problem in precise mathematical terms, we take Definition 25.1, and replace each interval $[x_i^-, x_i^+]$ from that definition by an infinite interval $(-\infty, \infty)$. Since the intervals are fixed, we do not need to describe them in the *GIVEN* part of the problem. Also, for such intervals, the condition $x_i \in (-\infty, \infty)$ is always satisfied, so, we do not need to consider it in the *CHECK* part of the problem. As a result, we arrive at the following definition:

Definition 25.2. *By the discrete analogue of the unconstrained optimization problem, we mean the following problem:*

GIVEN:

- *a positive integer n;*
- *a polynomial $f(x_1, \ldots, x_n)$;*
- *n positive rational numbers $q_1, \ldots, q_n$;*
- *a rational number y.*

CHECK: whether there exist real numbers $x_1, \ldots, x_n$ for which:

- *the ratio x_i/q_i is an integer for all $i = 1, \ldots, n$;*
- *$y = f(x_1, \ldots . x_n)$.*

Theorem 25.2. *The discrete analogue of unconstrained optimization problem is undecidable.*

	(Continuous) Interval Computations	Discrete Interval Computations
Linear f	Linear time	NP-hard
Quadratic f	NP-hard	NP-hard
Unconstrained Optimization	Decidable	Undecidable

Proofs

Proof of Theorem 25.1. To prove this result, we will reduce the *PARTITION* problem (known to be NP-hard) to this problem. In the *PARTITION* problem, we are given a sequence of integers $s_1, \ldots, s_n$, and we must check whether there exist values $y_1, \ldots, y_n$ for which $y_i \in \{-1, 1\}$ and $s_1 \cdot y_1 + \ldots + s_n \cdot y_n = 0$. In this problem, each variable y_i can take only two possible values: -1 and 1. Instead of these variables y_i, we can consider new variables $x_i = (y_i + 1)/2$. When $y_i \in \{-1, 1\}$, then $x_i \in \{0, 1\}$. To describe the condition $\sum s_i \cdot y_i = 0$ in terms of the new variables x_i, we will use the fact that $y_i = 2x_i - 1$; then, the desired condition takes the form $2\sum s_i \cdot x_i - \sum s_i = 0$. If we divide both sides of this equality by 2 and move the negative term to the other side of the equality, we get $\sum s_i \cdot x_i = y$, where $y = (1/2) \cdot (s_1 + \ldots + s_n)$. Hence, the *PARTITION* problem is equivalent to the following problem:

GIVEN:

- *a positive integer n;*
- *n non-negative integers $s_1, \ldots, s_n$;*

CHECK: whether there exist real numbers $x_1, \ldots, x_n$ for which:

- *$x_i \in \{0, 1\}$ for all $i = 1, \ldots, n$;*
- *$y = s_1 \cdot x_1 + \ldots + s_n \cdot x_n$, where $y = (1/2) \cdot (s_1 + \ldots + s_n)$.*

The only difference between this problem and the discrete interval problem for the linear function $f(x_1, \ldots, x_n)$ is that the condition $x_i \in \{0, 1\}$ is not in the desired form. But this condition can be easily represented in the desired form if we take $x_i^- = 0$, $x_i^+ = 1$, and $q_i = 1$: indeed, the only numbers from the interval $[0, 1]$ that are multiples of 1 are 0 and 1.

Thus, the *PARTITION* problem (which is known to be NP-hard) can indeed be reduced to a particular case of the basic problem of discrete interval computations for linear functions f; therefore, this basic problem is also NP-hard. The theorem is proven.

Proof of Theorem 25.2. For $q_1 = \ldots = q_n = 1$, the problem reduces to checking whether a given polynomial f has an integer solution. An equation in which we are looking for an integer solution is called *Diophantine.* It is known that there exists no algorithm for solving Diophantine equations (this result, originally proved by Matiyasevich [276] (see also [277, 86]), was a solution to the famous Hilbert's Tenth Problem [152]). Thus, the more general problem (discrete analogue of unconstrained optimization) is also undecidable. The theorem is proven.

26

ERROR ESTIMATION FOR INDIRECT MEASUREMENTS: INTERVAL COMPUTATION PROBLEM IS (SLIGHTLY) HARDER THAN A SIMILAR PROBABILISTIC COMPUTATIONAL PROBLEM

One of main applications of interval computations is estimating errors of indirect measurements. A quantity y is measured indirectly if we measure some quantities x_i related to y and then estimate y from the results $\tilde{x}_i$ of these measurements as $f(\tilde{x}_1, \ldots, \tilde{x}_n)$ by using a known relation f. Interval computations are used "to find the range of $f(x_1, \ldots, x_n)$ when x_i are known to belong to intervals $\mathbf{x}_i = [\tilde{x}_i - \Delta_i, \tilde{x}_i + \Delta_i]$," where Δ_i are guaranteed accuracies of direct measurements. It is known that the corresponding problem is intractable (NP-hard) even for polynomial functions f.

In some real-life situations, we know the probabilities of different value of x_i; usually, the errors $x_i - \tilde{x}_i$ are independent Gaussian random variables with 0 average and known standard deviations σ_i. For such situations, we can formulate a similar probabilistic problem: "given σ_i, compute the standard deviation of $f(x_1, \ldots, x_n)$". It is reasonably easy to show that this problem is feasible for polynomial functions f. So, for polynomial f, this probabilistic computation problem is easier than the interval computation problem.

It is not too much easier: Indeed, polynomials can be described as functions obtained from x_i by applying addition, subtraction, and multiplication. A natural expansion is to add division, thus getting rational functions. We prove that for rational functions, the probabilistic computational problem (described above) is NP-hard.

The results of this chapter appear in Kosheleva et al. [186].

26.1. Introduction

Interval computations solve a real-life problem. We are interested in some properties of known objects. Some of these properties we can simply measure (like a body temperature). No computation is necessary here. Such measurements are never absolutely accurate. Therefore, the result $\tilde{x}$ of measuring x can differ from the actual value of x. How large can an error $\Delta x = \tilde{x} - x$ be? Manufacturers of measuring instruments guarantee some *accuracy* Δ; this means that the error will never exceed Δ: $|\Delta x| \le \Delta$. So, if our measurement result is $\tilde{x}$, then the possible values of $x = \tilde{x} - \Delta x$ form an *interval* $[\tilde{x}-\Delta, \tilde{x}+\Delta]$.

Some quantities we can simply measure; but for many others quantities, it is either impossible or too costly to measure them directly. Such situations are very frequent. There is no ruler to measure the distance between us and a quasar, no scales to weigh the Earth or our Galaxy, no speedometer to measure directly the speed of elementary particles.

Since we cannot simply measure, we need to compute such values. All these values are measured *indirectly*: we measure several other quantities $x_1, \ldots, x_n$ that are related to the desired one y, and then we reconstruct the value of y from the results $\tilde{x}_i$ of measuring x_i. In other words, we have an algorithm f that takes the values $\tilde{x}_i$ and returns an estimate $\tilde{y} = f(\tilde{x}_1, \ldots, \tilde{x}_n)$. This estimate is called a *result of indirect measurement.* And here comes the problem: *to estimate the error of this estimate.* For example, in case we know the accuracies Δ_i with which we measured x_i (i.e., if we know the intervals $\mathbf{x}_i = [\tilde{x}_i - \Delta_i, \tilde{x}_i + \Delta_i]$ of possible values of x_i), then we would like to know the interval of possible values of y. This is the basic problem of interval computation with which the entire field started:

GIVEN:

- *a function* $f : R^n \to R$;
- n *intervals* $\mathbf{x}_1, \ldots, \mathbf{x}_n$.

COMPUTE:

the range

$$\mathbf{f}^u(\mathbf{x}_1, \ldots, \mathbf{x}_n) = \{f(x_1, \ldots, x_n) : x_1 \in \mathbf{x}_1, \ldots, x_n \in \mathbf{x}_n\}.$$

What does "compute" mean? For a continuous function f, the desired range is an interval, so, to compute a range means to compute the *endpoints* of this range interval, i.e., the values $\underline{y} = \inf \mathbf{f}^u(\mathbf{x}_1, \ldots, \mathbf{x}_n)$ and $\overline{y} = \sup \mathbf{f}^u(\mathbf{x}_1, \ldots, \mathbf{x}_n)$.

In the majority of the existing computers, only rational numbers are represented. It is therefore natural to assume that all the parameters in the definition of a function f are rational numbers:

- if f is a polynomial function, then f is a polynomial with rational coefficients;
- if f is a rational function, then it is a ratio of two polynomials with rational coefficients,

etc. For such functions, however, the (desired) endpoints of the resulting interval are not necessarily rational numbers. So, when we say that we want to compute these endpoints, we mean that we want, given any accuracy $\varepsilon > 0$, to be able to compute these endpoints with the given accuracy (i.e., to compute some numbers $\tilde{\underline{y}}$ and $\tilde{\overline{y}}$ that are ε–close to the desired endpoints).

If a function f is discontinuous for some x_i (e.g., if it is a rational function), then the bounds $\underline{y}$ and $\overline{y}$ may be infinite. In the majority of computers, there are upper bounds on the size of the numbers that can be represented. Therefore, to represent ∞, we can use the largest possible computer-represented number M (and we can use $-M$ to represent $-\infty$). In this situation, "to compute" each of the endpoints $\underline{y}$ or $\overline{y}$ means to be able, for any given $\varepsilon > 0$ and $M > 0$, to find a value $\tilde{\underline{y}}$ (correspondingly, $\tilde{\overline{y}}$) with the following properties:

- if $\underline{y} \geq M$, then $\tilde{\underline{y}} \geq M - \varepsilon$;
- if $\underline{y} \leq -M$, then $\tilde{\underline{y}} \leq -(M - \varepsilon)$;
- if $-M \leq \underline{y} \leq M$, then $|\tilde{\underline{y}} - \underline{y}| \leq \varepsilon$.

Interval computations are intractable. The basic problem of interval computations (formulated above) is known to be intractable (NP-hard) even for quadratic functions f (see previous chapters).

Probabilistic computational problem emerging from error estimation for indirect measurements: informal description. In many real-life situations, manufacturers of the measuring instruments provide us with the probabilities of different values of the error $\Delta x_i = \tilde{x}_i - x_i$.

The simplest (and the most frequent) case is when the errors Δx_i are assumed to be independent Gaussian random variables (see, e.g., Rabinovich [333]). To describe a Gaussian distribution, it is sufficient to describe the first two moments. The first moment (average) is usually assumed to be 0, because if for a given measurement instrument, the average is not 0, then we can *recalibrate* the measuring device to get rid of this *systematic* error. In this case, the second moment is simply the square of the standard deviation. So, we can assume that we know the standard deviation σ_i of each measurement error Δx_i.

The main fundamental motivation to use Gaussian distributions is that according to the central limit theorem, under reasonable assumptions, the distribution of the sum of several (N) independent small random variables tends to the Gaussian distribution as $N \to \infty$. Therefore, if we eliminate major error components in the measurement error, the resulting error Δx_i will be caused by the cumulative effect of many independent small factors, and hence, its distribution will be close to Gaussian (see, e.g., Wadsworth [422]).

If the errors Δx_i are independent Gaussian distributed random variables, then, for every box $\mathbf{B} = [a_1, b_1] \times \ldots \times [a_n, b_n] \subseteq R^n$, the probability that $(\Delta x_1, \ldots, \Delta x_n) \in \mathbf{B}$ is positive (this probability is positive for *any* random variable with unbounded joint density function). Therefore, in this Gaussian (probabilistic) case, for unbounded functions f, no *guaranteed* bounds for y are possible.

Instead, we would like to compute the probabilistic characteristics of the resulting random error

$$\Delta y = y - \tilde{y} = f(\tilde{x}_1 - \Delta x_1, \ldots, \tilde{x}_n - \Delta x_n) - f(\tilde{x}_1, \ldots, \tilde{x}_n).$$

The two simplest characteristics (the ones most widely used in engineering measurements; see Rabinovich [333]) are the first and the second moments, i.e., equivalently, the average $E(\Delta y)$ and the standard deviation $\sigma(\Delta y)$.

As a result, we arrive at the following problem:

Definition 26.1. *By a (E, σ)–probabilistic computational problem emerging from error estimation for indirect measurement* (or simply (E, σ)–*probabilistic computational problem, for short*), *we mean the following computational problem:*

GIVEN:

- *a function $f : R^n \to R$;*
- *n real numbers $\tilde{x}_1, \ldots, \tilde{x}_n$;*
- *n non-negative real numbers $\sigma_1, \ldots, \sigma_n$.*

COMPUTE:

the average value $E(\Delta y)$ and the standard deviation $\sigma(\Delta y)$ of $\Delta y = f(\tilde{x}_1 - \Delta x_1, \ldots, \tilde{x}_n - \Delta x_n) - f(\tilde{x}_1, \ldots, \tilde{x}_n)$, where Δx_i are independent random variables with Gaussian distribution, 0 average, and standard deviations σ_i.

How complicated is the (E, σ)–probabilistic computational problem? For a linear function $f(x_1, \ldots, x_n) = a_0 + a_1 x_1 + \ldots + a_n x_n$, both the interval and the (E, σ)–probabilistic computational problems can be explicitly solved:

- For an interval computation problem, the resulting interval for y is $\mathbf{y} = [\tilde{y} - \Delta, \tilde{y} + \Delta]$, where

$$\Delta = |a_1|\Delta_1 + \ldots + |a_n|\Delta_n.$$

- For a similar (E, σ)–probabilistic computational problem, the desired standard deviation $\sigma(\Delta y)$ is equal to

$$\sigma(\Delta y) = \sqrt{a_1^2 \sigma_1^2 + \ldots + a_n^2 \sigma_n^2}.$$

To compute Δ, we need n multiplications ($|a_i|\Delta_i$, $1 \le i \le n$) and $n - 1$ additions. To compute $\sigma^2(\Delta y)$, we need $3n$ multiplications ($a_i \cdot a_i \cdot \sigma_i \cdot \sigma_i$, $1 \le i \le n$). So, even if we discount the square root and the additions, the formula for $\sigma(\Delta y)$ requires more computation steps than the formula for Δ. Based on this simplest case, it may seem that our (E, σ)–probabilistic computational problem is more complicated than the corresponding interval computation problem. So, the natural question is: is this really so? In particular, is the (E, σ)–probabilistic computational problem (formulated above) intractable for polynomial functions f? And if it is not intractable, what if we enlarge the class of functions? These questions are answered in this chapter.

26.2. Main Results

Proposition 26.1. *There exists a polynomial-time algorithm that solves the (E,σ)−probabilistic computational problem (emerging from error estimation for indirect measurements) for polynomial functions f.*

Comment. In other words, for polynomial functions f, the (E,σ)−probabilistic computational problem described above is feasible. So, the interval computation problem emerging from error estimation for indirect measurements is harder than a similar (E,σ)−probabilistic computational problem. How harder? I.e., what operations can we add to polynomials and still retain feasibility of the (E,σ)−probabilistic computational problem?

Our *informal* answer to this question is: probably, none, because when we add the simplest possible operation, feasibility is lost.

The *precise* result can be formulated as follows: Polynomials can be described as functions obtained from x_i by applying addition, subtraction, and multiplication. A natural expansion is to add division, thus getting rational functions. Our second result is that already for rational functions, the (E,σ)−probabilistic computational problem is NP-hard.

Theorem 26.1. *For rational functions f, the (E,σ)−probabilistic computational problem (emerging from error estimation for indirect measurements) is intractable (NP-hard).*

The results of this chapter can be represented as a table:

$f(x_1,\ldots,x_n)$	Interval computations	(E,σ)-Probabilistic computations
Linear f	Linear time (slightly faster)	Linear time (slightly slower)
Quadratic f	NP-hard	Polynomial time
Polynomial f	NP-hard	Polynomial time
Rational f	NP-hard	NP-hard

26.3. Open Problem: How Complicated Is It to Compute Other Probabilistic Characteristics of Indirect Measurements (In Particular, Confidence Intervals)?

It is desirable to compute other characteristics. The first two moments do not describe the probability distribution completely. Therefore, to describe the distribution of Δy, we must also describe other probabilistic characteristics of Δy.

In particular, since in general, no *guaranteed* bounds for y are possible, it is natural to looks for an interval that contains y with a certain *probability* p_0 (close to 1), i.e., for a *confidence interval* for $f(x_1, \ldots, x_n)$.

For confidence intervals, traditional engineering methods do not give guaranteed estimates. In traditional engineering practice (see, e.g., Rabinovich [333]), confidence intervals for the result y of indirect measurement are estimated from the average $E(\Delta y)$ and the standard deviation $\sigma(\Delta y)$ on the assumption that the distribution of y is Gaussian (for Gaussian distribution, simple formulas are known for computing confidence intervals).

The actual probability distribution for Δy may be close to Gaussian, but for non-linear functions f, it is (in general) not exactly Gaussian. Therefore, these engineering estimates are only approximately true, and they do not *guarantee* that y belongs to the constructed interval with the desired probability p_0.

Chebyshev's inequalities give a guaranteed but too wide enclosure for the confidence interval. To get guaranteed bounds, commonly, the well known Chebyshev's theorem is used; this theorem states that for *any* random variable Δy, and for any positive real number k, the inequality

$$E(\Delta y) - k \cdot \sigma(\Delta y) \leq \Delta y \leq E(\Delta y) - k \cdot \sigma(\Delta y)$$

is true with probability $\geq 1 - 1/k^2$.

Chebyshev's inequality leads to a guaranteed interval, but for the typical case when the distribution of Δy *is* close to the Gaussian distribution, the resulting interval is often *too wide* (i.e., much wider than the smallest possible interval in which the measurement error Δy is located with a given probability). In other words, the interval that stems from the Chebyshev's inequality only gives a crude *enclosure* for the desired confidence interval.

It is desirable to compute the confidence interval exactly. Open problem. We, therefore, arrive at the following problem of estimating the confidence interval *exactly*:

GIVEN:

- *a function* $f : R^n \to R$;
- n *real numbers* $\tilde{x}_1, \ldots, \tilde{x}_n$;
- n *non-negative real numbers* $\sigma_1, \ldots, \sigma_n$;
- *a rational number* p_0 *from the open interval* $(0, 1)$.

COMPUTE:

an interval $\mathbf{y}$ *for which:*

- *the probability that* $y = f(\tilde{x}_1 - \Delta x_1, \ldots, \tilde{x}_n - \Delta x_n) \in \mathbf{y}$ *is* $\geq p_0$, *where* Δx_i *are independent random variables with Gaussian distribution, 0 average, and standard deviations* σ_i; *and*
- *no proper subinterval* $\mathbf{y}' \subset \mathbf{y}$ *has this property.*

Whether this probabilistic problem is feasible or not for polynomial and/or for rational functions f is an interesting *open problem.*

Proofs

Proof of Proposition 26.1. Let f be a polynomial, i.e.,

$$f(x_1, \ldots, x_n) = \sum c_{i_1,\ldots,i_n} x_1{}^{i_1} \ldots x_n{}^{i_n}.$$

Then,

$$\Delta y = f(\tilde{x}_1 - \Delta x_1, \ldots, \tilde{x}_n - \Delta x_n) - f(\tilde{x}_1, \ldots, \tilde{x}_n)$$

is also a polynomial. Our goal is to find $E[\Delta y]$ and $\sigma[\Delta y]$.

By definition, $\sigma[\alpha] = \sqrt{E[(\alpha - E(\alpha))^2]}$. So, $\sigma[\Delta y] = \sqrt{E[(\Delta y - E(\Delta y))^2]}$. Since Δy is a polynomial, the expression $(\Delta y - E(\Delta y))^2$ is also a polynomial ($E(\Delta y)$ is a constant). Therefore, if we can prove that computing $E[P]$ for an arbitrary polynomial P is feasible, then, as a result, we will be able to conclude that computing $\sigma[P]$ is feasible as well.

So, it is sufficient to show how to compute $E[P]$ for an arbitrary polynomial

$$P(\Delta x_1, \ldots, \Delta x_n) = \sum a_{i_1,\ldots,i_n} \Delta x_1{}^{i_1} \ldots \Delta x_n{}^{i_n}.$$

Since mathematical expectation $E[\alpha]$ is a linear operator, we conclude that

$$E[P] = E[\sum a_{i_1,\ldots,i_n} \Delta x_1{}^{i_1} \ldots \Delta x_n{}^{i_n}] = \sum a_{i_1,\ldots,i_n} E[\Delta x_1{}^{i_1} \ldots \Delta x_n{}^{i_n}].$$

Due to our assumption that the random variables Δx_i are independent, we have $E[\Delta x_1{}^{i_1} \ldots \Delta x_n{}^{i_n}] = E[\Delta x_1{}^{i_1}] \cdot \ldots \cdot E[\Delta x_n{}^{i_n}]$. Therefore,

$$E[P] = \sum a_{i_1,\ldots,i_n} E[\Delta x_1{}^{i_1}] \cdot \ldots \cdot E[\Delta x_n{}^{i_n}].$$

So, to compute $E[P]$, it is sufficient to be able to compute the values $E[\Delta x_j{}^k]$.

We know that Δx_i is distributed according to Gaussian distribution, i.e., with density

$$\rho_i(x) = \frac{1}{\sqrt{2\pi}\sigma_i} \exp\left(-\frac{x^2}{2\sigma_i^2}\right).$$

In terms of ρ,

$$E[(\Delta x_i)^k] = \int_{-\infty}^{\infty} x^k \rho_i(x) dx.$$

After we substitute the expression for ρ_i into this integral, we will get

$$E[(\Delta x_i)^k] = \frac{1}{\sqrt{2\pi} \cdot \sigma_i} \int_{-\infty}^{\infty} x^k \exp\left(-\frac{x^2}{2\sigma_i^2}\right) dx.$$

For odd k, we have an integral of an odd function over a symmetric interval, i.e., 0. For even $k = 2m$, we have an integral of an even function over a symmetric interval $(-\infty, \infty)$. This integral can be computed as two times the integral over $[0, \infty)$, and this integral is known (see, e.g., Beyer [43], formula 666):

$$\int_0^{\infty} x^{2m} \exp(-ax^2) dx = \frac{1 \cdot 3 \cdot 5 \cdot \ldots \cdot (2m-1)}{2^{m+1} a^m} \sqrt{\frac{\pi}{a}}.$$

We multiply this integral by 2, so we will get 2^m instead of 2^{m+1} in the denominator. Here, $a = 1/(2\sigma_i)^2$, so

$$\frac{1}{2^m a^m \sqrt{a}} = \frac{(2\sigma_i^2)^m \sqrt{2} \cdot \sigma_i}{2^m} = \frac{2^m \sigma_i^{2m+1} \sqrt{2}}{2^m} = \sigma_i^{2m+1} \sqrt{2}.$$

Hence,

$$E[(\Delta x_i)^{2m}] = \frac{1}{\sqrt{2\pi} \cdot \sigma_i} \cdot 1 \cdot 3 \cdot 5 \cdot \ldots \cdot (2m-1) \cdot \sigma_i^{2m+1} \sqrt{2} \sqrt{\pi}.$$

Cancelling equal terms in the numerator and the denominator, we arrive at the formula

$$E[(\Delta x_i)^{2m}] = 1 \cdot 3 \cdot 5 \cdot \ldots \cdot (2m-1) \cdot \sigma_i^{2m}.$$

This is an easily computable expression: it consists of the elementary computational steps (multiplications) whose total number is linear in m, and, therefore, limited by a linear function of the length of the description of a polynomial. The total number of such terms is also bounded by the total length of the description of a polynomial. Therefore, we can conclude that computing $E[P]$ is feasible (computable in quadratic time). Proposition is proven.

Proof of Theorem 26.1. To prove the theorem, we will reduce satisfiability problem for 3-CNF formulas to our problem. Let us assign to every 3-CNF formula $F = D_1 \& \ldots \& D_m$ with n Boolean variables $z_1, \ldots, z_n$ (and disjunctions D_j of the type $a \vee b$ or $a \vee b \vee c$, where a, b, c are variables or their negations), a rational function $f(x_1, \ldots, x_n)$ of n real variables $x_1, \ldots, x_n$. This assignment will be done in several steps:

- First, to every Boolean variable z_i, we put into correspondence an expression $C[z_i] = x_i^2$ (C stands for "correspondence").
- To every negative literal $\neg z_i$, we assign an expression $C[\neg z_i] = (1 - x_i)^2$.
- To every disjunction $D = a \vee b$, we assign an expression

  $$C[D] = C(a) \cdot C(b).$$

- To every disjunction $D = a \vee b \vee c$, we assign an expression

  $$C[D] = C(a) \cdot C(b) \cdot C(c).$$

- To the formula $F = D_1 \& \ldots \& D_m$, we assign an expression

  $$C[F] = C(D_1) + \ldots + C(D_m).$$

- Finally, we take $f = (C[F])^{-p}$, where p is an arbitrary integer that is greater than $n/2$.

We take $\tilde{x}_1 = \ldots = \tilde{x}_n = 0.5$, and $\sigma_1 = \ldots = \sigma_n = 0.5$.

For this function $f(x_1, \ldots, x_n)$, we are interested in computing $E[\Delta y]$, where

$$\Delta y = f(\tilde{x}_1 - \Delta x_1, \ldots, \tilde{x}_n - \Delta x_n) - f(\tilde{x}_1, \ldots, \tilde{x}_n),$$

and Δx_i are independent Gaussian random variables with 0 average and standard deviation σ_i.

We will prove the following two statements:

(a) *If a formula F is not satisfiable, then* $0 \leq |E[\Delta y]| \leq 64^p$.

(b) *If a formula F is satisfiable, then* $E[\Delta y] = \infty$.

So, if there exists an algorithm $\mathcal{U}$ that solves the (E, σ)−probabilistic computational problem for rational functions f, then for every 3-CNF formula F, by applying $\mathcal{U}$ to the above rational function f, we will be able to check whether F is satisfiable or not. Since this checking is known to be NP-hard, (E, σ)−probabilistic computational problem for rational functions f is also NP-hard.

To complete our proof, we must prove statements (a) and (b).

Proof of (a).

1. First, we will prove that if F is not satisfiable, then $C[F] \geq 1/64$ for all values of x_i. This statement is equivalent to the following one: "if the infimum $\inf C[F]$ of $C[F]$ over all $x_1, \ldots, x_n$ is smaller than 1/64, then a formula F is satisfiable". Let us prove it.

Since the infimum of $C[F]$ is smaller than 1/64, there exist values $x_1, \ldots, x_n$, for which $C[F](x_1, \ldots, x_n) < 1/64$. Let us take the following Boolean vector: for every i,

- $z_i =$"true" if $x_i < 1/2$, and
- $z_i =$"false" otherwise.

We will show that these Boolean values satisfy the formula F. Since $F = D_1 \& \ldots \& D_m$, it is sufficient to prove that all disjunctions D_j are satisfied by these Boolean variables z_i. Indeed, since $C[F] = C[D_1] + \ldots + C[D_n] < 1/64$, and $C[D_j] \geq 0$, we can conclude that for all j, $C[D_j] < 1/64$.

Let us now show that for every disjunction D_j, at least for one of its literals a, we have $C[a] < 1/4$. The proof will slightly differ depending on how many literals are there in this disjunction:

- If $D_j = a$, then $C[a] = C[D_i] < 1/64 < 1/4$.
- If $D_j = a \vee b$, then $C[D_i] = C[a] \cdot C[b] < 1/64$. Therefore, either $C[a] < 1/4$, or $C[b] < 1/4$, because otherwise, from $C[a] \geq 1/4$ and $C[b] \geq 1/4$, we would conclude that $C[a] \cdot C[b] \geq 1/16 > 1/64$.
- If $D_j = a \vee b \vee c$, then $C[D_i] = C[a] \cdot C[b] \cdot C[c] < 1/64$. Therefore, either $C[a] < 1/4$, or $C[b] < 1/4$, or $C[c] < 1/4$, because otherwise, from $C[a] \geq 1/4$, $C[b] \geq 1/4$, and $C[c] \geq 1/4$, we would conclude that $C[a] \cdot C[b] \geq C[c] \geq 1/64$.

Now, to prove that D_j is satisfied, we must consider two possible cases:

- $a = z_i$. In this case, $C[a] = C[z_i] = x_i^2$, so, from $C[a] < 1/4$, we conclude that $x_i < 1/2$. Therefore, z_i is "true", so a is true, and the entire disjunction $D_j = a \vee \ldots$ is satisfied.
- $a = \neg z_i$. In this case, $C[a] = (1 - x_i)^2 < 1/4$ hence, $1 - x_i < 1/2$ and so $x_i > 1/2$. Therefore, z_i is "false", $a = \neg z_i$ is "true", and the entire disjunction $D_j = a \vee \ldots$ is satisfied.

In all cases, all disjunctions D_j are satisfied and therefore, F is also satisfied. So, we have proved that if the infimum $\inf C[F]$ of $C[F]$ over all $x_1, \ldots, x_n$ is smaller than $1/64$, then a formula F is satisfiable.

2. Now, we can complete the proof of the statement (a). Indeed, if a formula F is not satisfiable, then, as we have proved in Part 1. of this proof, $C[F] \geq 1/64$. Therefore, $f = (C[F])^{-p} \leq (1/64)^{-p} = 64^p$. On the other hand, f is a positive function, so $0 \leq f(x_1, \ldots, x_n) \leq 64^p$ for all possible values $x_1, \ldots, x_n$.

By definition, $E[\Delta y] = E[y] - \tilde{y}$, where $\tilde{y} = f(\tilde{x}_1, \ldots, \tilde{x}_n)$, and

$$y = f(\tilde{x}_1 - \Delta x_1, \ldots, \tilde{x}_n - \Delta x_n).$$

From the inequality $0 \leq f \leq 64^p$, we can conclude that $0 \leq \tilde{y} \leq 64$, and that $0 \leq E[y] \leq 64^p$. If two numbers belong to an interval $[0, t]$, then the largest possible difference between them is t. Therefore, $|E[\Delta y]| = |E[y] - \tilde{y}| \leq 64^p$. Statement (a) is proved.

Proof of (b).

1. Let us first do some preliminary simplification. Suppose that a formula F is satisfiable. This means that it has a Boolean vector that satisfies it. Let us

fix one of these Boolean vectors $z_1, \ldots, z_n$. To simplify further computations, we will rename the variables:

- if z_i is true, then we take $t_i = z_i$;
- if z_i is false, then we take $t_i = \neg z_i$.

After this renaming, we get a new formula $G = D'_1 \& \ldots \& D'_m$ with new Boolean variables $t_1, \ldots, t_n$ that is satisfied by the values $t_1 = \ldots = t_n =$"true".

We can apply our procedure C to this new formula G, and get a polynomial $C[G]$ of n variables $u_1, \ldots, u_n$. The relationship between $C[G](u_1, \ldots, u_n)$ and $C[F](x_1, \ldots, x_n)$ is easy to trace: if we take

- $u_i = x_i$ if z_i is true, and
- $u_i = 1 - x_i$ if z_i is false,

then $C[G](u_1, \ldots, u_n) = C[F](x_1, \ldots, x_n)$.

Because of our choice of $\tilde{x}_i$, in terms of new variables, we will have either $\tilde{u}_i = 0.5$, or $\tilde{u}_i = 1 - \tilde{x}_i = 1 - 0.5 = 0.5$, i.e., $\tilde{u}_i = 0.5$ for all i. Therefore, the (E, σ)–probabilistic computational problem for $f = C[F]^{-p}$, $\tilde{x}_i = 0.5$ and $\sigma_i = 0.5$, is equivalent to the (E, σ)–probabilistic computational problem for $g = C[G]^{-p}$, with $\tilde{u}_i = 0.5$, and $\sigma_i = 0.5$.

2. To prove statement (a), we estimated $C[D_j]$ by a constant. To prove (b), we will need a better estimate. We will get such an estimate only for the case when $|u_i| \leq 1$ for all i. In this case:

- for every positive literal $a = t_i$, $C[a] = u_i^2 \leq 1$;
- and for every negative literal $a = \neg t_i$, $C[a] = (1 - u_i)^2$. Since $|u_i| \leq 1$, we can conclude that $|1 - u_i| \leq 1 + |u_i| \leq 2$, and $C[a] \leq 4$.

In both cases, $C[a] \leq 4$.

Let D'_j be an arbitrary disjunction from G. Since G is satisfied by the values $t_1 = \ldots = t_n =$"true", we can conclude that D'_j is also satisfied by these

Boolean values and therefore, one of its literals is positive. Without losing generality, we can denote this literal by $a = t_i$. For this literal, $C[a] = u_i^2 \leq S$, where we denoted

$$S = \sum_{i=1}^{n} u_i^2.$$

Depending on the number of literals in D'_j, we have three cases:

- $D'_j = a$. In this case, $C[D'_j] = C[a] \leq S$.
- $D'_j = a \vee b$. In this case, $C[D'_j] = C[a] \cdot C[b]$. Since $C[a] \leq S$, and $C[b] \leq 4$, we conclude that $C[D'_j] \leq 4S$.
- $D'_j = a \vee b \vee c$. In this case, $C[D'_j] = C[a] \cdot C[b] \cdot C[c]$. Since $C[a] \leq S$, $C[b] \leq 4$, and $C[c] \leq 4$, we conclude that $C[D'_j] \leq 16S$.

In all three cases, we have $C[D'_j] \leq 16S$.

Adding the inequalities for all disjunctions, we conclude that $C[G] = C[D'_1] + \ldots + C[D'_m] \leq 16mS$. Therefore, when $|u_j| \leq 1$, we have $C[G] \leq 16mS$. In particular, this inequality is true when $S \leq 1$, because then, for every j,

$$u_j^2 \leq \sum_{i=1}^{n} u_i^2 = S \leq 1,$$

and hence, $|u_i| \leq 1$. So, when $S \leq 1$, we have $C[G] \leq 16mS$. In this case, the function $g = C[G]^{-p}$ to which we apply (E, σ)–probabilistic computational problem, satisfies the inequality $g = C[G]^{-p} \geq (16m)^{-p} S^{-p}$.

We want to compute $E[\Delta y]$, where

$$\Delta y = g(\tilde{u}_1 - \Delta u_1, \ldots, \tilde{u}_n - \Delta u_n) - g(\tilde{u}_1, \ldots, \tilde{u}_n),$$

and Δu_i are Gaussian independent variables with average 0 and standard deviation σ_i. Therefore, $E[\Delta y] = E[y] - \tilde{y}$, where we denoted

$$y = g(\tilde{u}_1 - \Delta u_1, \ldots, \tilde{u}_n - \Delta u_n)$$

and $\tilde{y} = g(\tilde{u}_1, \ldots, \tilde{u}_n)$. Since $\tilde{y}$ is a finite number, in order to prove that $E[\Delta y] = \infty$, we must prove that $E[y] = \infty$. By definition,

$$E[y] = \int \ldots \int \rho(u_1, \ldots, u_n) \cdot g(u_1, \ldots, u_n) du_1 \ldots du_n,$$

where the integral is taken over the entire n–dimensional space, and

$$\rho(u_1, \ldots, u_n) = \prod_{i=1}^{n} \rho_i(u_i),$$

where

$$\rho_i(u_i) = \frac{1}{\sqrt{2\pi} \cdot \sigma_i} \exp\left(-\frac{(u_i - 0.5)^2}{2\sigma_i^2}\right),$$

and ρ is the probability density. Since g is non-negative, we can conclude that

$$E[y] = \int \ldots \int \rho(u_1, \ldots, u_n) \cdot g(u_1, \ldots, u_n) du_1 \ldots du_n \geq I_1,$$

where we denoted

$$I_1 = \int \ldots \int_B \rho(u_1, \ldots, u_n) \cdot g(u_1, \ldots, u_n) du_1 \ldots du_n,$$

and B is the set of all values $(u_1, .., u_n)$ for which $S \leq 1$. So, to prove that $E[\Delta y] = \infty$, it is sufficient to prove that $I_1 = \infty$.

To prove that, let us estimate ρ. We have chosen $\sigma_i = 0.5$, so $2\sigma_i^2 = 0.5$, $1/(2\sigma_i^2) = 2$, and $1/(\sqrt{2\pi} \cdot \sigma_i) = 2/(\sqrt{2\pi}) = \sqrt{2/\pi}$. For $(u_1, \ldots, u_n) \in B$, we have $|u_i| \leq 1$, therefore, we have $|u_i - 0.5| \leq |u_i| + 0.5 \leq 1 + 0.5 = 1.5$, $(u_i - 0.5)^2 \leq 2.25$, and

$$\frac{(u_i - 0.5)^2}{2\sigma_i^2} = 2(y - 0.5)^2 \leq 4.5.$$

Therefore,

$$\exp\left(-\frac{(u_i - 0.5)^2}{2\sigma_i^2}\right) \geq \exp(-4.5),$$

and

$$\rho_i(u_i) = \frac{1}{\sqrt{2\pi} \cdot \sigma_i} \exp\left(-\frac{(u_i - 0.5)^2}{2\sigma_i^2}\right) \geq \sqrt{\frac{2}{\pi}} \exp(-4.5) > 0.$$

If we denote

$$\sqrt{\frac{2}{\pi}} \exp(-4.5)$$

by ρ_0, we can describe the resulting inequality as $\rho_i(u_i) \geq \rho_0 > 0$. Hence, the function ρ that is the product of m such terms, is bounded from below by a positive number ρ_0^m. Therefore,

$$I_1 = \int \ldots \int_B \rho(u_1, \ldots, u_n) \cdot g(u_1, \ldots, u_n) du_1 \ldots du_n \geq$$

$$\rho_o^m \cdot \int \ldots \int_B g(u_1, \ldots, u_n) du_1 \ldots du_n.$$

So, to prove that I_1 is infinite, it is sufficient to prove that the integral

$$I_2 = \int \ldots \int_B g(u_1, \ldots, u_n) du_1 \ldots du_n$$

is infinite.

For $(u_1, \ldots, u_n) \in B$, we have $g(u_1, \ldots, u_n) \geq (16m)^{-p} S^{-p}$. Therefore,

$$I_2 = \int \ldots \int_B g(u_1, \ldots, u_n) du_1 \ldots du_n \geq (16m)^{-p} I_3,$$

where we denoted

$$I_3 = \int \ldots \int_B S^{-p} du_1 \ldots du_n.$$

So, to prove that I_2 is infinite, it is sufficient to prove that the integral I_3 is infinite. But this integral can be easily expressed in spherical coordinates $(r, \theta_1, \theta_2, \ldots \theta_{n-1})$, where $r = \sqrt{u_1^2 + \ldots + u_n^2}$ and θ_i are angles such that

$$u_1 = r \cdot \cos(\theta_1),$$

$$u_2 = r \cdot \sin(\theta_1) \cdot \cos(\theta_2),$$

$$\ldots$$

$$u_i = r \cdot \sin(\theta_1) \cdot \ldots \cdot \sin(\theta_{i-1}) \cdot \cos(\theta_i),$$

$$\ldots$$

$$u_{n-1} = r \cdot \sin(\theta_1) \cdot \ldots \cdot \sin(\theta_{n-2}) \cos(\theta_{n-1}),$$

$$u_n = r \cdot \sin(\theta_1) \cdot \ldots \cdot \sin(\theta_{n-2}) \cdot \sin(\theta_{n-1}).$$

Indeed, the area B is spherically symmetric (it is actually the ball of radius 1), the integrand is symmetric because $S = r^2$, $du_1 \ldots du_n = A(\theta_1, \ldots, \theta_n) r^{n-1} dr d\theta_1 \cdot \ldots \cdot d\theta_{n-1}$ for some expression $A(\theta_1, \ldots \theta_n)$. So, when we integrate over $\theta_1, \ldots, \theta_n$, the integral takes the form

$$I_3 = \text{const} \cdot \int_0^1 \frac{r^{n-1} dr}{r^{2p}} = \text{const} \cdot \int_0^1 r^{-2p+n-1} dr =$$

$$\left(\frac{r^{-2p+n}}{-2p+n} \right) |_{r=1} - \left(\frac{r^{-2p+n}}{-2p+n} \right) |_{r=0}.$$

Since we have chosen $p > n/2$, this expression is infinite (for $r = 0$), so $I_3 = \infty$, and hence, $I_2 = \infty$, $I_1 = \infty$, $E[y] = \infty$, and $E[\Delta y] = \infty$. The statement (b) is proved, and so is the theorem.

A

IN CASE OF INTERVAL (OR MORE GENERAL) UNCERTAINTY, NO ALGORITHM CAN CHOOSE THE SIMPLEST REPRESENTATIVE

When we only know the interval of possible values of a certain quantity (or a more general set of possible values), it is desirable to characterize this interval by supplying the user with the "simplest" element from this interval, and by characterizing how different from this value we can get. For example, if, for some unknown physical quantity x, measurements result in the interval $[1.95, 2.1]$ of possible values, then, most probably, the physicist will publish this result as $y \approx 2$. Similarly, a natural representation of the measurement result $x \in [3.141592, 3.141593]$ is $x \approx \pi$.

In this appendix, we show that the problem of choosing the simplest element from a given interval (or from a given set) is, in general, not algorithmically solvable.

The results presented in this appendix first appeared in Heindl et al. [144].

A.1. In Case of Interval (or More General Set) Uncertainty, a User Would Like to Have a Representative Value from This Interval (Set)

The value of a physical quantity y is usually obtained either by a direct measurement, or by an *indirect measurement*, i.e., by processing the results of some related measurements $x_1, \ldots, x_n$. Since measurements are normally not 100% precise, their results may differ from the actual values of the measured quantities. As a result, after the measurement (direct or indirect), we do *not* get the *exact value* of the desired quantity, we only get a *set* Y of its possible values.

In many cases, this set Y is an interval, but more complicated sets are also possible: e.g., if we know that $y^2 = x_1$, and that $x_1 \in [1,4]$, then the set of possible values of y is the union of two intervals $[-2,-1] \cup [1,2]$.

For each set, it would be very convenient for the users of this information, if we could select a *representative* element of this set and describe the possible deviations from this value.

For example, if the set of possible values is the interval $[3,5]$, it is natural to take the midpoint of this interval (i.e., the number 4) as the desired representative, and describe the possible positive and negative deviations from 4 as 4^{+1}_{-1}.

For more complicated intervals, a midpoint may not be the best choice. For example, for an interval $[1.95, 2.1]$, the natural representative is 2, so the natural representation of this interval is $2^{+0.1}_{-0.05}$. By a "natural representation" we mean that if, say, a physicist tries to measure an unknown quantity y, and as a result of the measurement, he gets the interval $[1.95, 2.1]$ of possible values, then, most probably, he will publish this result as $y \approx 2$. The reason for choosing 2 is that the hypothesis $y = 2$ seems to be the *simplest possible hypothesis*, i.e., 2 *seems to be the simplest possible number from this interval.*

This "simplest" number is not always an integer or a rational number: e.g., for an interval $[3.141592, 3.141593]$, the natural representative is, most likely, π.

The natural question is: is it possible to design an algorithm that would, given a set of real numbers, choose its simplest element?

A.2. How to Formalize "The Simplest"?

To answer this question, we must first define what "simple" means. Intuitively, a number is simple if it is easy to describe; in other words, a number is simple if the length of its description is small. It is natural to use this intuitive idea to give a precise definition.

Traditionally, foundations of mathematics are based on set theory. So, in principle, we can fix some standard version of set theory (e.g., Zermelo-Fraenkel theory ZF), and consider definitions within this theory. Such a formalization would have made the definitions (and maybe proofs) slightly shorter. However, these shorter-to-prove results will only apply to ZF. What is we use another version of set theory? What if we use alternative foundations of mathematics (e.g., based on categories instead of sets)? We will see that our results do not depend on which version of set theory we choose, and do not even depend on whether we use set theory or any other formalization of mathematics. To convey this generality, we will formulate our results in the most general form.

We want to be able to have variables that run over real numbers (i.e., whose possible values are real numbers), variables that run over integers, and maybe some other types of variables (e.g., variables that run over intervals, and/or variables that run over arbitrary sets). So, the natural language to use is ***multi-sorted first order logic***. For the convenience of the readers who may not be well familiar with this notion, let us give sketchy definitions here; readers who are interested in technical details can look, e.g., in Barwise [19], Enderton [101], and Schoenfield [380].

Definition A.1.

- *Let a finite set A be fixed. This set will be called an **alphabet**, and elements of this set will be called **symbols**. We assume that this set does not contain symbols (,), &, $\vee$, $\neg$, $\rightarrow$, $\forall$, $\exists$, and symbols with subscripts.*
- *By a **multi-sorted first order language**, we mean the tuple $L = (\mathcal{S}, \mathcal{P}, \mathcal{F}, ar)$, where*
  - *$\mathcal{S}$, $\mathcal{P}$, and $\mathcal{F}$ are subsets of the set A that have no common elements (i.e., $\mathcal{S} \cap \mathcal{P} = \mathcal{S} \cap \mathcal{F} = \mathcal{P} \cap \mathcal{F} = \emptyset$);*
    - *elements of the set $\mathcal{S}$ will be called **sorts**;*
    - *elements of the set $\mathcal{P}$ will be called **predicate symbols**;*
    - *elements of the set $\mathcal{F}$ will be called **function symbols**.*
  - *ar is a function that transforms every element from the set $\mathcal{P} \cup \mathcal{F}$ into a non-empty finite sequence of sorts (i.e., of elements of $\mathcal{S}$).*
  - *for every predicate symbol $P \in \mathcal{P}$, the number of elements in a sequence $ar(P)$ is called the **arity** of this predicate; if this number is 1, the predicate is called **unary**; if it is 2, the predicate is called **binary**, etc.;*
  - *for every function symbol $f \in \mathcal{F}$, its **arity** is defined as the number of elements in $ar(f)$ minus 1; the last symbol is the sequence $ar(f)$ is called its **output type**; 0-ary functions are called **constants** of this output type.*

- *Let $s \in \mathcal{S}$ be a sort. By a variable of sort s, we mean an expression of the type x_n^s, where n is a natural number.*
- *The notion of the term and its type is defined as follows:*
  - *every variable x_n^s is a term of type s;*
  - *if $t_1, \ldots, t_m$ are terms of types $s_1, \ldots, s_m$, and if $f \in \mathcal{F}$ is a function symbol for whom $ar(f) = s_1 \ldots s_m s$, then the expression $f(t_1, \ldots, t_m)$ is a term of type s.*
- *The notion of an elementary formula is defined as follows: If $t_1, \ldots, t_m$ are terms of types $s_1, \ldots, s_m$, and $P \in \mathcal{P}$ is a predicate for which $ar(P) = s_1 \ldots s_m$, then $P(t_1, \ldots, t_m)$ is an elementary formula.*
- *The notion of a formula is defined as follows:*
  - *Every elementary formula is a formula.*
  - *If F and G are formulas, then the expressions (F), $F \& G$, $F \vee G$, $\neg F$, and $F \rightarrow G$ are formulas.*
  - *If F is a formula and v is a variable, then expressions $\forall v F$ and $\exists v F$ are formulas.*

In a standard manner, we can now define *closed* formulas, formulas with one free variable, etc.

Comment. In the following text, we will consider languages in which the list of sorts $\mathcal{S}$ contains two symbols: "integer" and "real", and which contain standard arithmetic predicates and function symbols such as 0, 1, +, −, ·, /, =, <, ≤, both for integers and for reals.

For reader's convenience:

- we will denote variables that run over real numbers by $x, y, \ldots$ (instead of $x_1^{\text{"real"}}, x_2^{\text{"real"}}, \ldots$), and, correspondingly, variables that run over natural numbers by $m, n, \ldots$; and
- for terms containing standard functions like $+$, we will use traditional notations $x + y$ (with a function symbol inside the expression) instead of a more precise expression $+(x, y)$ following from Definition A.1.

Definition A.2. *Let a (multi-sorted first order) language L be fixed. By a theory T, we mean a finite set of closed formulas.*

Comment. In the following text, we will always assume that a theory T is *consistent.*

In a standard (and natural) manner, we can now define the notion of an *interpretation* of a language L, in which, crudely speaking:

- to every sort $s \in \mathcal{S}$, we assign a set (whose elements will be called objects of this sort);
- to every predicate symbol, we assign a predicate;
- to every function symbol, a corresponding function.

For each interpretation, we can then interpret terms and formulas, and we say that an interpretation is a *model* of the theory T if all formulas from T are true in this interpretation.

We say that a formula F is *deducible* from the theory T (and denote it by $T \vdash F$) if this formula F is true in every model of the theory T.

Comment. In the following text, we will assume that a theory T is fixed. We will assume that this theory contains both the standard first order theory of integers (Peano arithmetic (Barwise [19], Enderton [101], Schoenfiled [380]) and a standard first order theory of real numbers (Tarksi [408], Seidenberg [388], Ben-Or *et al.* [31], Canny [60]).

As we have already mentioned, one of the possibilities is to consider, as the theory T, axiomatic set theory (e.g., ZF), together with explicit definitions of integers, real numbers, and standard operations and predicates in terms of set theory. In this case, the set of sorts consists of three elements: "integer", "real", and "set". However, as we have also mentioned, our definitions and results apply to other theories as well.

Now, we are ready to define "definability".

Definition A.3. *Let a language L (whose set of sorts includes the sort of real numbers) and a theory T be fixed. By a **definable real number**, we mean a formula $F(y)$ with one free variable for real numbers for which*

$$T \vdash \exists y F(y) \& \forall x \forall y (F(x) \& F(y) \to x = y).$$

*We will also say that a formula $F(y)$ **defines a real number**.*

Comment. Informally, a real number x_0 is definable if there exists a formula $F(x)$ that is true for this real number x_0 that is not true for any other real number $x \neq x_0$.

Examples.

- A formula $y \cdot y = 1 + 1 \& y \geq 0$ satisfies the Definition A.3; thus, it defines a unique real number ($\sqrt{2}$).
- A formula $\forall x(x \cdot y = x + x + x)$ defines a real number 3.
- If the language L contains all symbols for standard mathematical functions, including the sine function $\sin(x)$, and if the theory T contains ZF + definitions of these mathematical functions in terms of set theory, then the formula $\sin(y) = 0 \ \& \ 3 \leq y \leq 4$ defines a real number: actually, this number is π.

Definition A.4. *Let $F(y)$ and $F'(y)$ be definable real numbers. We say that they define the same real number if*

$$T \vdash \forall x \forall y (F(x) \& F'(y) \rightarrow x = y).$$

We say that $F(y)$ and $F'(y)$ define different numbers if

$$T \vdash \forall x \forall y (F(x) \& F'(y) \rightarrow x \neq y).$$

Definition A.5.

- *By a length of a formula F, we mean its total length that is counted as follows:*
  - *every symbol from the alphabet A, every parenthesis (,), and every logical symbol ($\&, \vee, \neg, \rightarrow, \forall, \exists$) is counted as one symbol;*
  - *every variable x_n^s is counted as 2 symbols + as many symbols as there are bits in the binary representation of the integer n.*
- *Let $F(y)$ be a definable real number. By its complexity $D(F)$, we mean the length of the shortest formula that defines the same real number.*

Comments.

- This definition is similar to the so-called *Kolmogorov complexity* $C(x)$ (invented independently by Chaitin, Kolmogorov, and Solomonoff), which is defined as the smallest length of the program that *computes* x (for a current survey on Kolmogorov complexity, see, e.g., Li *et al.* [254]). In our case, however, we do not care that much about how to compute: computing 3.141592 may be easier than computing π; we are more interested in how easy it is to *describe* x. Due to this difference, we had to modify Kolmogorov's definition.

- The above-defined complexity of a number depends on the theory. Correspondingly, the choice of the simplest number from an interval may also depend on the theory. For example, if this interval is $[0.0625, 0.15625]$, then we can have at least two different situations:

  - On one hand, if we are preparing a publication, then the simplest element of this interval is, most probably, 0.1, because 0.1 is, probably, the simplest possible decimal number on this interval.
  - On the other hand, if we must choose a number for a further computer processing, it makes more sense to choose a number 1/8 that has the simplest *binary* representation (0.001_2).

  These two different situations correspond to two different theories:

  - in the theory that correspond to the first situation, we allow decimal numbers as constants but not binary numbers;
  - in the theory that correspond to the second situation, we allow binary numbers as constants but not decimal numbers.

A.3. First Result: It Is Impossible to Algorithmically Choose the Simplest Element of a Finite Set

Comment. When we say that a real number is given, we mean that we are given a *formula* $F(y)$ that defines this number. So, the question becomes: suppose that we are given several numbers. Can we choose the one with the the smallest complexity? We will prove that the answer is negative even for the simplest sets that consist of two real numbers.

Definition A.6. *By the problem of choosing the simplest representative from a finite set, we mean the following problem:*

GIVEN:

- *an integer n, and*
- *a set of n definable real numbers, i.e., n formulas $F_1(y), \ldots, F_n(y)$ that define real numbers.*

FIND:

the value i for which the corresponding real number is the simplest, i.e., for which $D(F_i) = \min(D(F_1), \ldots, D(F_n))$.

Proposition A.1. *Even for $n = 2$, no algorithm is possible that, given a finite set with n elements, chooses the simplest representative from this set.*

A.4. Second Result: It Is Impossible to Algorithmically Choose the Simplest Element of an Interval

Definition A.7a. *By a definable interval, we mean a pair of formulas $\underline{F}(y)$ and $\overline{F}(y)$ that define real numbers and for which*

$$T \vdash \forall x \forall y (\underline{F}(x) \& \overline{F}(y) \rightarrow x \leq y).$$

Definition A.7b. *We say that a definable real number $F(y)$ belongs to the definable interval $[\underline{F}(y), \overline{F}(y)]$ if*

$$T \vdash \forall x \forall y \forall z (\underline{F}(x) \& F(z) \& \overline{F}(y) \rightarrow (x \leq z \& z \leq y).$$

Definition A.8. *By the problem of choosing the simplest representative from an interval, we mean the following problem:*

GIVEN:

a definable interval $[\underline{F}(y), \overline{F}(y)]$.

FIND:

- *the formula* $F(y)$ *that defines the simplest definable real number from the interval* $[\underline{F}(y), \overline{F}(y)]$*; and,*
- *in case one of the endpoints* $\underline{F}(y), \overline{F}(y)$ *is the simplest definable number on this interval, the value* $-$ *or* $+$ *indicating, correspondingly, whether the lower endpoint* $\underline{F}(y)$ *or the upper endpoint* $\overline{F}(y)$ *is the simplest.*

Proposition A.2. *No algorithm is possible that, given a definable interval, would return the simplest representative from this interval.*

Comment. The same impossibility result holds if we fix one of the endpoints. To be more precise, this result holds for *almost all* possible choices of the endpoint ("almost all" in some natural sense).

Proposition A.3.

- *If* $F(y)$ *is the simplest possible definable real number, then:*

  There exists an algorithm that, given any definable real number $F'(y)$ *that defines a different number, chooses the simplest representative from the corresponding interval* $[F(y), F'(y)]$ *or* $[F'(y), F(x)]$.

- *If* $F(y)$ *is not the simplest possible real number, then:*

  No algorithm is possible that, given any definable interval with $F(y)$ *as one of the endpoints, would choose the simplest representative from this interval.*

Definition A.9. *We say that a property* $\tilde{P}(x)$ *holds for almost all definable real numbers if there exists finitely many definable real numbers* $F_1(y), \ldots, F_n(y)$ *such that: if the definable number* $F(y)$ *is different from each of them, then the property* $\tilde{P}(x)$ *holds for the number that is defined by the formula* $F(y)$.

Comment. Informally, we can say that a property holds for almost all definable real numbers if it holds for all definable real numbers, except, maybe, finitely many of them.

Corollary. *For almost all definable real numbers $F(y)$, the following property holds:*

* *No algorithm is possible that, given a definable interval with $F(y)$ as one of its endpoints, would choose the simplest representative from this interval.*

Comment. Similar results are true if we restrict ourselves to intervals in which the given number $F(y)$ is the lower endpoint (or, correspondingly, the upper endpoint).

Proposition A.4.

- *For any definable real number $F(y)$, the following two properties are equivalent to each other:*
  - *The number defined by the formula $F(y)$ is the simplest of all definable real numbers that are $\geq$ that this number.*
  - *There exists an algorithm that, given any definable interval with $F(y)$ as its lower endpoint, chooses the simplest representative from this interval.*
- *For any definable real number $F(y)$, the following two properties are equivalent to each other:*
  - *The number defined by the formula $F(y)$ is the simplest of all definable real numbers that are $\leq$ that this number.*
  - *There exists an algorithm that, given any definable interval with $F(y)$ as its upper endpoint, chooses the simplest representative from this interval.*

A.5. Similar Results Hold for Computable Real Numbers

For the cases when the intervals (from which we are choosing the simplest numbers) come from computations, it is reasonable not to consider arbitrary *definable* real numbers, but to restrict ourselves to *computable* real numbers, i.e., real numbers that can be computed with an arbitrary accuracy (see, e.g., Bishop [48], Bridges [58], Beeson [26], Bishop *et al.* [49]):

Definition A.10. *A real number x is called constructive if there exists an algorithm (program) that transforms an arbitrary integer k into a rational number x_k that is 2^{-k}–close to x. It is said that this algorithm computes the real number x.*

Comment. Every constructive real number is uniquely determined by the corresponding algorithm and is, therefore, definable.

Comment. When we say that a constructive real number is given, we mean that we are given an algorithm that computes this real number.

Definition A.11. *By the problem of choosing the simplest representative from a constructive interval, we mean the following problem:*

> *GIVEN:*
>
> > *a constructive interval, i.e., algorithms $\underline{U}$ and $\overline{U}$ that compute real numbers $\underline{x} < \overline{x}$.*
>
> *FIND:*
>
> > *the simplest (in the sense of $D(F) \rightarrow \min$) constructive real number from the interval $[\underline{x}, \overline{x}]$.*

Proposition A.5. *No algorithm is possible that, given a constructive interval, returns the simplest representative from this interval.*

Proposition A.6.

- *If x is the simplest possible constructive real number, then:*

  There exists an algorithm that, given any other constructive real number $y \neq x$, chooses the simplest representative from the corresponding interval $[x, y]$ or $[y, x]$.

- *If x is not the simplest possible constructive real number, then:*

  No algorithm is possible that, given any other constructive real number $y \neq x$, would choose the simplest representative from the corresponding interval $[x, y]$ or $[y, x]$.

Proposition A.7.

- *For any constructive real number x, the following two properties are equivalent to each other:*
  - *The number x is the simplest of all constructive real numbers $\geq x$.*
  - *There exists an algorithm that, given any constructive real number $y > x$, chooses the simplest constructive real number from the interval $[x, y]$.*
- *For any definable real number x, the following two properties are equivalent to each other:*
  - *The number x is the simplest of all constructive real numbers $\leq x$.*
  - *There exists an algorithm that, given any constructive real number $y < x$, chooses the simplest constructive real number from the interval $[y, x]$.*

Proofs

General comment. The results of this appendix are mainly based on results from mathematical logic.

Proof of Proposition A.1. We will prove our result by reduction to a contradiction. Let us assume that there exists an algorithm U that for every two defining properties F_1 and F_2 tells whether the first or the second one defines the simplest number.

Since we have assumed, in effect, that the theory T contains formal (= first order) arithmetic, we can use the famous Gödel's theorem and conclude that this theory is *undecidable*, i.e., that there exists no algorithm that, given a formula F from this languages, would tell whether this formula is deducible from T or not (see, e.g., Barwise [19], Enderton [101], Schoenfield [380]).

Moreover, no algorithm is possible, that is applicable to an arbitrary arithmetic formula F and that would return "yes" if F is deducible from T and "no" if the negation $\neg F$ of the formula F is deducible from T (see, e.g., Schoenfield [380], Chapter 6, Ex. 13(c)); see also Rogers [362], Sections 7.7–7.9). We will show that our hypothetic algorithm U leads exactly to such an impossible algorithm.

Indeed, let us take an arbitrary definable number and denote it by F^-.

Since the language L contains the formal arithmetic, all integers are defined in this language. Therefore, there are infinitely many definable numbers. Since for every length l, there are only finitely many formulas of this length, these formulas can only define finitely many different numbers. Thus, for every length l, there exists a definable number that cannot be defined by any formula of length $\le l$, and for which, therefore, the complexity is $> l$. In particular, there exists a definable number whose complexity is greater than $l = D(F^-)$. Let us pick one such number and denote it by $\tilde{F}(y)$. Similarly, there exists a definable number whose complexity is $> D(\tilde{F})$. Let us pick one such number and denote it by F^+.

So, we have three definable numbers $F^-(y)$, $\tilde{F}(y)$, and $F^+(y)$, for which $D(F^-) < D(\tilde{F}) < D(F^+)$.

Let us now consider the following formula:

$$(F \rightarrow F^-(y))\&(\neg F \rightarrow F^+(y)).$$

We will denote this formula by $F'(y)$. Let us first prove that this formula indeed defines a real number:

- if F is deducible from T, then this formula $F'(y)$ clearly defines a real number (namely, the same real number as $F^-(y)$);
- similarly, if $\neg F$ is deducible from T, then this formula $F'(y)$ also defines a real number (namely, the same real number as $F^+(y)$);

- since in classical logic, we have $T \vdash F \vee \neg F$, we can thus conclude that this formula *always* defines a real number.

In particular, if either F or $\neg F$ is deducible from the theory T, then:

- If F is deducible from T, then this new formula is equivalent to $F^-(y)$ and therefore, it defines the same number as $F^-(y)$.
- If F is not deducible from T, then this formula is equivalent to $F^+(y)$ and thus, it defines the same number as $F^+(y)$.

Let us now apply our hypothetic algorithm U to the formulas $F'(y)$ and $\tilde{F}(y)$:

- If F is deducible from T, then U will select the number defined by $F'(y)$, because this number $(F^-(y))$ is simpler than the number defined by $\tilde{F}(y)$.
- If $\neg F$ is deducible from T, then U will select the number defined by the formula $\tilde{F}(y)$, because in this case, this number is simpler than the number $(F^+(y))$ defined by the formula $F'(y)$.

Thus, simply by looking at the output of the algorithm U, we get an algorithm that returns "yes" if F is deducible from T and "no" if its negation $\neg F$ is deducible from T. We already know that such an algorithm is impossible.

This contradiction shows that our initial assumption — that the problem of choosing the representative from a finite set is algorithmically solvable — is false. Hence, this problem is not algorithmically solvable. Proposition A.1 is proven.

Comments.

- This result is similar to the known result that Kolmogorov complexity is not computable (Li *et al.* [254]). In effect, our result sounds slightly stronger because we have proven that not only computing the actual values of complexity is impossible, but even deciding which of the values has larger complexity is also impossible.

- Simplicity seems to be a natural criterion for choosing a representative, but we can also look for other ways in which a number can be representative. For example, we may want a number that is the most "typical" of the elements of the given set; this approach is outlined, for different notions of "typicality", in Heindl *et al.* [145, 146, 147], Friedman *et al.* [112], Bouchon-Meunier *et al.* [55].

Proof of Propositions A.2–A.4. Let us first prove Proposition A.3. Then, we will show that the Corollary (and hence, Proposition A.2) is also true.

Case of the Simplest Possible Definable Real Number $F(y)$. Let $F(y)$ be the simplest possible definable real number. This means that its complexity $D(F)$ is the smallest possible complexity that a real number can have. In other words, the number $F(y)$ is defined by a formula of length $l_{\min}$ that is the shortest possible formula defining a real number. For such $F(y)$, it is easy to describe the desired algorithm: from every interval $[F(y), F'(y)]$ or $[F'(y), F(y)]$ that has $F(y)$ as one of its endpoints, we can return this very definable number $F(y)$ as the desired simplest representative.

Case of a Definable Real Number $F(y)$ That Is Not the Simplest Possible. Let now $F(y)$ be *not* the simplest possible real number. For such $F(y)$, we will prove the impossibility of an algorithm by reduction to a contradiction. Let us assume that there exists an algorithm U that, given any other definable real number $F'(y)$:

- chooses the simplest representative s from the corresponding interval $[F(y), F'(y)]$ or $[F'(y), F(y)]$; and
- if this simplest representative coincides with one of the endpoints, returns $-$ or $+$ depending on whether s is the left or the right endpoint.

The fact that $F(y)$ is not the simplest possible number means that there exist other definable real numbers whose complexity is smaller than $D(F)$, i.e., that are defined by formulas shorter than $D(F)$. We have already shown in the proof of Proposition A.1 that for every length l, there exist finitely many definable real numbers of complexity l. Thus, there exist finitely many definable real numbers that are simpler than $F(y)$. From these numbers, let us pick the formula $G(y)$ for which the number defined by it is the closest to $F(y)$ (if there are two such numbers, let us pick the one that is greater than the number defined by $F(y)$).

Without loss of generality, we can assume that the number x_F defined by the formula $F(y)$ is smaller than the number x_G defined by the formula $G(y)$ (the case $x_G < x_F$ can be considered similarly). Now, let $f(n)$ be any algorithmic function from natural numbers to natural numbers. It is known that every algorithmic sequence is definable in Peano arithmetic, and therefore, since out theory T includes Peano arithmetic, $f(n)$ is definable in T as well.

For every such function, we can define a new definable number z_f as follows:

- If $\forall n(f(n) = 0)$, then $z_f = x_G$.
- If $\exists n(f(n) \neq 0)$, then $z_f = x_G - 2^{-n_{\min}} \cdot (x_G - x_F)$, where $n_{\min}$ is the smallest natural number n for which $f(n) \neq 0$.

(We have used words to define z_f, but this definition can be easily reformulated in terms of formulas, so, the number z_f is indeed definable.)

For each function f, it is easy to see which element from the interval $[x_F, z_f]$ is the simplest:

- If $\exists n(f(n) \neq 0)$, then $x_F < z_f < x_G$. Since we have chosen x_G as the closest of all definable real numbers that are simpler than x_F, and since all the elements of the semi-open interval $(x_F, z_f]$ are closer to x_F than x_G, we can conclude that none of the real numbers from the interval $(x_F, z_f]$ is simpler than x_F. Thus, x_F is the simplest of all real numbers from the interval $[x_F, z_f]$.
- If $\forall n(f(n) = 0)$, then $z_f = x_G$. Since we have chosen x_G as the closest of all definable real numbers that are simpler than x_F, and since all the elements of the open interval (x_F, x_G) are closer to x_F than x_G, we can conclude that none of the real numbers from the open interval (x_F, x_G) is simpler than x_F. Thus, x_G is the simplest of all real numbers from the interval $[x_F, x_G] = [x_F, z_f]$.

In both cases, the simplest element coincides with one of the endpoints, so, the algorithm U will return either $-$ or $+$:

- If $\exists n(f(n) \neq 0)$, then the lower endpoint (x_F) is the simplest, and hence, the algorithm U will return $-$.

- If $\forall n(f(n) = 0)$, then the upper endpoint (z_f) is the simplest, and hence, the algorithm U will return $+$.

Thus, by checking whether the sign returned by the algorithm U is $-$ or $+$, we will be able to check, for a given computable function f, whether $\forall n(f(n) = 0)$ is true or not.

However, it is known (see, e.g., Lewis *et al.* [251], Martin [274], Papadimitriou [323]) that there exists *no* algorithm for deciding whether a program (to be more precise, a program that always finishes its computations) always returns 0. In other words, there exists no algorithm, that, given an algorithmic (everywhere defined) function $f(n)$ from natural numbers to natural numbers would check whether $\forall n(f(n) = 0)$. This contradiction shows that our initial assumption — that the problem of choosing the representative from an interval is algorithmically solvable — is false. Hence, this problem is not algorithmically solvable. Proposition A.3 is proven.

Proof of the Corollary. Let us now prove the Corollary (and thus, Proposition A.2). In the proof of Proposition A.1, we have already shown that for every length l, there exist finitely many definable real numbers of complexity l. In particular, this means that there exist finitely many definable real numbers of the smallest possible complexity $l_{\min}$. Thus, every property (including the property $*$) that holds for all definable real numbers, except for the simplest ones, is thus true for almost all definable real numbers. Corollary is proven.

Proofs of Proposition A.4. Proposition A.4 can be proven similarly to the proof of Proposition A.3.

Proof of Propositions A.5–A.7. If x is the simplest possible constructive real number, then we can always return x.

If x is not the simplest possible constructive real number, then we can use the same construction as in the proof of Proposition A.3. To complete the proof, we must now prove only the following two additional statements:

- First, we need to prove that z_f is a constructive real number (and that, given a program f, we can construct a program (algorithm) for computing z_f).
- Second, in our definition, we no longer require the algorithm to return $-$ or $+$. Therefore, to complete the proof, we must show that if an algorithm

returns a constructive real number s that is equal to one of the endpoints (i.e., to x or to z_f), then we can algorithmically check whether this constructive real number coincides with the left or with the right endpoint.

Both statements are (relatively) easy to prove:

- To compute z_f with an accuracy $(z - x) \cdot 2^{-k}$, it is sufficient to compute first k values of f, and take:

    - If $\forall n(n \leq k \rightarrow f(n) = 0)$, then $a_k = z$.
    - If $\exists n(n \leq k \& f(n) \neq 0)$, then $a_k = z - 2^{-n_{\min}} \cdot (z - x)$, where $n_{\min}$ is the smallest natural number $n \leq k$ for which $f(n) \neq 0$.

  Then, as one can easily see, $|a_k - z_f| \leq 2^{-k} \cdot |z_f - x| \leq 2^{-k} \cdot |z - x|$. From these values $a_1, a_2, \ldots, a_k, \ldots$, we can easily compute the desired rational approximations z_{fk} to z_f.

- If an algorithm returns a constructive real number s that coincides with one of the constructive endpoints of the interval $[x, z_f]$, then, by computing x, z_f, and s with sufficient accuracy (namely, with accuracy $\varepsilon < (z_f - x)/4$), and comparing the corresponding rational numbers, we will be able to check whether $s = x$ or $s = z_f$. Indeed, in this case, from $|s_k - s| \leq \varepsilon$, and $|z_{fk} - z_f| \leq \varepsilon$, we can conclude that

$$|z_{fk} - s_k| \geq |z_f - s| - |s_k - s| - |z_{fk} - z_f| >$$

$$|z_f - s| - 2 \cdot (1/4) \cdot |z_f - s| > (1/2) \cdot |z_f - x|.$$

  Hence:

    - If $s = x$, then, similarly, $|s_k - z_{fk}| > (1/2) \cdot |z_f - x|$. On the other hand, in this case, $|s_k - x_k| \leq |s_k - s| + |x_k - x| \leq 2\varepsilon < (1/2) \cdot (z_f - x)$. Therefore, in this case, $|s_k - x_k| < |s_k - z_{fk}|$.
    - Similarly, if $s = z_f$, then $|s_k - x_k| > |s_k - z_{fk}|$.

  Thus, comparing two rational numbers $|s_k - x_k|$ and $|s_k - z_{fk}|$, we can tell with which of the endpoints s coincides.

Propositions are proven.

B

ERROR ESTIMATION FOR INDIRECT MEASUREMENTS: CASE OF APPROXIMATELY KNOWN FUNCTIONS

In the main text, we analyzed the error estimation problem for indirect measurements with a precisely known $f(x_1, \ldots, x_n)$. In particular, we showed that for a linear function $f(x_1, \ldots, x_n) = a_0 + a_1 \cdot x_1 + \ldots + a_n \cdot x_n$, this problem can be solved in linear time. In real life, we sometimes know the function f only approximately. For example, we may know that f is linear, but we do not know the exact values of the coefficients a_i; these values must be determined from measurements. In this appendix, we show that in this situation, even for linear functions, the error estimation problem for indirect measurements becomes computationally intractable (depending on the formulation, this problem is either NP-hard or exponential time).

B.1. Introduction to the problem

Indirect measurements: brief reminder. One of the main problems to which interval computations are applied is the problem of *error estimation for indirect measurements*. In indirect measurements, we are interested in the value of a physical quantity y that is difficult to measure directly. To overcome this difficulty, we:

- measure some other physical quantities $x_1, \ldots, x_n$ that uniquely determine y (i.e., for which $y = f(x_1, \ldots, x_n)$ for some function f), and then
- use the results $\tilde{x}_1, \ldots, \tilde{x}_n$ of these measurements to estimate the value of y as $\tilde{y} = f(\tilde{x}_1, \ldots, \tilde{x}_n)$.

Measurements are not absolutely precise; hence, the measurement results $\tilde{x}_i$ may differ from the actual values x_i of the measured quantities. In some cases, we know the *probabilities* of different measurement errors $\Delta x_i = \tilde{x}_i - x_i$, but in many cases, we only know the *upper bounds* Δ_i on the errors Δx_i. In such cases, from the result $\tilde{x}_i$ of each direct measurement, we can conclude that the (unknown) actual value x_i of the measured physical quantity belongs to the *interval* $\mathbf{x}_i = [\tilde{x}_i - \Delta_i, \tilde{x}_i + \Delta_i]$. Because of the potential inaccuracy of direct measurements ($x_i \neq \tilde{x}_i$), the result $\tilde{y} = f(\tilde{x}_1, \ldots, \tilde{x}_n)$ of indirect measurements can also be inaccurate, i.e., different from the (unknown) actual value $y = f(x_1, \ldots, x_n)$. For a given measurement, possible values of y form an *interval* $\mathbf{y}$. It is, therefore, desirable to estimate this interval. This is one of the *basic* problems of *interval computations.*

We have shown that for the simplest case of *linear* functions f, there exist a linear-time algorithm for computing $f(x_1, \ldots, x_n)$, while for the next-simplest quadratic functions $f(x_1, \ldots, x_n)$, the problem becomes, in general, computationally intractable (NP-hard).

In indirect measurements, we do not always know the function $f(x_1, \ldots, x_n)$ precisely. In the main text, we considered the idealized situation when we know the *exact* expression for the function $f(x_1, \ldots, x_n)$. This expression usually comes from a theory that describes the relation between x_i and y. Sometimes, such a theory gives an *exact* dependence, but often, we only get a formula with coefficients that needs to be experimentally determined. (For example, Newton's gravitation theory includes a gravitation constant that must be determined from experiments; special relativity contains the speed of light c, and quantum mechanics contains Planck's constant $\hbar$.) In such situations, error estimation for indirect measurements becomes a more difficult problem than for the case of precisely known f.

What is the *computational complexity* of this more difficult problem? Is it feasible or intractable?

For quadratic (and more complicated) functions f, the problem is NP-hard, so we will only consider linear functions f. For quadratic functions $f(x_1, \ldots, x_n)$, the above problem of interval computation is NP-hard even if we know the *exact* function f. Therefore, in any reasonable formulation, a more complicated problem of error estimation for indirect measurements for the case of *approximately known* f is intractable (NP-hard) as well.

Therefore, to answer the question about computational complexity, it is sufficient to consider the situations when the actual (unknown) function $f(x_1, \ldots, x_n)$ is *linear*.

How can we determine f? How can we determine the function f experimentally? Since we assumed that the quantity y depends on the quantities x_i, a natural idea is to generate some values of x_i and measure the resulting value of y. After these measurements, we get several *patterns* $(\tilde{x}_1^{(p)}, \ldots, \tilde{x}_n^{(p)}, y^{(p)})$, $1 \leq p \leq P$, from which we determine f.

Usually, a function f is used for many indirect measurements. Each pattern measurement (that leads to determining f) will benefit all these measurements. Therefore, when we plan these pattern measurements, we can afford to spend more and thus get better sensors and better accuracy than in further (one-time) measurement of x_i. Hence, we will assume that the patterns are measured with *higher accuracy* that the values x_i in the follow-up measurement.

Passive and active formulations. In the main text, we assumed that we *first* somehow determine the function f, and *then* use this function to compute the desired interval $\mathbf{y}$. We can use a similar *two-stage* approach for a new situation as well; in other words:

- on the first stage, we perform several (highly accurate) measurements to determine the function $f(x_1, \ldots, x_n)$; and then
- on the second stage, we get the measurement results $\mathbf{x}_i$, and use the information obtained on the first stage to compute the desired interval $\mathbf{y}$ for different input intervals $\mathbf{x}_i$.

It is possible that for a particular problem, the information that we have gathered on the first stage is not sufficient to compute $\mathbf{y}$. In this case, it is desirable to perform new experiments to recover the missing information about f. (For example, if we are planning an automatic mission to a distant planet, and we cannot, based on the known values of the masses and gravitational constant, predict whether the spaceship will successfully reach the planet, then we must undertake new measurements to get better values of the constants.) In other words, instead of the traditional *passive* two-stage approach, it is reasonable to use an *active* approach, in which, after measuring $\mathbf{x}_i$, we may (if necessary) return to the first stage and make new measurements to get more information about f.

From the viewpoint of the accuracy, this *active* process is ideal. But realistically, extra measurements cost time and money, so, in applications in which time and/or cost is limited (e.g., in real-time control where time is severely limited, and in manufacturing where cost is severely limited) we will not be able to afford this active process. Since such applications are quite frequent, in the following, we will consider *two* possible formulations: *passive* (two-stage) and *active*.

Measurement errors corresponding to measuring f. To formulate the problem of error estimation for indirect measurement in precise terms, we must describe all measurement errors that influence this problem.

In the basic problem as described in the main text, the only measurement errors that we had to consider where the errors in measuring x_i (as described by their bounds Δ_i). In the new problem, the function f itself comes from measurements, so, we have to take into consideration the errors with which this function is measured, i.e., errors with which we get each of the patterns.

To get a pattern $(\tilde{x}_1^{(p)}, \ldots, \tilde{x}_n^{(p)}, y^{(p)})$, we must set up x_i (and maybe also measure x_i to check how accurately we set this value), and then measure y. So, instead of a *single* type of measurement errors, we have now *three* types of measurement errors:

- errors Δx_i related to measuring x_i in the indirect measurement;
- errors $\Delta x_i^{(p)}$ related to measuring x_i when forming a pattern;
- errors $\Delta y^{(p)}$ related to measuring y (when forming a pattern).

In general, all three errors are present, and our general formulation of the error estimation problem will take all three types of errors into consideration.

Possibility of simplification. The general error estimation problem (with all three types of measurement errors present) turns out to be computationally *intractable* (see below). Therefore, we have to look for *simplifications* that make this problem feasible. There is a natural way to simplify this general problem: namely, as we will show, some of these measurement errors are much smaller than the others and therefore, these smaller errors can be often safely neglected (i.e., assumed to be equal to 0).

- We have already remarked that the *patterns* are usually measured with much *higher accuracy* than in the follow-up measurements of x_i, i.e., that $|\Delta x_i^{(p)}| \ll |\Delta x_i|$ and $|\Delta y^{(p)}| \ll |\Delta x_i|$.
- The entire necessity for indirect measurements is caused by the situation in which it is much *more difficult to* accurately *measure* y directly *than to measure* x_i; therefore, $|\Delta x_i^{(p)}| \ll |\Delta y^{(p)}|$.

Hence, $|\Delta x_i^{(p)}| \ll |\Delta y^{(p)}| \ll |\Delta x_i|$. In view of this relation, we can have two consequent simplifications:

- First, we can *neglect* the smallest possible measurement errors $\Delta x_i^{(p)}$, i.e., assume that, in patterns, the *values of* $x_i^{(p)}$ *are measured* absolutely *precisely*.
- Second, we can also *neglect* the errors of the second smallest type ($\Delta y^{(p)}$), i.e., assume that in each pattern not only the values $x_i^{(p)}$, but also the *values* $y^{(p)}$ *are measured with* absolute *precision*.

In this appendix, we will analyze each of the two problems (passive and active) in all three settings: in the most general setting and in these two simplified settings.

Comment. Theoretically, we can go one more step further and consider the situation in which measurement errors of *all three* types are negligible, but that assumption would simply mean that $\tilde{x}_i$ are the precise values of x_i and therefore, no interval computations are needed at all.

B.2. Precise formulation of the problem: the simplest case

Let us start with the simplest setting, in which in each pattern, we can measure both $x_i^{(p)}$ and $y^{(p)}$ precisely. This setting will be the easiest to analyze.

We know that the actual dependence f is linear, i.e., that $f(x_1, \ldots, x_n) = a_0 + a_1 \cdot x_1 + \ldots + a_n \cdot x_n$ for some (initially unknown) coefficients a_i. Therefore:

- In the *active* problem, we can easily compute all the coefficients:
  - First, we take the pattern with the values $x_1^{(0)} = \ldots = x_n^{(0)} = 0$. For this pattern, $y^{(0)} = a_0$, and thus, after measuring $y^{(0)}$, we get a_0.
  - Next, for all $i = 1, \ldots, n$, we take the pattern with $x_i^{(i)} = 1$ and $x_j^{(i)} = 0$ for all $j \neq i$. For this pattern, $y^{(i)} = a_0 + a_i$. Since we already know a_0, we can thus determine a_i as $y^{(i)} - a_0$.

  Thus, after $n+1$ pattern measurements, we get all $n+1$ coefficients, and so, we can compute the desired interval $\mathbf{y}$ in linear time. Thus, if we count both the measurement and the computations, we still get the time that is linear in n.

- In the *passive* problem, we cannot choose the patterns. Instead, we are given the patterns $(x_1^{(p)}, \ldots, x_n^{(p)}, y^{(p)})$, $1 \leq p \leq P$. These patterns mean that the (unknown) coefficients a_i satisfy the system of linear equations:

$$a_0 + a_1 \cdot x_1^{(p)} + \ldots + a_n \cdot x_n^{(p)} = y^{(p)}.$$

  There are two possibilities here:
  - One possibility is that this system of linear equations is *under-determined.* In this case, the set A of all possible values of $\vec{a} = (a_0, a_1, \ldots, a_n)$ is an (infinite) *plane* (of dimension ≥ 1), and, therefore, for any non-degenerate interval $\mathbf{x}_1, \ldots, \mathbf{x}_n$, the set of possible values of $y = a_0 + a_1 \cdot x_1 + \ldots + a_n \cdot x_n$ (for all $x_i \in \mathbf{x}_i$ and $\vec{a} \in A$) coincides with the entire real line R. In this case, we get *no information* at all about the actual value of y, and therefore, there is *no indirect measurement* of y.
  - The only case when there *is* an indirect measurement of y is when the corresponding system of linear equations is *determined.* In this case, we can *uniquely* determine the coefficients a_i by solving the corresponding system of linear equations.
    * Solving a linear system requires polynomial time ($O(n^{2.376})$; see, e.g., Cormen *et al.* [76], Chapter 31).
    * After we have computed all the coefficients a_i, computing $\mathbf{y}$ requires linear time.

  Thus, totally, for passive situation, we need *polynomial time.*

B.3. Precise formulation of the problem: the second simplest case

Passive case. Let us first consider the passive case.

Definition B.1.

- *Let a positive integer n be given; it will be called the number of inputs.*
- *By a linear function (or a coefficient vector), we mean a sequence of $n+1$ real numbers $\vec{a} = (a_0, a_1, \ldots, a_n)$.*
- *By a pattern, we mean a sequence of $n+2$ real numbers $\mathcal{P} = (x_1, \ldots, x_n, y, \Delta)$, with $\Delta \geq 0$.*
- *We say that a pattern $\mathcal{P}$ is consistent with the linear function (coefficient vector) $\vec{a}$ is the following inequality holds: $|l(x_1, \ldots, x_n) - y| \leq \Delta$, where $l(x_1, \ldots, x_n) = a_0 + a_1 \cdot x_1 + \ldots + a_n \cdot x_n$.*
- *We say that a sequence of patterns $\mathcal{P}^{(p)}$, $1 \leq p \leq P$, is consistent if there exists a coefficients vector that is consistent with all the patterns.*

Definition B.2. *By a problem of estimating error of an approximately linear indirect measurement (in the passive setting), we mean the following problem:*

GIVEN:

- *n intervals $\mathbf{x}_1, \ldots, \mathbf{x}_n$;*
- *a consistent sequence of patterns $\mathcal{P}^{(1)}, \ldots, \mathcal{P}^{(P)}$;*
- *a positive real number δ.*

COMPUTE:

rational numbers that are $\delta-$close to the endpoints of the interval $\mathbf{y}$ of all possible values of $y = l(x_1, \ldots, x_n) = a_0 + a_1 \cdot x_1 + \ldots + a_n \cdot x_n$ for all $x_i \in \mathbf{x}_i$ and for all linear function $l = (a_0, a_1, \ldots, a_n)$ that are consistent with all the patterns.

Theorem B.1. *The problem of estimating error of an approximately linear indirect measurement (in the passive setting) is NP-hard.*

Active case. In the active case, in addition to the purely *computation* steps, we can also make a *measurement step*: namely, we can take an arbitrary sequence $(x_1, \ldots, x_n)$, and perform an experiment that will determine the appropriate value of $f(x_1, \ldots, x_n)$ (with the accuracy corresponding to pattern-measuring y). In this section, we will denote the (approximate) measured value of y that corresponds to the given $x_1, \ldots, x_n$, by $\tilde{f}(x_1, \ldots, x_n)$.

Both computational and measurement steps take time; therefore, to estimate the total running time of the algorithm, we will count *both* computational and measurement steps. In the language of theory of computing, these "computations + measurements" are called *computations with an oracle* $\tilde{f}(x_1, \ldots, x_n)$.

We will show that the resulting problem requires exponential time.

One last comment before we describe the precise formalization: In real life, we *cannot* simulate arbitrary *large* values of a physical quantity, there is usually an *upper bound* on its physically possible values. So, to make our formulation realistic, we will assume that we can get the values $\tilde{f}(x_1, \ldots, x_n)$ not for *arbitrary* vectors $\vec{x} = (x_1, \ldots, x_n)$ of rational numbers, but only for vectors $\vec{x}$ that are *sufficiently close* to the measurement vector $\vec{\tilde{x}} = (\tilde{x}_1, \ldots, \tilde{x}_n)$, i.e., for which the Euclidean distance $d(\vec{\tilde{x}}, \vec{x}) = \sqrt{(\tilde{x}_1 - x_1)^2 + \ldots + (\tilde{x}_n - x_n)^2}$ does not exceed some given value $r > 0$. We must, of course, make sure that all *possible* values x_i (i.e., all values for which $|\Delta x_i| = |x_i - \tilde{x}_i| \leq \Delta_i$, where Δ_i is an upper bound on the error of i-th direct measurement) are still sufficiently close to $\vec{\tilde{x}}$ in this sense. Thus, we must require that $\sqrt{\Delta_1^2 + \ldots + \Delta_n^2} \leq r$.

Definition B.3.

- *Let a positive integer n be given; n will be called the **number of inputs**.*
- *By a **coefficient vector**, we mean a sequence of $n + 1$ real numbers $\vec{a} = (a_0, a_1, \ldots, a_n)$.*
- *By an **oracle**, we mean a function $\tilde{f}(\tilde{x}_1, \ldots, \tilde{x}_n)$ of n variables.*
- *By a **measurement result**, we mean two vectors $\vec{\tilde{x}} = (\tilde{x}_1, \ldots, \tilde{x}_n)$ and $\vec{\Delta} = (\Delta_1, \ldots, \Delta_n)$, where $\Delta_i \geq 0$. The value $\tilde{x}_i$ is called the **result of i-th direct measurement**, and the value Δ_i is called the **accuracy of i-th direct measurement**.*
- *Let a positive real number $\varepsilon > 0$ be give, It will be called the **accuracy of pattern measurement**.*

- *We say that an oracle* $\tilde{f}$ *is* $(\tilde{\vec{x}}, r, \varepsilon)$*-consistent with a linear function* $l(\vec{x}) = a_0 + a_1 \cdot x_1 + \ldots + a_n \cdot x_n$ *if* $|\tilde{f}(\vec{x}) - l(\vec{x})| \le \varepsilon$ *for all vectors* $\vec{x} = (x_1, \ldots, x_n)$ *that are r-close to* $\tilde{\vec{x}}$ *(i.e., for which* $d(\tilde{\vec{x}}, \vec{x}) \le r$*).*
- *We say that an oracle* $\tilde{f}$ *is* $(\tilde{\vec{x}}, r, \varepsilon)$*-consistent if there exists a linear function* $l(x)$ *that is* $(\tilde{\vec{x}}, r, \varepsilon)$*-consistent with* $\tilde{f}$*.*

Definition B.4. *By a problem of estimating error of an approximately linear indirect measurement (in the active setting), we mean the following problem:*

GIVEN:

- *n real numbers* $\tilde{x}_1, \ldots, \tilde{x}_n$*;*
- *n non-negative real numbers* $\Delta_1, \ldots, \Delta_n$*;*
- *positive real numbers* $r \ge \sqrt{\Delta_1^2 + \ldots + \Delta_n^2}$ *and* ε*.*
- *a* $(\tilde{\vec{x}}, r, \varepsilon)$*-consistent oracle* $\tilde{f}(x_1, \ldots, x_n)$*;*
- *a positive real number* δ*.*

COMPUTE:

rational numbers that are δ*−close to the endpoints of the interval*

$$\mathbf{y} = [\underline{y}, \overline{y}] = \{y = l(x_1, \ldots, x_n) | x_i \in \mathbf{x}_i \text{ for all } i, \text{ and a linear function}$$

$$l(x_1, \ldots, x_n) \text{ is } (\tilde{\vec{x}}, r, \varepsilon)\text{-consistent with the given oracle } \tilde{f}.\}$$

Comments.

- If the measurements of x_i are absolutely accurate (i.e., if $\Delta_i = 0$ and $\tilde{x}_i = x_i$), and if the patterns are measured absolutely accurately (i.e., $\varepsilon = 0$ and $\tilde{f}(x_1, \ldots, x_n) = f(x_1, \ldots, x_n)$), then we can compute the desired value y as $\tilde{y} = \tilde{f}(\tilde{x}_1, \ldots, \ldots x_n)$. In the general case of non-zero errors, it is still reasonable to take this value $\tilde{y} = \tilde{f}(\tilde{x}_1, \ldots, \tilde{x}_n)$ as the numerical estimate for y. From this viewpoint, it makes sense to ask for the *largest possible deviation* Δ between this estimate $\tilde{y}$ and all possible values $y \in \mathbf{y}$ (i.e., to be more precise, the largest possible value of $|y - \tilde{y}|$). (Of course, if we can compute the endpoints of the interval $\mathbf{y}$, then we can easily compute this value Δ as well.)

- The only way to use the oracle function $\tilde{f}$ is to generate some rational values $x_1, \ldots, x_n$ and to return $\tilde{f}(x_1, \ldots, x_n)$. By the *computational complexity* of an algorithm that solves our problem, we will mean the *total number* of elementary computational operations plus the number of oracle calls. (Of course, by counting the call as *one* computational step, we *underestimate* the computational complexity of an algorithm, but since we are going to prove an *exponential lower bound* for the number of calls, we will thus get an exponential lower bound for *any other* reasonable definition of a computational complexity.)

Theorem B.2. (Kreinovich [206]) *If an algorithm solves the problem of estimating errors of approximately linear indirect measurements (in the active setting), then its worst-case computational complexity is* $\geq 2^{n-1}$.

Comment. This theorem means that whatever algorithm solves our problem, this algorithm will require exponential time on *some* instances. Of course, the worst-case exponential complexity does not mean that we *always* have exponential time: for some oracles, faster computations are possible (see an example in the Proofs section).

B.4. Precise formulation of the problem: the general case

Since our problem is computationally intractable even in the simplified case, when the errors $\Delta_i^{(p)}$ are negligible, it is intractable in the general case as well.

The complexity of different problems (related to error estimation for indirect measurements for approximately known linear functions f) can be represented as a table:

Accuracy of experiments for measuring f	Active setting (we can make new experiments)	Passive setting (we can only use results of given experiments)
x_i and y measured precisely	Polynomial time	Linear time
x_i measured precisely, y measured with error	Exponential time (or worse)	NP-hard
x_i and y measured with error	Exponential time (or worse)	NP-hard

We can use another table to compare the computational complexity of the problems with *precisely* known and *approximately* known f:

	Precisely known f	Approximately known f
Linear f	Linear time	Exponential time (or worse)
Quadratic f	NP-hard	Exponential time (or worse)
Polynomial f	NP-hard	Exponential time (or worse)

B.5. Proofs

Proof of Theorem B.1.

We already know, from the main text, that for *bilinear* functions $y = \sum a_{ij} x_i y_j$, the basic problem of interval computations is NP-hard. To be more precise, the problem of computing the largest possible value of a bilinear function $\sum a_{ij} x_i y_j$, when x_i and y_j run over given intervals $\mathbf{x}_i$ and $\mathbf{y}_j$, is NP-hard. To show that *our* problem is NP-hard, let us reduce this bilinear problem to ours. Namely, if we:

- rename y_j as a_{n+j},
- introduce a new coefficient a_0 that is equal to 0, and
- denote $\sum_j a_{ij} y_j$ by a_i,

then we can re-formulate the original bilinear problem in the following equivalent form:

$$a_0 + a_1 \cdot x_1 + \ldots + a_n \cdot x_n + a_{n+1}x_{n+1} + \ldots + a_{2n} \cdot x_{2n} \to \max$$

under the conditions that $x_1 \in \mathbf{x}_1, \ldots, x_n \in \mathbf{x}_n$, $x_{n+1} \in \mathbf{x}_{n+1} = [0,0]$, ..., $x_{2n} \in \mathbf{x}_{2n} = [0,0]$, and the values a_i satisfy the following system of $2n+1$ two-sided linear inequalities:

$$0 \le a_i - \sum_{j=1}^{n} a_{ij} a_{n+j} \le 0, \;\; 1 \le i \le n;$$

$$\underline{y}_j \le a_{n+j} \le \overline{y}_j, \;\; 1 \le j \le n;$$

$$0 \le a_0 \le 0.$$

The last inequality can be represented as consistency condition between the coefficient vector $\vec{a}$ and the all-zeroes pattern $(0, \ldots, 0, 0, 0)$. This condition guarantees that $a_0 = 0$.

Since $a_0 = 0$, each of remaining $2n$ double-sided *inequalities*

$$\underline{c} \le k_1 \cdot a_1 + \ldots + k_{2n} \cdot a_{2n} \le \overline{c},$$

can, in its turn, be represented as a *consistency condition* between a vector $\vec{a}$ and a pattern $(k_1, \ldots, k_{2n}, y, \Delta)$, where $y = (\underline{c} + \overline{c})/2$ and $\Delta = (\overline{c} - \underline{c})/2$. Thus, the original bi-linear problem is equivalent to a particular case of our problem.

Hence, any method of solving *our* problem in polynomial time would lead to solving the *bilinear* problem in polynomial time; since the bilinear problem is NP-hard, our problem is, thus, also NP-hard. The theorem is proven.

Proof of Theorem B.2. We will prove this theorem by reduction to a contradiction. Assume that we have an algorithm $\mathcal{U}$ that solves this problem and whose worst-case computational complexity for some n is $< 2^{n-1}$. This means that this algorithm will always produce the answer in $\le 2^{n-1} - 1$ computational and measurement steps. In particular, it will produce the desired answer with $\le 2^{n-1} - 1$ calls to the oracle $\tilde{f}(\vec{x})$. Let us show that this assumption leads to a contradiction.

To show this, we will take $\tilde{x}_i = 0$, $\Delta_i = 1$, $\varepsilon = 1$, and $r = \sqrt{n}$ (we will choose $\delta > 0$ later). As an oracle $\tilde{f}(\vec{x})$, we will take a function that is identically 0. (For this function, $\tilde{y} = \tilde{f}(\tilde{x}_1, \ldots, \tilde{x}_n)$ is always equal to 0.) This oracle $\tilde{f}(\vec{x})$ is itself a linear function, so it is $(\vec{\tilde{x}}, r, \varepsilon)$-consistent for all $\vec{\tilde{x}}$, r, and ε.

If a linear function $l(\vec{x})$ is $(\vec{\tilde{x}}, r, \varepsilon)$-consistent with this oracle, then for all $x_i \in \mathbf{x}_i$, we have $|l(\vec{x}) - \tilde{f}(\vec{x})| \le \varepsilon = 1$. Since $\tilde{f}(\vec{x}) = 0$, we have $|l(\vec{x})| \le 1$. Thus, all possible values $y = l(\vec{x})$ are between -1 and 1, and therefore, the deviation $|y - \tilde{y}|$ between y and $\tilde{y} = 0$ cannot exceed 1. Hence, the largest possible value Δ of this deviation is also ≤ 1: $\Delta \le 1$.

The algorithm $\mathcal{U}$ computes the value $\tilde{\Delta}$ that is δ–close to Δ. Therefore, from $\Delta \le 1$, we conclude that $\tilde{\Delta} \le \Delta + \delta \le 1 + \delta$.

This algorithm $\mathcal{U}$ uses $\le 2^{n-1} - 1$ values of the oracle $\tilde{f}(\vec{x})$. So, if we use another oracle $\tilde{g}(\vec{x})$ which is also $(\vec{\tilde{x}}, r, \varepsilon)$-consistent, and which has the same values in all tested points $\vec{x}^{(p)}$ (i.e., $\tilde{f}(\vec{x}^{(p)}) = \tilde{g}(\vec{x}^{(p)})$), then the algorithm $\mathcal{U}$ will not notice the difference and thus produce the same estimate $\tilde{\Delta}$ for the new largest deviation $\Delta(\tilde{g})$. To get the desired contradiction, we will produce an oracle $\tilde{g}(\vec{x})$ for which this estimate will be incorrect.

For each of $\le 2^{n-1} - 1$ patterns $\vec{x}^{(p)} = (x_1^{(p)}, \ldots, x_n^{(p)})$ for which the algorithm $\mathcal{U}$ calls the oracle, we can find a vector of signs $(\text{sign}(x_1^{(p)}), \ldots, \text{sign}(x_n^{(p)}))$, where the sign function is defined as usual: $\text{sign}(x) = 1$ if $x > 0$, $\text{sign}(x) = -1$ if $x < 0$, and $\text{sign}(0) = 0$. These $\le 2^{n-1} - 1$ patterns lead to $\le 2^{n-1} - 1$ sign vectors. If we add, for each pattern, a vector of its *negative* signs $(-\text{sign}(x_1), \ldots, -\text{sign}(x_n))$, we still get no much than twice the number of patterns, i.e., no more than $2^n - 2$ different sign vectors.

There are totally 2^n sign vectors that consist of ± 1. Therefore, at least one of these vectors is not appearing neither as the sign vector of a pattern, nor as a negative sign vector of a pattern. Let us denote one of these vectors by $\vec{\varepsilon} = (\varepsilon_1, \ldots, \varepsilon_n)$; let us then take

$$l_0(\vec{x}) = \frac{1}{\sqrt{n(n-1)}} \sum_{i=1}^{n} \varepsilon_i \cdot x_i,$$

and define the oracle $\tilde{g}(\vec{x})$ as follows:

- $\tilde{g}(\vec{x}) = l_0(\vec{x}) - 1$ if $l_0(\vec{x}) > 1$;
- $\tilde{g}(\vec{x}) = 0$ if $-1 \le l_0(\vec{x}) \le 1$;
- $\tilde{g}(\vec{x}) = l_0(\vec{x}) + 1$ if $l_0(\vec{x}) < -1$.

This oracle is $(\vec{\tilde{x}}, r, \varepsilon)$-consistent, because it is $(\vec{\tilde{x}}, r, \varepsilon)$-consistent with the linear function $l_0(\vec{x})$.

Let us prove that this new oracle $\tilde{g}(\vec{x})$ attains the same values (identically 0) for all the patterns $\vec{x}^{(p)}$ (to which $\tilde{f}$ has been applied while the algorithm $\mathcal{U}$ was running).

Indeed, let $\vec{x}^{(p)}$ be one of such patterns. Because of our choice of signs ε_i, not all the signs of $x_i^{(p)}$ coincide with ε_i. For those i for which $x_i^{(p)}$ and ε_i have different signs, the product $\varepsilon_i \cdot x_i^{(p)}$ is non-positive. Therefore,

$$\sum_{i=1}^{n} \varepsilon_i \cdot x_i^{(p)} \le \sum_{i \in P_p} \varepsilon_i \cdot x_i^{(p)},$$

where by P_p, we denoted the set of all i for which the sign of $x_i^{(p)}$ coincides with ε_i. Since not all of these signs coincide, the number of elements $|P_p|$ in a set P_p is smaller than n, i.e., $|P_p| \le n-1$.

It is well known that the scalar (dot) product $\vec{a} \cdot \vec{b} = \sum a_i \cdot b_i$ of arbitrary two vectors cannot exceed the product of their lengths: $\vec{a} \cdot \vec{b} \le \sqrt{\sum a_i^2} \cdot \sqrt{\sum b_i^2}$. Therefore,

$$\sum_{i \in P_p} \varepsilon_i \cdot x_i^{(p)} = \sum_{i \in P_p} 1 \cdot |x_i| \le \sqrt{\sum_{i \in P_p} 1} \cdot \sqrt{\sum_{i \in P_p} |x_i^{(p)}|^2}.$$

But

$$\sum_{i \in P_p} 1 = |P_p| \le n-1,$$

and

$$\sum_{i \in P_p} |x_i^{(p)}|^2 = \sum_{i \in P_p} \Delta_i^2 \le \sum_{i=1}^{n} \Delta_i^2 \le r^2 = n.$$

Therefore,

$$\sum_{i=1}^{n} \varepsilon_i \cdot x_i^{(p)} \le \sum_{i \in P_p} \varepsilon_i \cdot x_i^{(p)} \le \sqrt{n-1} \cdot \sqrt{n} = \sqrt{n(n-1)},$$

and

$$l_0(\vec{x}^{(p)}) = \frac{1}{\sqrt{n(n-1)}} \sum_{i=1}^{n} \varepsilon_i \cdot x_i^{(p)} \le 1.$$

Similarly, one can prove that $l_0(\vec{x}^{(p)}) \ge -1$. Therefore, according to our definition of the oracle $\tilde{g}(\vec{x})$, we have $\tilde{g}(\vec{x}^{(p)}) = 0$. So, on all the patterns $\vec{x}^{(p)}$, the new oracle $\tilde{g}(\vec{x})$ indeed attains the same value ($= 0$) as the original oracle $\tilde{f}(\vec{x})$.

Let us now estimate the largest possible deviation Δ for this new oracle. This oracle is $(\vec{\tilde{x}}, r, \varepsilon)$-consistent with the linear function $l_0(\vec{x})$. The values $x_i = \varepsilon_i$ are possible values of i–th quantity, because $|\tilde{x}_i - x_i| = |0 - \varepsilon_i| = |\varepsilon_i| = 1 \le \Delta_i$. So, the largest possible deviation $\Delta(\tilde{g})$ is greater than or equal to the deviation $|y - \tilde{y}| = |l_0(\vec{\varepsilon}) - \tilde{g}(\vec{\tilde{x}})|$. For $\tilde{x}_i = 0$, we have $l_0(\vec{\tilde{x}}) = 0$ and therefore, $\tilde{y} = \tilde{g}(\vec{\tilde{x}}) = 0$. On the other hand,

$$y = l_0(\vec{\varepsilon}) = \frac{1}{\sqrt{n(n-1)}} \sum_{i=1}^{n} \varepsilon_i^2 = \frac{n}{\sqrt{n(n-1)}} = \sqrt{\frac{n}{n-1}}.$$

Hence, $|y - \tilde{y}| = \sqrt{n/(n-1)}$, and $\Delta(\tilde{g}) \ge \sqrt{n/(n-1)}$.

We have already shown that for this oracle $\tilde{g}(\vec{x})$, the algorithm $\mathcal{U}$ returns the estimate $\tilde{\Delta} \le 1 + \delta$. So, if we choose $\delta < 1/2(\sqrt{n/(n-1)} - 1)$, then this estimate cannot be δ-close to the value $\Delta(\tilde{g}) \ge \sqrt{n/(n-1)}$ and will, therefore, be *incorrect.*

This contradiction shows that our initial assumption that we can have an algorithm $\mathcal{U}$ (that solves the problem of estimating errors of approximately linear indirect measurements) with $< 2^{n-1}$ computational and measurement steps is wrong. The theorem is proven.

Proof of the comment after Theorem B.2. Let us show an example of problems that can be solved fast. Let us take arbitrary $\varepsilon > 0$, $r \ge 3 \cdot \sqrt{n}$, $\tilde{x}_i \le 1$, $\Delta_i \le 1$, and the following oracle $\tilde{f}$:

- $\tilde{f}(0, 0, \ldots, 0) = -\varepsilon$;
- $\tilde{f}(1, 0, \ldots, 0) = \tilde{f}(0, 1, 0, \ldots, 0) = \ldots = \tilde{f}(0, \ldots, 0, 1, 0, \ldots, 0) = \ldots = \tilde{f}(0, \ldots, 0, 1) = \varepsilon$;
- $\tilde{f}(-1, 0, \ldots, 0) = \tilde{f}(0, -1, 0, \ldots, 0) = \ldots = \tilde{f}(0, \ldots, 0, -1, 0, \ldots, 0) = \ldots = \tilde{f}(0, \ldots, 0, -1) = \varepsilon$;
- $\tilde{f}(x_1, \ldots, x_n) = 0$ for all other vectors $\vec{x} = (x_1, \ldots, x_n)$.

This oracle is $(\vec{\tilde{x}}, r, \varepsilon)$-*consistent* because it is consistent with the linear function $l_0(x_1, \ldots, x_n) \equiv 0$. Let us show that $l_0(\vec{x})$ is the *only* linear function $l(\vec{x}) = a_0 + a_1 \cdot x_1 + \ldots + a_n \cdot x_n$ that is consistent with this oracle.

Indeed, consistent means that $|\tilde{f}(\vec{x}) - l(\vec{x})| \le \varepsilon$ for all $\vec{x}$ that are r-close to $\vec{\tilde{x}}$. In particular, this inequality must be true for $\vec{x} = (0, 0 \ldots, 0)$, $\vec{x} = (1, 0, \ldots, 0)$, and $\vec{x} = (-1, 0, \ldots, 0)$. Substituting the values of the oracle and the general expression for the linear function $l(x_1, \ldots, x_n)$ into these inequalities, we get the following three inequalities: $|a_0 + \varepsilon| \leq \varepsilon$, $|a_0 + a_1 - \varepsilon| \leq \varepsilon$, and $|a_0 - a_1 - \varepsilon| \leq \varepsilon$. These inequalities can be re-formulated in the following two-sided form:

$$-2\varepsilon \leq a_0 \leq 0;$$

$$0 \leq a_0 + a_1 \leq 2\varepsilon;$$

$$0 \leq a_0 - a_1 \leq 2\varepsilon.$$

Adding the second and the third of these inequalities, and dividing all three sides of the resulting two-sided inquality by two, we conclude that $0 \leq a_0 \leq 2\varepsilon$, i.e., that a_0 is non-negative. Since from the first inequality, we know that a_0 is non-positive, we conclude that $a_0 = 0$.

Since $a_0 = 0$, the second inequality leads to $a_1 \geq 0$, and the third leads to $-a_1 \geq 0$, i.e., to $a_1 \leq 0$. Thus, $a_1 = 0$. Similarly, we can show that $a_i = 0$ for $i = 2, \ldots, n$, and $l(\vec{x}) = 0 = l_0(\vec{x})$.

Since $l_0(\vec{x}) = 0$ is the only linear function that is consistent with the oracle, the desired set $\mathbf{y}$ of all possible values of $y = l(\vec{x})$ consists of a single value 0, and this degenerate interval is easy to compute.

C

FROM INTERVAL COMPUTATIONS TO MODAL MATHEMATICS

In this appendix, we describe the computational complexity and feasibility of another natural generalization of interval computations: modal mathematics.

This appendix was written in collaboration with B. Bouchon-Meunier. The results presented in this appendix first appeared in Bouchon-Meunier et al. [54].

C.1. Formulation of the Problem

Traditional interval mathematics: a brief reminder. Before we start explaining why we need to go beyond interval computations, let us briefly recall our motivation for the use of interval computations in data processing.

Traditional data processing methods of numerical mathematics are based on the assumptions that we know the exact values of the input quantities. In reality, the data come from measurements, and measurements are never 100% precise; hence, the actual value x of each input quantity may differ from its measurement result $\tilde{x}$. In some cases, we know the probabilities of different values of error $\Delta x = \tilde{x} - x$, but in most case, we only know the guaranteed upper bound Δ for the error; in these cases, the only information we have about the (unknown) actual value x is that x belongs to the *interval* $\mathbf{x} = [\tilde{x}-\Delta, \tilde{x}+\Delta]$.

One of the basic problems of interval mathematics is, therefore, as follows: given a data processing algorithm $f(x_1, \ldots, x_n)$ and n intervals $\mathbf{x}_1, \ldots, \mathbf{x}_n$, compute the range $\mathbf{y}$ of possible values of $y = f(x_1, \ldots, x_n)$ when $x_i \in \mathbf{x}_i$.

Non-traditional interval problems: control, design, optimization, etc. The above formulation makes perfect sense when we estimate the value of a certain physical quantity y that is related to the directly measured quantities x_i. In this case, the goal is to find all real numbers that *can* be the values of this quantity y (for the given measurement results).

The goal of data processing is, however, often more complicated. For example, we may want to *control* a certain system; in this case, we must find a control value that, e.g., guarantees stability of the system for all possible values of the parameters x_i (i.e., for all $x_i \in \mathbf{x}_i$). In this case, we want to find all real numbers y for which stability *must* occur. Similarly, many real-life problems of design, control, and optimization lead to complicated mathematical formulations.

Shary's approach: successes and limitations. For the case when the relationship between different variables is described by a system of (linear or non-linear) equations, different possible problems have been described by Shary (see, e.g., [393]). Shary distinguishes between different possible formulations by using different quantifiers for different variables: e.g., if we are interested in the set of *possible* values of y, we are interested in values y for which $\exists x_i$ such that the given equations are true; if we want a control y that leads to stability for *all* possible values x_i, we use a universal quantifier $\forall x_i$.

Successes. Shary's classification contains practically all known problems, and it seems to be sufficient for describing *objective* knowledge, that is usually described in terms of equations.

Limitations. The main limitation of Shary's approach is that an essential part of our knowledge comes from experts, and experts often describe their knowledge not in terms of equations, but in terms of logical statements (e.g., in terms of "rules" of the type "if A then B").

Modal logic: a way of describing "can" and "must". In case of measurement uncertainty, expert statements cannot be formulated in terms of pure logic; we also need some formalization of the words like "can" and "must" that were used in the above descriptions of our objectives. Logics that formalize these terms are called *modal logics* (see, e.g., Reyes *et al.* [339], Mints [284]); in these logics, "A can happen" is usually described as $\Diamond A$, and "A must happen" as $\Box A$.

Our problem. Our problem is thus: *to add these modal operators to mathematics, and thus, go from interval mathematics to modal mathematics.*

Since our intended application is data processing, this generalization only makes sense while its results are still computable (and feasibly computable). So, among the first problems to handle are the problems of *computability* and *computational complexity* of the resulting modal mathematics. These problems will be handled in the chapter.

Historical comment. The fact that formulas and relations of interval mathematics can be described in terms of modal logic is not new: it was first noticed in Gardeñes [120] (see also Kreinovich *et al.* [212] and [236]).

C.2. Example, and an Idea of the General Description

Example. Let us use capital letters (e.g., X_i) to describe variables that are not necessarily uniquely determined by our knowledge, i.e., that can still take different values (e.g., values from an interval). In these terms, if a measurement leads us to a conclusion that the value of this variable belongs to an interval $[\underline{x}_i, \overline{x}_i]$, then we can express this knowledge as: $\Box(\underline{x}_i \le X_i \le \overline{x}_i)$. In these terms, the basic problem of interval computations can be, crudely speaking, reformulated as follows: given the values $\underline{x}_i$, $\overline{x}_i$, and y, check whether the following formula is true:

$$(\Box(\underline{x}_1 \le X_1 \le \overline{x}_1)\& \ldots \& \Box(\underline{x}_n \le X_n \le \overline{x}_n)) \to$$
$$\Box(f(X_1, \ldots, X_n) \le y). \quad \text{(C.1)}$$

Actually, by checking the validity of this formula, we check whether the upper bound of the range $f([\underline{x}_1, \overline{x}_1], \ldots, [\underline{x}_n, \overline{x}_n])$ is greater than a given number y or not. If we can do that for all numbers y, then, by applying bisection, we can compute the actual upper endpoint of the range interval with greater and greater accuracy.

Similarly, we can compute the lower endpoint of the range interval if we check a formula

$$(\Box(\underline{x}_1 \le X_i \le \overline{x}_1)\& \ldots \& \Box(\underline{x}_n \le X_n \le \overline{x}_n)) \to$$
$$\Box(f(X_1, \ldots, X_n) \ge y).$$

The idea of a general description. Let us denote by $\mathcal{X} \subseteq R^n$ the (unknown) set of possible values of tuples $(X_1, \ldots, X_n)$. For each tuple $(X_1, \ldots, X_n)$, it is easy to define truth values of elementary formulas (e.g., equations, inequalities, etc.), and formulas of first order logic (that are obtained from elementary ones by using propositional connectives "and", "or", "not", and quantifiers). If a formula A is well defined, then:

- we say that $\Diamond A$ is true if and only if A is true for *at least one* tuple from $\mathcal{X}$, and
- that $\Box A$ is true if A is true for *all* tuples from $\mathcal{X}$.

We then say that a formula is true if it is true for all sets $\mathcal{X} \subseteq R^n$.

In particular, the validity of formula (C.1) means the following:

- The formula $\Box(\underline{x}_i \le X_i \le \overline{x}_i)$ means that for every tuple from $\mathcal{X}$, the value X_i belongs to the interval $[\underline{x}_i, \overline{x}_i]$. In other words, this formula is true if and only if the set $\mathcal{X}$ is a subset of the box $B = [\underline{x}_1, \overline{x}_1] \times \ldots \times [\underline{x}_n, \overline{x}_n]$.
- The formula $\Box(f(X_1, \ldots, X_n) \le y)$ means that for every tuple $(X_1, \ldots, X_n) \in \mathcal{X}$, the value $f(X_1, \ldots, X_n)$ does not exceed y. In other words, this formula means that the range $f(\mathcal{X})$ of the function f on the set $\mathcal{X}$ belongs to the semi-line $(-\infty, y]$.
- Finally, the formula (C.1) itself means that if $\mathcal{X} \subseteq B$, then $f(\mathcal{X}) \subseteq (-\infty, y]$. To check this implication, it is sufficient to check it for $\mathcal{X} = B$, because from $F(B) \subseteq (-\infty, y]$ and $\mathcal{X} \subseteq B$, it follows that $f(\mathcal{X}) \subseteq f(B) \subseteq (-\infty, y]$. Thus, the formula (C.1) is equivalent to $f(B) \subseteq (-\infty, y]$.

Comment. The possibility of quantification over all possible *sets* (not only numbers from intervals) is what distinguishes this formalism from Shary's. In particular, this formalism leads to a new description of the so-called *Kaucher arithmetic* (see, e.g., [227]).

This description can also be generalized to describe unknown *functions* [236]: For example, we often know that a physical quantity y depend on some other quantity x (i.e., that $y = f(x)$ for an unknown function f); as a result of the measurements, we have intervals $[\underline{y}_i, \overline{y}_i]$ of possible values of $y_i = f(x_i)$ at given points $x_1 < \ldots < x_n$. A natural question is: is this information consistent with the assumption that the function f is monotonic?

For example, if $f(x)$ describes the dependency of the brightness y of the astronomical source on the coordinate x, then this question has a precise physical meaning: does this course constitute a single component, or it consists of several components? (There exist several other applications of this problem; these applications and a feasible algorithm for solving this problem are described in Villaverde *et al.* [420, 421].) In terms of modal logic, this problem can be formulated as follows: if we are ***sure*** that $f(x_i) \in [\underline{y}_i, \overline{y}_i]$ for all i, then is it ***possible*** that $f(x)$ is monotonic? In other words, is it true that

$$(\Box(\underline{y}_1 \leq f(x_1) \leq \overline{y}_1) \& \ldots \& \Box(\underline{y}_n \leq f(x_n) \leq \overline{y}_n)) \rightarrow$$
$$\Diamond \forall x \forall y (x < y \rightarrow f(x) \leq f(y)).$$

C.3. Definitions and the Main Result

Let us define the language of modal mathematics. Let us first describe the alphabet of the designed language. The formulas of the desired language will be formed from the following symbols:

- ***constants*** for all rational numbers;
- ***variables*** $x_1, \ldots$ (denoted by small letters) that run over all real numbers;
- ***modal variables*** $X_1, \ldots$ (denoted by capital letters);
- ***arithmetic operations*** $+$, $-$, $\cdot$, and $/$; and
- ***relations*** $=, <, \leq, >, \geq$.

Definintion C.1. *A term is defined as follows:*

- *every constant and every variable is a term;*
- *if t and t' are terms, then (t), $t + t'$, $t - t'$, $t \cdot t'$, and t/t' are terms;*
- *nothing else is a term.*

Definition C.2. *An* ***elementary formula*** *is defined as a formula of the type $t \circ t'$, where t and t' are terms, and $\circ$ is one of the relations (i.e., $=$, $<$, $\leq$, $>$, or $\geq$).*

Definition C.3. *A formula is defined as follows:*

- *every elementary formula is a formula;*
- *if F and F' are formulas, then (F), $\neg F$, $F \& F'$, $F \vee F'$, and $F \rightarrow F'$ are formulas;*
- *if F is a formula, and x_i is a variable, then $\forall x_i F$ and $\exists x_i F$ are formulas;*
- *if F is a formula, then $\Diamond F$ and $\Box F$ are formulas;*
- *nothing else is a formula.*

A formula is called quantifier-free if it does not use quantifiers $\forall$ and $\exists$, and modal-free if it does not use modalities $\Diamond$ and $\Box$. A formula is called closed if every variable x_i is within the scope of some quantifier, and every modal variable X_i is within the scope of some modality.

Comment. Modal-free formulas are exactly formulas of first order theory of real numbers described in Tarski, Seidenberg [408, 388].

To define the *truth value* of an arbitrary formula F of modal mathematics, we must select:

- some values $x_1, \ldots, x_m$ for all free numerical variables from this formula;
- some values $X_1, \ldots, X_f$ for all free modal variables from this formula, and
- a set $\mathcal{X} \subseteq R^n$ (where n is the total number of non-free modal variables in a formula F).

For this choice, the truth value of a formula F is defined as follows:

Definition C.4. *Let F be a modal formula. By a selection, we mean a tuple $\langle x_1, \ldots, x_m, X_1, \ldots, X_f, \mathcal{X} \rangle$, where m is the total number of free numerical variables, f and n are, correspondingly, the total numbers of free and non-free modal variables, x_i and X_j are real numbers, and $\mathcal{X} \subseteq R^n$.*

Definition C.5. *Let F be a formula, and let $\langle x_1, \ldots, x_m, X_1, \ldots, X_f, \mathcal{X} \rangle$ be a selection. Then, the truth value of a formula is defined as follows:*

- *The value of a term is defined in a straightforward way (as the result of the corresponding computations).*
- *Correspondingly, the truth value of an elementary formula $t \circ t'$ depends on whether the values of the terms t and t' are indeed connected by a relation $\circ$.*
- *The truth value of a composite formula $F\&F'$, $\forall x F$, etc., is defined according to the normal understanding of the logical operations &, $\forall$, etc.*
- *The formula $\Diamond A$ is true if and only if there exists a tuple $(X_1, \ldots, X_n) \in \mathcal{X}$ for which A is true.*
- *The formula $\Box A$ is true if and only if A is true for all tuples $(X_1, \ldots, X_n)$ from the set $\mathcal{X}$.*

Comment. In particular, for closed formulas, a selection consists of only the set $\mathcal{X}$.

Definition C.6. *We say that a closed formula is valid (in modal mathematics) if it is true for all sets $\mathcal{X} \subseteq R^n$.*

In terms of this definition, the basic computation problem of modal mathematics is to check whether a given closed formula is valid or not.

Theorem C.1.

- *For closed quantifier-free formulas:*
  - *Validity of modal-free formulas can be checked in polynomial time.*
  - *Checking validity of formulas that use modalities is an algorithmically decidable but NP-hard problem.*
- *For formulas with quantifiers:*
  - *Checking validity of arbitrary modal-free formulas is an algorithmically decidable problem.*
  - *Checking validity of arbitrary formulas with modality is an algorithmically undecidable problem.*

This result can be represented as a table:

	Modal-free formulas	Formulas with modalities
Quantifier-free formulas	Polynomial time	Decidable, NP-hard
Formulas with quantifiers	Decidable, $\geq$ exponential time	Undecidable

Proof of Theorem C.1.

Modal-free formulas: general comment. According of our definition of a closed formula, every modal formula must be within a scope of some modality. Therefore, if a closed formula is modal-free (i.e., contains no modalities), then it does not contain any modal variables at all.

Modal-free quantifier-free formulas. If a closed modal-free formula F is also quantifier-free, this means that F contains no variables at all, only constants, and the formula itself is a propositional combination of elementary formulas of the type $t \circ t'$, where t and t' are terms made composed from constants by applying elementary arithmetic operations. Computing the values of these terms step-by-step takes linear time (i.e., time that is bounded by a linear function of the size of the formula); comparing these values and applying propositional connectives to the Boolean-valued results of this comparison is also linear-time. So, as a result, we can check whether a formula is valid or not in linear (hence, in polynomial) time.

Modal-free formulas: general case. In general, if a closed modal-free formula is not necessarily quantifier-free, it is, as we have mentioned, a first order formula from the theory of real numbers described in Tarski and Seidenberg [408, 388]. For this theory, there are algorithms that test whether a given closed formula is valid or not; the first such algorithm was proposed by Tarski and Seidenberg [408, 388]; for more practical algorithms, see, e.g., Collins and Hong [72, 154, 73]. Therefore, for the class of modal-free formulas, the problem of checking whether a given formula is valid or not is algorithmically decidable.

Exponential (actually, doubly exponential) lower bounds for this problem were proven in Davenport [85].

Quantifier-free formulas with modalities. Let us first show that this problem is NP-hard. Indeed, we know (from the previous chapters) that the problem of computing the range of an interval function is NP-hard; actually, in our proofs, we have shown that the problem of checking whether, say, one of the endpoints of the resulting range interval is $\geq$ a given number, is NP-hard. But this problem, as we have shown, can be reformulated as a quantifier-free formula of modal mathematics. Thus, checking validity of such formulas is an NP-hard problem.

Let us now prove that checking validity of quantifier-free closed formulas F of modal mathematics is algorithmically decidable. For this, we will use the deciding algorithm from Tarski and Seidenberg [408, 388] or from Collins and Hong [72, 154, 73]. This algorithm transforms each first-order formula from the theory of real numbers (i.e., in our terms, each modal-free formula) into an equivalent quantifier-free modal-free formula (in particular, a closed formula is transformed into its Boolean value "true" or "false").

For each quantifier-free modal-free formula, we can describe the set of all tuples that make it true. Such a set is called *semi-algebraic*. By this definition, the union, complement, and intersection of semi-algebraic sets are also semi-algebraic (because they correspond to the disjunction $\vee$, conjunction &, and the negation $\neg$ of the corresponding formulas).

We will first show that if F is a quantifier-free formula with modalities, then each subformula of the formula F is equivalent to the propositional combination of quantifier-free modal-free formulas and formulas of the type $\mathcal{X} \subseteq A$ for semi-algebraic sets A. We will show this by induction over the length of the formula:

Induction base: Elementary formulas are quantifier-free modal-free, and therefore, they are themselves of the right type.

Induction step: Let us assume that all formulas shorter than a formula G are equivalent to formulas of the desired type. This means, in particular, that all subformulas of G are equivalent to formulas of the given type. Then, depending on the structure of the formula G, we have three possible situations:

- If G is a propositional combination (i.e., if $G = G'\&G''$, $G' \vee G''$, etc.), then, since each of its subformulas G', G'' is equivalent to a propositional combination of the formulas of the right type, G is also equivalent to such a propositional combination; so, for this case, the induction step is proven.

- Let us now consider the case when G has the form $\Box H$ for some formula H. We know that H is equivalent to a propositional combination H' of quantifier-free formulas and finitely many formulas of the type $\mathcal{X} \subseteq A$; let us denote the total number of such sets A by m, and the sets themselves by $A_1, \ldots, A_m$. Depending on whether $\mathcal{X} \subseteq A_j$ for each $j = 1, \ldots, m$, we have 2^m possible situations. Each situation will be denoted by s_k, where k is an m–digit number, in which 1 in i–th place means that $\mathcal{X} \subseteq A_i$, and 0 that $\mathcal{X} \not\subseteq A_i$. (For example, s_0 means that $\mathcal{X} \not\subseteq A_1 \& \ldots \& \mathcal{X} \not\subseteq A_p$.) For each of these situations s_k, H' can be reduced to a quantifier-free modal-free formula; we will denote such a formula by H_k. Hence, for every set $\mathcal{X}$ that is characterized by this situation s_k, the formula G of the type $\Box H$ is equivalent $\Box H'$, which, in its turn, is equivalent to $\forall X_1, \ldots, \forall X_n((X_1, \ldots, X_n) \in \mathcal{X} \to H_k)$. If we denote the set of all tuples that satisfy the condition H_k by A'_k, then this condition is equivalent to $\mathcal{X} \subseteq A'_k$. So, for each situation s_k, the formula $\Box H$ is equivalent to $\mathcal{X} \subseteq A'_k$. Therefore, in general, the formula $\Box H$ is equivalent to the following propositional combination

$$(s_1 \& (\mathcal{X} \subseteq A'_1)) \vee \ldots \vee (s_{2^m-1} \& (\mathcal{X} \subseteq A'_{2^m-1})).$$

  Each of the conditions s_k is already in the desired form, so H is also in the desired form.

- The case when G is of the type $\Diamond H$ can be reduced to the previous one, because, as one can easily check, $\Diamond H$ is equivalent to $\neg\Box(\neg H)$.

The statement is proven. In particular, it is applicable to the original closed formula F. Since this formula is closed, it has no modal variables left, and therefore, F is equivalent to a propositional combination of the formulas f_j of the type $\mathcal{X} \subseteq A$. This propositional combination can be reduced to a conjunctive normal form (CNF), i.e., to a formula $C_1 \& \ldots \& C_p$, where each subformula C_k (called *conjunction*) is of the type $a \vee \ldots \vee b$, and each of the subformulas $a, \ldots, b$ is either f_j, or $\neg f_j$. So, the validity of the formula F is equivalent to the fact that for every set $\mathcal{X}$, the conjunction $C_1 \& \ldots \& C_p$ is true. This means that for each set $\mathcal{X}$, each conjunction must be true. So, to check validity, it is sufficient to be able to check, for $k = 1, \ldots, p$, that each conjunction C_k is true for all sets $\mathcal{X} \subseteq R^n$.

If we take into consideration that each conjunction C_k is a conjunction of the formulas f_j and $\neg f_j$, and f_j is of the type $\mathcal{X} \subseteq A_j$, then (after, if necessary, reordering the terms inside C_k) we get the formula of the type

$$(\mathcal{X} \subseteq B_1) \vee \ldots (\mathcal{X} \subseteq B_q) \vee (\mathcal{X} \not\subseteq C_1) \vee \ldots \vee (\mathcal{X} \not\subseteq C_r)$$

for some semi-algebraic sets B_k and C_l. This formula is equivalent to

$$(\mathcal{X} \subseteq C_1 \& \ldots \& \mathcal{X} \subseteq C_r) \rightarrow (\mathcal{X} \subseteq B_1 \vee \ldots \mathcal{X} \subseteq B_q).$$

The condition of this implication is that $\mathcal{X}$ is a subset of r sets $C_1, \ldots, C_r$; this is equivalent to $\mathcal{X}$ being a subset of the intersection $C_1 \cap \ldots \cap C_r$. This intersection of semi-algebraic sets (we will denote it by C) is also semi-algebraic. So, for a given set $\mathcal{X}$, the truth of the formula F is equivalent to the truth of the following formula:

$$\mathcal{X} \subseteq C \rightarrow (\mathcal{X} \subseteq B_1 \vee \ldots \mathcal{X} \subseteq B_q). \tag{C.2}$$

The validity of F means that this implication must be true for all sets $\mathcal{X}$. Let us show that this formula is true for all $\mathcal{X}$ if and only if

$$C \subseteq B_1 \vee \ldots \vee C \subseteq B_q. \tag{C.3}$$

Indeed:

- If (C.2) is true for all sets $\mathcal{X}$, it must also be true for $\mathcal{X} = C$. For this set, the condition of (C.2) is true, and therefore, the conclusion must be true, and this conclusion is exactly (C.3).
- Vice versa, let (C.3) be true. Then, if a set $\mathcal{X}$ satisfies the condition of the implication (C.2), then $\mathcal{X}$ is a subset of C, and C (by (C.3)) is a subset of one of the sets B_k. Hence, $\mathcal{X}$ is a subset of one of the sets B_k, which is exactly the conclusion of the formula (C.2). So, the implication that constitutes the formula (C.2) is true.

So, validity of F is equivalent to the validity of a formula (C.3) with semi-algebraic sets C and B_k. The fact that these sets are semi-algebraic means that each of these sets is a set of all tuples $(X_1, \ldots, X_n)$ that satisfy a certain modal-free formula; we will denote the corresponding formulas by $F_C(X_1, \ldots, X_n)$ and $F_{B_k}(X_1, \ldots, X_n)$. In terms of these formulas, the condition $C \subseteq B_k$ becomes a first order (modal-free) statement $\forall x_1 \ldots \forall x_n (F_C(x_1, \ldots, x)n \rightarrow F_{B_k}(x_1, \ldots, x_n))$, and first order modal-free statement are algorithmically decidable (we can use Tarski-Seidenberg algorithm [408, 388] or a more modern algorithm from Collins and Hong [72, 154, 73]).

General case: formulas that contain both quantifiers and modalities. Let us now prove that the problem of checking validity of general formulas is algorithmically undecidable. To prove this statement, we will use the known fact that there exists no algorithm for checking whether a given *Diophantine equation* has a solution, i.e., whether for a given polynomial $P(x_1, \ldots, x_n)$ with integer coefficients, there exist natural numbers (non-negative integers) $x_1, \ldots, x_n$ for which $P(x_1, \ldots, x_n)$ Matiyasevich, Davis *et al.* [276, 86]. Let us show that checking whether such integers exist is equivalent to checking the validity of the following formula from our language:

$$(Z_1 \& \ldots Z_n \& C) \rightarrow \Diamond P(X_1, \ldots, X_n) = 0, \tag{C.4}$$

where Z_i stands for

$$\Diamond(X_i = 0) \& \forall x (\Diamond(X_i = x) \rightarrow \Diamond(X_i = x + 1)),$$

and C for

$$\forall x_1 \ldots \forall x_n ((\Diamond(X_1 = x_1) \& \ldots \& \Diamond(X_n = x_n)) \leftrightarrow \Diamond(X_1 = x_1 \& \ldots \& X_n = x_n)).$$

Indeed, according to our definitions, the validity of the statement Z_i means that the set $\mathcal{X}_i$ of all possible values of X_i (i.e., of all values X_i from the tuples $(X_1, \ldots, X_i, \ldots, X_n) \in \mathcal{X}$) contains 0 and contains $x+1$ with each its element x. This means, in particular, that it contains $0, 1, 2, \ldots$, i.e., all natural numbers.

The condition C means that if for each $i = 1, \ldots, n$, X_i is possible (i.e., $X_i \in \mathcal{X}_i$), then the tuple $(X_1, \ldots, X_n)$ is also possible (i.e., $(X_1, \ldots, X_n) \in \mathcal{X}$). In particular, since each natural number x_i is a possible value of X_i, an arbitrary tuple of natural numbers $(x_1, \ldots, x_n)$ belongs to the set $\mathcal{X}$.

The conclusion of the implication (C.4) means that $P(X_1, \ldots, X_n) = 0$ for some tuple $(X_1, \ldots, X_n) \in \mathcal{X}$.

If the equation $P(x_1, \ldots, x_n)$ has a solution $x_1^{(0)}, \ldots, x_n^{(0)}$ in which all values $x_i^{(0)}$ are natural numbers, then for every set $\mathcal{X}$, for which the conditions of the implication (C.4) are satisfied, the tuple $(x_1^{(0)}, \ldots, x_n^{(0)})$ corresponding to this solution is an element of $\mathcal{X}$, and therefore, the conclusion of the implication is true. Thus, if the original Diophantine equation has a solution, then the formula (C.4) is valid.

Vice versa, if the formula (C.4) is valid, then it is valid for an arbitrary set $\mathcal{X}$, in particular, for the set $\mathcal{X} = N^n$ (where N is the set of all integers). For this set, the conditions of the implication (C.4) are true, and therefore, the conclusion must also be true. Hence, the equation $P(X_1, \ldots, X_n)$ must have a solution $(X_1, \ldots, X_n) \in \mathcal{X}$. Since $\mathcal{X} = N^n$, this means that the original equation has a solution in natural numbers.

So, a Diophantine equation has a solution if and only if the corresponding formula (C.4) of modal mathematics is valid. Since it is impossible to algorithmically check whether a given Diophantine equation has a solution, it is thus impossible to check whether a given formula of modal mathematics is valid or not. Hence, in the general case, the problem of checking validity of formulas of modal mathematics is algorithmically undecidable. The theorem is proven.

D

BEYOND NP: TWO ROOTS GOOD, ONE ROOT BETTER

One of the main objectives of theoretical research in computational complexity and feasibility is to explain experimentally observed difference in complexity. In some cases, this experimental difference can be theoretically explained by proving that the experimentally harder problem is NP-hard, while the experimentally easier problem is computationally feasible. But sometimes, both problems are NP-hard. In this case, we need to find a way to compare NP-hard problems. This is what we will do in this appendix.

*Specifically, in this appendix, we explain why finding a **unique** root is easier than finding **multiple** roots. This chapter contains our joint results with R. B. Kearfott; these results were announced in Kreinovich and Kearfott [192, 193, 194, 195, 196, 201, 218].*

D.1. Formulation of the Problem: An Empirical Fact Needs to Be Explained

Experimental fact. The main objective of this appendix is to explain the following experimental facts (see, e.g., Kearfott [171, 172, 173, 174, 175]):

- In general, it is *easier* to find a solution $(x_1, \ldots, x_n)$ to a system of equations $f_1(x_1, \ldots, x_n) = 0, \ldots, f_k(x_1, \ldots, x_n) = 0$ when this system has a *unique* solution *than* when this system has *several* solutions.
- In general, it is *easier* to find a point $(x_1, \ldots, x_n)$, in which a given function $f(x_1, \ldots, x_n)$ attains its maximum, when there is only *one* such point, and much *harder* when there are *several*.

Two possible explanations. In principle, there are two possible explanations:

- this is a drawback of the existing methods; for other methods, finding non-unique solutions is as easy as finding unique ones;
- it actually *is* harder to find a non-unique solution.

The first explanation is quite possible: there are examples when a similar problem turned out to be a method's fault – e.g., Newton's method works *faster* and *better* if the root is in the middle of the domain, and *worse* if the root is close to the border. However, other methods find near-the-border roots much easier, almost as easily as the roots in the middle of the domain.

Both problems are, crudely speaking, NP-hard. We want to compare the complexity of two problems:

- solving systems of equations with *arbitrarily many* solutions, and
- solving the system of equations with a *unique* solution.

We already know that the general problem of solving a system of polynomial equations (without any limitations on the number of solutions) is NP-hard (it is even NP-hard for quadratic equations). We cannot exactly prove that finding the *unique* solution is NP-hard, but we can prove that it is "almost" NP-hard in the following precise sense: namely, using the reduction described in the proof of Theorem 3.1, one can show that the problem of finding the *unique* solution to a system of equations (or the unique point where the maximum is attained) is as complicated as the problem of finding the *unique satisfying vector* for a given propositional formula. The latter problem (it is usually denoted by *USAT*, from *u*nique *sat*isafiability) is known to be "almost" NP-hard in the sense that every other problem from the class NP can be *probabilistically* reduced to *USAT*; so if we were able to solve all the instances of *USAT* in polynomial time, we would have a *probabilistic* polynomial-time algorithm that solves almost all instances of all problems from the class NP. (Exact definition are somewhat complicated so, due to the lack of space, we refer the interested reader to Johnson [165] and Valiant *et al.* [417]. Note that in Beigel *et al.* [27], arguments are given that this problem may not be NP-hard.)

How can we compare these two situations? Since both problems seem to be equally difficult, how can we compare them?

D.2. Our Idea, and the Resulting Theoretical Explanation

Our idea. For polynomials whose coefficients are *rational* numbers, both problems are of approximately the same complexity (both are, crudely speaking, NP-hard). A natural idea is to consider polynomials whose coefficients are *computable* real numbers (see Appendix A for precise definitions). Such polynomials are called *computable*. For computable polynomials, we already get a clear distinction between unique and non-unique cases.

Indeed, for the case of the unique solutions, the following results are known in constructive mathematics (see, e.g., Kreinovich [193, 195], Kohlenbach [179, 180]):

Theorem D.1. *There exists an algorithm that is applicable to an arbitrary system of polynomial equations* $f_1(x_1,\ldots,x_n)=0,\ldots,f_k(x_1,\ldots,x_n)=0$ *(with computable* $f_i(x_1,\ldots,x_n)$*) that has a unique solution, and computes this solution.*

Theorem D.2. *There exists an algorithm that is applicable to an arbitrary computable polynomial* $f(x_1,\ldots,x_n)$ *on a computable box* $\mathbf{X}=\mathbf{x}_1\times\ldots\times\mathbf{x}_n$ *that attains its maximum on* $\mathbf{X}$ *at exactly one point* $x=(x_1,\ldots,x_n)$, *and computes this point* x.

If we allow two or more roots, the complexity of the problem changes drastically:

Theorem D.3. *No algorithm is possible that is applicable to any polynomial function* $f(x)$ *with exactly two roots, and returns these two roots.*

Theorem D.4. *No algorithm is possible that is applicable to any polynomial function* $f(x)$ *that attains its maximum at exactly two points, and returns these two points.*

These theorems explain why it is much more difficult to find all the roots if there are two of them than if there is a unique root. In other words, we *have a theoretical explanation* of the above-mentioned experimental fact.

	Unique solution	Two solutions
System of equations	Decidable, "almost" NP-hard	Algorithmically undecidable
Optimization	Decidable, "almost" NP-hard	Algorithmically undecidable

D.3. Philosophical Comment: These Results Form the Basis for Optimism

The problem of finding a unique solution is algorithmically solvable not only for computable *polynomials* $f_k(x_1, \ldots, x_n)$, but also for *arbitrary computable functions* $f_k(x_1, \ldots, x_n)$. On the other hand, if we have at least two solutions, no general algorithm is possible. This result can be viewed as a foundation of *optimism* (see, e.g., [200]).

Let us illustrate this idea on the example of *historical processes*. It often happens in history that a country is in a bad situation; it may be that a tyrant is ruining it, it may be that the enemies are devastating it. As times goes by, the situation gets gloomier and gloomier; it seems that there are fewer and fewer chances of survival, and then, in the darkest hour, when there seems to be the only easy-to-miss chance of survival, a miracle happens and this only possible way out is indeed followed.

It often happens that the army wins only after it was almost defeated. It often happens that the personal enlightment comes only after a person has plunged into despair. It often happens that a proof of the theorem comes only after unsuccessful attempts has almost led to despair. The above results explain these "miracles": according to these results, in general, it is much easier to find a way out when there is the only way out left, than to find it when there were still several possible.

This result teaches us to be *optimistic*: if the situation gets gloomier and gloomier we should not despair but try harder, and then, hopefully, when there will be only one way out left, we will find it!

Proofs

We will only give the proofs of Theorems D.3 and D.4. Both proofs use the fact that it is algorithmically impossible to tell whether a real number is equal to 0 or not (see, e.g., [191], or Bishop [48], Bridges [58], Beeson [26], Bishop *et al.* [49]; in Appendix A, we had, in effect, proved this impossibility).

Proof of Theorem D.3. We will prove this Theorem by reduction to a contradiction. Assume that such an algorithm U exists. So, U is applicable to an arbitrary polynomial with exactly two roots, and returns exactly these roots. As an example of such a polynomial, let's take

$$f_\alpha(x) = (x - 1 - \alpha^2) \cdot (x - 1 + \alpha^2) \cdot ((x + 1)^2 + \alpha^2),$$

where α is some constructive real number.

It is easy to check that for every α, this polynomial has exactly two roots. Indeed, $f_\alpha(x)$ is the product of three factors, so $f_\alpha(x) = 0$ if and only if one of these factors is equal to 0. We will consider two cases:

- If $\alpha = 0$, then $f_\alpha(x) = (x - 1)^2 \cdot (x + 1)^2$, so $f_\alpha(x) = 0$ if either $x = 1$, or $x = -1$.
- If $\alpha \neq 0$, then the third factor is positive, so for $f_\alpha(x)$ to be 0, one of the first two factors must be equal to 0. In other words, the roots are $x = 1 - \alpha^2$ and $x = 1 + \alpha^2$.

Now, we can get the desired contradiction: for every constructive number α, we can apply U to the polynomial $f_\alpha(x)$ and get the roots with an arbitrary accuracy. Let's compute them with the accuracy 1/4. Depending on whether $\alpha = 0$ or not, we have two cases:

- If $\alpha = 0$, then one of the roots is -1, so the (1/4)-approximation to this root will be a negative rational number.
- If $\alpha \neq 0$, then both roots are $\geq 1 - (1/2)^2 = 3/4$, hence, their (1/4)–approximations are greater than 0.

So, if one of the approximations is negative, then $\alpha = 0$, else $\alpha \neq 0$. Hence, based on U, we can construct the following algorithm V that would check whether a constructive real number is equal to 0 or not:

- apply U to $f_\alpha(x)$, and compute both roots with accuracy 1/4;
- if both resulting approximations are positive, return the answer "$\alpha \neq 0$", else return the answer "$\alpha = 0$".

But we have already mentioned that such an algorithm is impossible. So, our initial assumption (that an algorithm U exists) was wrong. The theorem is proven.

Proof of Theorem D.4. The proof is similar to the proof of Theorem D.3, with the only difference that as $f_\alpha(x)$, we now take

$$f_\alpha(x) = -(x - 1 - \alpha^2)^2 \cdot (x - 1 + \alpha^2) \cdot ((x + 1)^2 + \alpha^2).$$

This function is always non-negative, and since it attains 0 (in exactly the same points as the function from the proof of Theorem 1), its maximum is equal to 0. So, the polynomial $f_\alpha(x)$ attains its maximum for some x if and only if $f_\alpha(x) = 0$. Then, similar arguments complete the proof. The theorem is proven.

E

DOES "NP-HARD" REALLY MEAN "INTRACTABLE"?

Most computer scientists believe that NP-hard problem are really computationally intractable. This belief is well justified for traditional computers, but there are non-traditional physical and engineering ideas that may make NP-hard problem easily solvable. These ideas are briefly overviewed in this appendix.

Parallelism is desirable for interval computations. Many problems of interval computations and data processing are NP-hard. This means that in the worst case, computations take a very long time. To speed up these computations, it is desirable to have several computers working in *parallel.*

Parallelism is useful. Parallelization has indeed been successfully used in interval computations: see, e.g., Bernat *et al.* [215, 214, 216, 34, 35, 36, 37], Bhamidipati [44], Caprani *et al.* [61], Cooke [74], Deboeck *et al.* [87], Doser *et al.* [95], Eriksson *et al.* [102], Henriksen *et al.* [150], Hu *et al.* [158, 159], Kreinovich *et al.* [213, 237], Leclerc [250], Morgenstein *et al.* [294], Nemir *et al.* [299], Nguyen *et al.* [304, 307], Schaefer *et al.* [375], Schnepper *et al.* [381], Villa *et al.* [419], Villaverde *et al.* [421], Wolff von Gudenberg [432] and references therein.

Even with the fastest parallel machines, we still cannot solve some problems. Even on the fastest parallel computers, some problems cannot be solved. There have been several attempts to design hardware specifically tailored towards parallel interval computations (see, e.g., Schulte *et al.* [384], Cooke *et al.* [75], Nguyen *et al.* [306]), but even for specifically tailored hardware, the worst-case computation time remains unrealistically exponential.

What can we do?

Within Newtonian physics, NP-hardness does seem to mean "intractable". The common belief that NP-hard means intractable is based on the abilities of the physical processes that are used in the existing computers; it has been proven that is we only use processes from Newtonian physics, then we do not add additional ability to the computational devices (for exact formulations and proofs, see, e.g., Gandy [117, 118]).

Within traditional (Newtonian) physical and engineering solutions, NP-hard seems to indeed mean "intractable". Indeed, the existing computations schemes describe (more or less accurately) the ability of the modern computers. The only thing that is missing from the standard algorithms is *randomness*, i.e., the ability to input *truly random* data and use them in computations. In the language of theory of computation, the outside source of data is called an *oracle*. As early as 1981, Bennet *et al.* has shown [32] that if we allow a random sequence as an oracle, and correspondingly reformulate the definitions of the classes P and NP, then we can *prove* that P≠NP [32].

What if we use non-traditional physical and engineering ideas in computer design? Since we seem not to be able to avoid the unrealistic exponential time with traditional, Newtonian-physics-based computers, a question naturally appears: what if in the future, we will find *non-Newtonian* processes; will then NP-hard problems still be intractable? This question was first formulated by G. Kreisel [238].

In this appendix, we will show that by using some of these processes, we will be able to compute NP-processes fast.

Classification of possible non-Newtonian physical processes. Traditional computers use discrete-oriented deterministic processes in normal space and time. In reality, physical processes are (1) continuous, (2) non-deterministic (as described by *quantum mechanics*), and (3) they occur in non-traditional (*curved*) space-time. So, to describe how using additional physical processes will help in computations, we must consider how these three factors (adding non-determinism and taking curvature into consideration) change our computational abilities.

Non-Newtonian processes of first type: Use of physical fields. For a physical field, the value $f(\vec{x}, t)$ of the field f in a future moment of time t can be expressed in terms of the current state of this field $f(\vec{x}, 0)$ by an explicit integral formula. This formula is usually computable on existing computers, and therefore, the evolution described by the fields is *recursive* (relevant theorems are proved, e.g., in Pour-El *et al.* [332]).

In some cases, however, $f(\vec{x},t)$ is described as an integral in terms of the function $f(\vec{x},t)$ and its spatial derivatives. So, if we start with a function $f(\vec{x},0)$ that is recursive, but whose (spatial) derivatives are not recursive, we may end up with a *non-recursive value* $f(\vec{x},t)$. This was shown by Pour-El *et al.* in [331] (see also Beeson [26], Ch. 15, and Pour-El *et al.* [332]). This result generalizes a theorem proved by Aberth in 1971 and rediscovered in Pour-El *et al.* [330].

Another possible way of using fields to speed up computations is described in Beltran *et al.* [29].

Comment. This result does not necessarily mean that we have found a way to compute a function that is not computable on a normal computer (see, e.g., Kreisel [239]), because for that, we would need to find a way to implement the initial conditions with a non-recursive derivative. A more definite possibility of solving NP-hard problem fast comes from the other two aspects of physical processes.

Non-Newtonian processes of second type: Quantum processes (adding non-determinism). In [90], Deutch has shown that the use of quantum processes can potentially speed up computations (see also Penrose [325], Ch. 4, p. 146). The fact that quantum processes can speed up computations follows from the fact that the prediction problem in quantum mechanics is NP-hard [235] (see also Kosheleva *et al.* [184]), and therefore, if we can simply wait to see what will happen, we will thus get (in time t) the values whose computation on a normal computer would require a time that grows exponentially with t.

Specifically, in Stannett [403], it is shown that in some *versions of quantum mechanics, relativistic physics, and algebraic field theory*, it is possible not only to solve NP-hard problems fast, but even to compute functions that are not computable on regular computers at all. For several approaches to *quantum gravity*, a similar property is described in Geroch *et al.* [125] and in Penrose [325], Ch. 8, and Ch. 10, p. 431.

Comment 1. Penrose in [325], Ch. 9, pp. 400–402, provides arguments that *quantum processes* may be essential for the functioning of the brain (in particular, he cites experimental data from Hecht *et al.* [142] and Baylor *et al.* [21]). He also remarks that the well-known fact that our brain easily solves problems that are still difficult for modern computers, may be an indication that our brain uses non-Newtonian processes for computations (and thus. it has additional computational abilities).

Comment 2. Rosen in [367], Ch. 11, p. 257, argues that biological processes cannot be explained by the existing physical processes, and thus concludes that some (***not yet known) physical processes*** must be responsible for these processes.

Comment 3. In Brasher *et al.* [57], it is proved that quantum processes can not only ***speed up*** computations, they can also ***decrease the energy consumption*** of the computation processes.

Non-Newtonian processes of third type: Using curved space-time for computations. If we allow heavily curved space (e.g., semi-closed ***black holes***), we can get the results faster if we stay in the area where the curvature is strong and time goes slower, and let the computations be done outside (see, e.g., Morgenstein *et al.* [200, 287]); then, we will even be able to compute NP-hard problems in polynomial time.

Some physical theories describe ***acausal processes*** (e.g., closed timelike curves that enable us to send information from the future to the past). The possibility of such processes is seriously discussed in physics (see, e.g., Tipler, Thorne, *et al.* [411, 412, 413, 295, 296, 310, 132, 319, 409, 311, 11]). Rosen in [365] provides ***biological*** motivations for the existence of such processes: namely, he suggests that the living beings can use physical processes that influence the current events depending on the future ones (he calls such acausal processes ***anticipatory***; see also [364, 366, 367, 368]).

If we allow such processes, then talking about computation time may make no sense: we can spend all the time we want on solving a problem, and then simply send the result back in time to the moment of time when we asked for a solution. In this case, many computational steps pass, but we get the solution exactly at the same moment when we ask for it (i.e., we do not feel any waiting at all). This idea was described in Kosheleva *et al.* [185] (see also Maslov [275] and Kreinovich [205]), Moravec [292], and Nahin ([298], pp. 202–203).

An even more unusual idea has been proposed by Penrose: using his analysis of how new ideas come to mind (sometimes we kind of "see" them), he argues that the platonic world of ideas may physically exist, and communication with this world is part of our conscience [325], Ch. 10, p. 427. If we assume that there exists a platonic world of ideas in which time is not defined, then we can simply read the solution from that world without bothering about time at all.

F

BRIGHT SIDES OF NP-HARDNESS OF INTERVAL COMPUTATIONS II: FREEDOM OF WILL?

In this book, we have proven that many computational problems of data processing and interval computations are NP-hard. The immediate conclusion of these results is negative: *one cannot expect an algorithm that solves all the problems of data processing and interval computations in reasonable time. However, as we will mention in this chapter, the NP-hardness results also have their* bright sides. *The first bright side was described in Chapter 15: NP-hardness interval computations enables us to apply efficient heuristic methods, originally developed for interval computations, to other complicated problems, and thus, get new heuristics.*

In this appendix, we present a more speculative idea: namely, we show that this NP-hardness may help us to somewhat patch the seeming contradiction between the determinism of modern physics and the notion of freedom of will.

F.1. Freedom of Will

Determinism of modern physics. Physical theories are usually described in terms of differential equations; in these theories, the *current* state of the world is uniquely determined by the *initial* state of the world. So, if we know the initial state of the world, then we can uniquely determine the state of the world at any future moment of time.

A more accurate description of this statement is as follows: if we know the initial state of the *model* of a part of the world, then we can determine the future state of the model.

This prediction is not an abstract possibility: physical equations often lead to reasonably efficient predictions.

Comment. A comment is in order regarding quantum physics:

- In classical (non-quantum) physics, the state of the world is uniquely determined by the corresponding differential equations, and it uniquely determines the results of all possible experiments.
- In quantum mechanics, the state ψ is still uniquely determined by Schroedinger's equations, but this state does not determine the results of the experiments: it only determines the *probabilities* of different experimental results.

Freedom of will. This *determinism* contradicts to the idea of *freedom of will*, i.e., to the fact that we humans feel ourselves capable of making free decisions – decisions that cannot be predicted. According to the idea of the freedom of will, not only the decisions cannot be predicted, but we cannot even always predict what decisions will be possible.

At first glance, there seems to be a contradiction. At first glance, there seems to be a contradiction between determinism and the freedom of will, and philosophers have been viewing it as such.

Contradiction disappears if we take interval uncertainty into consideration. Let us show that the seeming contradiction between physics and freedom of will almost disappear if we take into consideration that we never know the initial state of the world precisely:

Indeed, we get this state from measurements, and measurements are never absolutely precise. As a result, for every measured physical quantity x, we do not know its exact value, we only know the result $\tilde{x}$ of measuring x, together with the guaranteed accuracy Δ of the measuring device. From this measurement, we can only conclude that the actual value x is somewhere on the *interval* $[\tilde{x} - \Delta, \tilde{x} + \Delta]$. So, instead of knowing the *exact* values of the parameters of the initial state, we only know *intervals* of possible values of these parameters. For intervals, as we have seen, the problem of computing the exact range is NP-hard, and therefore, it is not possible to compute, in reasonable time, the interval of possible values of parameters that describe the future state. This impossibility means that we cannot predict which decisions will be possible and which decision will not. In other words, this impossibility leaves room for freedom of will.

Disclaimer: this is not yet a solution of the freedom of will problem, but it may be a step towards the solution. Our remarks, of course, do not explain whether freedom of will exists of not, which is *the* basic question argued by theologians and philosophers over the centuries; our remarks only say that in spite of a seeming contradiction with physical determinism, there seems to be a room for freedom of will.

Historical comments.

- The relationship between freedom of will and computational complexity was mentioned by Penrose in [325]; see also Balogh [15] and Alefeld *et al.* [3, 4].
- For *quantum* systems, a similar result (that prediction is NP-hard) was proved in Kreinovich *et al.* [235]; for *gravitational* systems, it was proven in Kreinovich [203]. We have shown that this result is true even for simple linearized equations of *classical physics.*

F.2. Possible (Speculative) Consequences

Our conclusion about freedom of will has the following two (even more speculative) consequences. We fully realize that these consequences are speculative and that we are not specialists in this field. We hope however that these comments will be useful to the corresponding specialists:

Is absolute totalitarianism possible? L. A. Levin, one of the authors of the notion of NP-hardness, remarked that the impossibility to always predict the exact consequences of a person's behavior means that *absolute totalitarianism is impossible.* Indeed, the idea of a totalitarian control is based on the assumption that dictators can always predict the results of their pressure, and choose the strategy that guarantees their power in the future. Hopefully, such a prediction is impossible, and therefore, the dictators will always make an erroneous move with unpredicted results that they were trying to avoid.

Towards solving the sociobiological dilemma. The impossibility of prediction may also help in resolving the following *sociobiological dilemma* (see, e.g., Wilson [430, 431], Lumsden *et al.* [260, 261], MacDonald [264], Ruse [373]):

- On one hand, many biologists believe that *genes determine* many important human characteristics, including many characteristics related to *intelligence.*
- However, on the other side, they abhor the (seemingly inevitable) conclusion of this belief, that *people with certain genetic characteristics are intellectually inferior* to others.

The computational intractability of prediction helps us *avoid* the inevitability of *this conclusion*: genes may *theoretically* determine our behavior, but *in practice*, it may be computationally intractable to make any behavior predictions based on the genetic characteristics.

G

THE WORSE, THE BETTER: PARADOXICAL COMPUTATIONAL COMPLEXITY OF INTERVAL COMPUTATIONS AND DATA PROCESSING

In this book, we have presented many results about computational complexity and feasibility of interval computations and data processing.

*From the **technical** viewpoint, **most important** practical **problems**, related to data processing and interval computations, are either known to be computational intractable, or known to be computationally feasible. There are only a few (relatively minor) problems for which it is not known whether a problem is feasible or not, so the reader may get an impression that this area of research is almost finished.*

*To avoid creating this false impression, we decided to finish this book with this appendix that describes the **fundamental challenge**: many important technical results about computational complexity and feasibility of interval computations and data processing, with all their technical sound proofs, are actually intuitively **paradoxical**. Namely, problems that should intuitively be more complicated actually turn out to be computationally easier, and vice versa.*

We believe that if somebody can come up with a reasonable explanation of this paradoxical complexity, this explanation will greatly enhance our understanding of the fundamental problems of computing. With this grand challenge, we finish the book.

(This appendix was written in collaboration with V. Nesterov. Its main ideas first appeared in Nesterov et al. [301].*)*

Intuitive complexity of a computational problem: three main factors. Intuitively, the complexity of the computational problem depends on the following three factors:

- The first factor is the *size* of the area where solution has to be found. This size describes the total number of objects that we need to analyze in order to find a solution; so, the larger this size, the more complicated the problem.

  In general, this conclusion is *true*: e.g., the more variables a problem has, the more difficult it is to solve it. However, we will see that this intuitive conclusion is *not always true*.

- Another factor is the *number of solutions*. Intuitively, the more objects are solutions to our problem, the easier it must be to find one. For example, if half of the possible objects are solutions, then we can find one by several successive random number simulations.

  Again, we will show that this is not always the case, and sometimes, the unique solution is easier to find than a multiple one.

- Finally, the complexity of the problem is determined by the *complexity of the condition* that we want the desired solution to satisfy.

  We will see that sometimes, problems with more complicated conditions are easier to solve.

In this appendix, we will briefly describe the corresponding basic examples of paradoxical computational complexity.

Decreasing the size of the area, in which a solution has to be found, can drastically increase the computational complexity. Traditionally, problems of interval computations are solved under the assumption that the corresponding variables x_i take arbitrary values from the given intervals $[\underline{x}_i, \overline{x}_i]$. A typical problem is the problem of *range estimation*:

GIVEN:

- an integer n;
- a polynomial $f(x_1, \ldots, x_n)$ with rational coefficients;
- n intervals $\mathbf{x}_i = [\underline{x}_i, \overline{x}_i]$ $(1 \leq i \leq n)$ with rational endpoints; and
- a rational number y.

CHECK whether y belongs to the range of possible values of $f(x_1, \ldots, x_n)$ (i.e., whether y can be represented as $f(x_1, \ldots, x_n)$ for some $x_i \in \mathbf{x}_i$).

For *linear* functions $f(x_1, \ldots, x_n)$, this problem is *easy to solve*; for *polynomial* functions $f(x_1, \ldots, x_n)$, it is NP-hard but still *algorithmically solvable* (even if we allow infinite intervals $\mathbf{x}_i$).

Here comes a paradox. In many real-life situations, we know that a quantity x_i takes only values proportional to a certain quantum q_i (electric charge is a typical example). If we impose this additional condition that the ratio x_i/q_i is an integer, then, at first glance, we *drastically decrease the size of the area* where the solution can be found. However, as we have seen, the *computational complexity* immediately *increases*:

- even for *linear* functions $f(x_1, \ldots, x_n)$, the problem becomes *NP-hard* (i.e., crudely speaking, not solvable by a feasible algorithm), and
- in the general case of *arbitrary polynomials* and possibly infinite intervals, the problem becomes *algorithmically undecidable*.

The fewer solutions, the ... easier to find them. Intuitively, as we have mentioned, the fewer solutions, the more difficult it is to find one. However, it is well known in numerical computations that numerical methods converge faster if there is a *unique* solution (e.g., the unique root). Since there is a clear contradiction between the intuitive expectations and the existing numerical techniques, it is natural to ask a question:

- is this easy-when-few-roots phenomenon caused by the inadequacy of the known methods, or
- is a general property of all possible methods?

Surprisingly, the second answer is correct, and intuition is all wrong here; as we have mentioned earlier in the book:

- *there exists* a general *algorithm* for finding roots of computable functions that have *only one* root, but
- it is *algorithmically impossible* to find a root of all functions who have *one or two roots*.

Sometimes, problems involving more complicated properties are easier to solve. Finally, *computational complexity of a problem* "find an object x that satisfies a given property $P(x)$" (or "check whether a given object x satisfies a given property $P(x)$") is influenced by the *computational complexity of the* corresponding *property* $P(x)$. In particular, the range problem of interval computations can be reformulated as the problem of checking the formula

$$\exists x_1 \ldots \exists x_n (\underline{x}_i \le x_i \le \overline{x}_i \& f(x_1, \ldots, x_n) = y).$$

This formula has a *quantifier complexity 1* (meaning that it contains only one type of quantifiers), and it may seem, at first glance, that it is easier to check than any problem with higher quantifier complexity.

However, if we restrict ourselves to *monotonic* functions, i.e., if we check formulas

$$\exists x_1 \ldots \exists x_n (\underline{x}_i \le x_i \le \overline{x}_i \& f(x_1, \ldots, x_n) = y) \&$$
$$\forall x_i, x_i' (x_i < x_i' \rightarrow$$
$$f(x_1, \ldots, x_{i-1}, x_i, x_{i+1}, \ldots, x_n) \le f(x_1, \ldots, x_{i-1}, x_i', x_{i+1}, \ldots, x_n))$$

of quantifier complexity 2, the range problem becomes easy to solve, *much easier* than the general range problem.

Summarizing:

The results about
computational complexity and feasibility
of interval computations and data processing
are often paradoxical and counter-intuitive.

To explain
these paradoxical results
is a fundamental challenge.

With this challenge, we finish the book.

Thank you for being with us!

REFERENCES

[1] C. Abdallah, P. Dorato, R. Liska, S. Steinberg, and W. Yang, "Applications os quantifier elimination theory to control system design", *4th IEEE Mediterranean Symposium on Control and Automation*, 1996.

[2] J. Abello, V. Kreinovich, H. T. Nguyen, S. Sudarsky, and J. Yen, "Computing an appropriate control strategy based only on a given plant's rule-based model is NP-hard", In: L. Hall, H. Ying, R. Langari, and J. Yen (eds.), *NAFIPS/IFIS/NASA'94, Proceedings of the First International Joint Conference of The North American Fuzzy Information Processing Society Biannual Conference, The Industrial Fuzzy Control and Intelligent Systems Conference, and The NASA Joint Technology Workshop on Neural Networks and Fuzzy Logic, San Antonio, December 18–21, 1994*, IEEE, Piscataway, NJ, pp. 331–332.

[3] G. Alefeld, M. Koshelev, and G. Mayer, "Fixed Future and Uncertain Past: Theorems Explain Why It Is Often More Difficult To Reconstruct the Past Than to Predict the Future", *Proceedings of the NASA University Research Centers Conference*, Albuquerque, New Mexico, February 16–19, 1997, pp. 23–27.

[4] G. Alefeld, M. Koshelev, and G. Mayer, "Why is it computationally harder to reconstruct the past than to predict the future?", *International Journal of Theoretical Physics*, 1997, Vol. 36, No. 8, pp. 1709–1715.

[5] G. Alefeld, V. Kreinovich, and G. Mayer, "Symmetric Linear Systems with Perturbed Input Data", In: Götz Alefeld and Jürgen Herzberger (eds.), *Numerical Methods and Error Bounds. Proceedings of the IMACS-GAMM International Symposium on Numerical Methods and Error Bounds, Oldenburg, Germany, July 9–12, 1995*, Akademie Verlag, Berlin, 1996, pp. 16–22.

[6] G. Alefeld, V. Kreinovich, and G. Mayer, "The shape of the symmetric solution set", In: R. B. Kearfott et al (eds.), *Applications of Interval Computations*, Kluwer, Dordrecht, 1996, pp. 61–79.

[7] G. Alefeld, V. Kreinovich, and G. Mayer, "The Shape of the Symmetric, Persymmetric, and Skew–Symmetric Solution Set", *SIAM Journal on Matrix Analysis and Applications (SIMAX)*, 1997, Vol. 18, No. 3, pp. 693–705.

[8] G. Alefeld, V. Kreinovich, and G. Mayer, "The Shape of the Solution Set for Systems of Interval Linear Equations with Dependent Coefficients", *Mathematische Nachrichten* (to appear; preliminary version appeared as University of Texas at El Paso, Department of Computer Science, Technical Report UTEP-CS-95-39a, 1995; available by ftp from **cs.utep.edu**, login **anonymous**, file **pub/reports/tr95-39a.tex**).

[9] G. Alefeld and G. Mayer, "On the symmetric and unsymmetric solution set of interval systems," *SIAM J. Matr. Anal. Appl.*, 1995, Vol. 16, pp. 1223–1240.

[10] G. Alefeld and J. Herzberger, *Introduction to Interval Computations*, Academic Press, N. Y. 1983.

[11] B. Allen and J. Simon, "Time travel on a string", *Nature*, 1992, Vol. 357, May 7, pp. 19–21.

[12] E. Allender, "Applications of time-bounded Kolmogorov complexity in complexity theory", In: O. Watanabe (ed.), *Kolmogorov complexity and computational complexity*, Springer-Verlag, Berlin, Heidelberg, 1992, pp. 4–22.

[13] K. Ambos-Spies, H. Fleischhack, and H. Huwig, "Diagonalizations over polynomial-time computable sets", *Theoretical Computer Science*, 1987, Vol. 51, pp. 177–204.

[14] T. V. Arak and A. Yu. Zaitsev, *Uniform limit theorems for sums of independent random variables* (Proceedings of the Steklov Institute of Mathematics, Vol. 174), American Mathematical Society, Providence, RI, 1988.

[15] I. Balogh et al., "Responses to 'Computationalism'", *Social Epistemology*, 1990, Vol. 4, No. 2, pp. 155–199.

[16] B. R. Barmish, *New tools for robustness of linear systems*, McMillan, N.Y., 1994.

[17] B. R. Barmish, M. Fu, and S. Saleh, "Stability of a polytope of matrices: Counterexamples", *IEEE Transactions on Automatic Control*, 1988, Vol. 33, pp. 569–572.

[18] B. R. Barmish and C. V. Hollot, "Counter–example to a recent result on the stability of interval matrices by S. Białas", *International Journal of Control*, 1984, Vol. 39, pp. 1103–1104.

[19] J. Barwise (ed.). *Handbook of Mathematical Logic*. North-Holland, Amsterdam, 1977.

[20] M. Baumann, "A regularity criterion for interval matrices", In J. Garloff et al,, ed., *Collection of Scientific Papers Honouring Prof. Dr. K. Nickel on Occasion of his 60th Birthday, Part I*, Freiburg University, Freiburg, 1984, pp. 45–50.

[21] D. A. Baylor, T. D. Lamb, and K.-W. Yau, "Responses of retinal rods to single photons", *Journal of Physiology*, 1979, Vol. 288, pp. 613–634.

[22] O. Beaumont, *Evaluation of Polynomials through the Simplex Algorithm*, April 1996, Institut de Recherche en Informatique et Systèmes Aléatoires, Publication Interne No. 1007.

[23] H. Beeck, "Zur Problematik der Hüllenbestimmung von Intervallgleichungssystemen", In K. Nickel, ed., *Interval Mathematics*, Lecture Notes in Computer Science, Vol. 29, Springer–Verlag, Berlin, 1975, pp. 150–159.

[24] R. Beigel, H. Burman, and L. Fortnow, "NP might not be as easy as detecting unique solutions", *Complexity Conference Abstracts 1997*, June 1997, Abstract No. 97–5, p. 8; full version is available from http://www.cs.uchicago.edu/ fortnow/papers.

preliminary version appeared as University of Texas at El Paso, Department of Computer Science, Technical Report UTEP-CS-95-46a, 1995; available by ftp from **cs.utep.edu**, login **anonymous**, file **pub/reports/tr95-46a.tex**).

[25] R. Bellman, *Introduction to matrix analysis*, McGraw Hill Co., N.Y., 1970.

[26] M. J. Beeson, *Foundations of constructive mathematics*, Springer-Verlag, N.Y., 1985.

[27] R. Beigel, H. Burnham, and L. Fortnow, "NP might not be as easy as detecting unique solutions", *Complexity Conference Abstracts 1997*, June 1997, Abstract No. 97–5, p. 8 (full text is available from http://www.cs.uchicago.edu/ fortnow/papers).

[28] G. Belforte and B. Bona, "An improved parameter identification algorithm for signal with unknown-but-bounded errors", *Proceeding of the 7th IFAC Symposium on Identification and Parameter Estimation*, York, U.K., 1985.

[29] A. Beltran, V. Kreinovich, and L. Longpré, *QFT + NP = P Quantum Field Theory (QFT): A Possible Way of Solving NP-Complete Problems in Polynomial Time*, University of Texas at El Paso, Department of Computer Science, Technical Report UTEP-CS-96-45, 1996; available by ftp from `cs.utep.edu`, login `anonymous`, file `pub/reports/tr96-45.tex`.

[30] M. Beltran, G. Castillo, and V. Kreinovich, "Algorithms That Still Produce a Solution (Maybe Not Optimal) Even When Interrupted: Shary's Idea Justified", *Reliable Computing*, 1998, Vol. 4, No. 1 (to appear; preliminary version appeared as University of Texas at El Paso, Department of Computer Science, Technical Report UTEP-CS-95-46a, 1995; available by ftp from `cs.utep.edu`, login `anonymous`, file `pub/reports/tr95-46a.tex`).

[31] N. Ben-Or, D. Kozen, and J. Reif, "The complexity of elementary algebra and geometry", *Journal of Computer and System Sciences*, 1986, Vol. 32, pp. 251–264.

[32] G. G. Bennet and J. Gill, "Relative to a random oracle A, $P^A \neq NP^A \neq$co-NP^A with probability 1", SIAM Journal of Computer Science, 1981, Vol. 10, pp. 96–113.

[33] A. Berman and R. J. Plemmons, *Nonnegative Matrices in the Mathematical Sciences*, SIAM, Philadelphia, 1994.

[34] A. Bernat, L. Cortes, V. Kreinovich, and K. Villaverde. "Intelligent parallel simulation – a key to intractable problems of information processing." *Proceedings of the Twenty-Third Annual Pittsburgh Conference on Modelling and Simulation*, Pittsburgh, PA, 1992, Part 2, pp. 959–969.

[35] A. Bernat and V. Kreinovich (eds.), Special issue on parallel interval computations, *Interval Computations*, 1994, No. 3.

[36] A. Bernat and V. Kreinovich (eds.), Special issue on parallel algorithms for interval computations, *Reliable Computing*, 1995, Vol. 1, No. 2.

[37] A. Bernat, E. Villa, K. Bhamidipati, V. Kreinovich, "Parallel interval computations as a background problem: when processors come and go", *International Conference on Interval and Computer-Algebraic Methods in Science and Engineering (Interval'94), St. Petersburg, Russia, March 7-10, 1994, Abstracts*, pp. 51–53.

[38] M. Berz, "Differential algebraic treatment of beam dynamics to very high orders including applications to spacecharge", *AIP Conference Proceedings*, Vol. 177, 1988, p. 275.

[39] M. Berz, "COSY INFINITY Version 6", In: M. Berz, S. Martin and K. Ziegler (Eds.), *Proc. Nonlinear Effects in Accelerators*, IOP Publishing, 1992, p. 125.

[40] M. Berz and G. Hoffstätter, "Exact bounds of the long term stability of weakly nonlinear systems applied to the design of large storage rings", *Interval Computations*, 1994, No. 2, pp. 68–89.

[41] M. Berz and G. Hoffstätter, "Computation and Application of Taylor Polynomials with Interval Remainder Bounds", *Reliable Computing*, 1998, Vol. 4, No. 1 (to appear).

[42] M. Berz, G. Hoffstätter, W. Wan, K. Shamseddine, and K. Makino, "COSY INFINITY and its applications to nonlinear dynamics", In: M. Berz, C. Bischof, A. Griewank, and G. Corliss (Eds.), *Computational Differentiation: Techniques, Applications, and Tools*, SIAM, Philadelphia, 1996.

[43] W. H. Beyer, *CRC standard mathematical tables and formulae*, CRC Press, Boca Raton, FL, 1991.

[44] K. Bhamidipati, "PVM estimates errors caused by imprecise data", In: *Proceedings of the 1994 PVM Users' Group Meeting, Oak Ridge, TN, May 19–20*, Center for Research on Parallel Computations, Session 2A, 1994.

[45] S. Białas, "A necessary and sufficient condition for the stability of interval matrices", *International Journal of Control*, 1983, Vol. 37, pp. 717–722.

[46] S. Białas and J. Garloff, "Intervals of P–matrices and related matrices", *Linear Algebra and Its Applications*, 1984, Vol. 58, pp. 33–41.

[47] P. J. Bickel and E. L. Lehmann, "Descriptive Statistics for Nonparametric Models. 1. Introduction", *Ann. Statist.*, 1975, Vol. 3, pp. 1045–1069.

[48] E. Bishop, *Foundations of Constructive Analysis*, McGraw-Hill, 1967.

[49] E. Bishop, D. S. Bridges, *Constructive Analysis*, Springer, N.Y., 1985.

[50] L. Blum, M. Shub, and S. Smale, "On a theory of computation and complexity over the real numbers: NP-completeness, recursive functions, and universal machines", *Bull. Amer. Math. Soc.*, 1989, Vol. 21, pp. 1–46.

[51] M. Blum and R. Impagliazzo, "Generic oracles and oracle classes", In: *Proceedings of 28th IEEE Symposium on Foundations of Computer Science*, 1987, pp. 118–126.

[52] B. Bouchon-Meunier, O. Kosheleva, V. Kreinovich, and H. T. Nguyen, "Fuzzy Numbers are the Only Fuzzy Sets That Keep Invertible Operations Invertible", *Proceedings of the International Conference on Information Processing and Management of Uncertainty in Knowledge-Based Systems (IPMU'96)*, Granada, Spain, July 1–5, 1996, Vol. 2, pp. 1049–1054.

[53] B. Bouchon-Meunier, O. Kosheleva, V. Kreinovich, and H. T. Nguyen, "Fuzzy numbers are the only fuzzy sets that keep invertible operations invertible", *Fuzzy Sets and Systems*, 1997, Vol. 91, No. 2 (to appear).

[54] B. Bouchon-Meunier and V. Kreinovich, *From Interval Computations to Modal Mathematics: Applications and Computational Complexity*, University of Texas at El Paso, Department of Computer Science, Technical Report UTEP-CS-96-47, 1996; available by ftp from **cs.utep.edu**, login **anonymous**, file **pub/reports/tr96-47.tex**.

[55] B. Bouchon-Meunier, M. Rifqi, and S. Bothorel, "Towards general measures of comparison of objects", *Fuzzy Sets and Systems*, 1996 (to appear).

[56] R. D. Braatz, P. M. Young, J. C. Doyle, and M. Morari, "Computational complexity of mu calculation", *IEEE Transactions on Automatic Control*, 1994, Vol. 39, pp. 1000–1002.

[57] J. D. Brasher, C. F. Hester, and H. J. Caulfield, "Virtually-deterministic quantum computing of nondeterministic polynomial problems", International Journal of Theoretical Physics, 1991, pp. 973–977.

[58] D. S. Bridges, *Constructive Functional Analysis*, Pitman, London, 1979.

[59] V. Brimkov, B. Codenotti, M. Leoncini, and G. Resta, "Strong NP-completeness of a matrix similarity problem", *Theoretical Computer Science*, 1996, Vol. 165, pp. 483–490.

[60] J. Canny, "Improved algorithms for sign determination and existential quantifier elimination", *The Computer Journal*, 1993, Vol. 36, No. 5, pp. 409–418.

[61] O. Caprani, B. Godthaab, and K. Madsen "Use of a Real-Valued Local Minimum in Parallel Interval Global Optimization", *Interval Computations*, 1993, No. 2, pp. 71–82.

[62] N. Cartwright, *How the laws of physics lie*, Oxford, 1983.

[63] L. Chee, *Computing the Value of a Boolean expression with intervals is NP-hard*, Master Thesis, University of Texas at El Paso, Department of Computer Science, 1996.

[64] L. S. Chee, "Computing the Value of a Boolean Expression with Interval Inputs is NP-Hard", *Reliable Computing*, 1997, Vol. 3, pp. 155–172.

[65] F. L. Chernousko, *Estimation of the phase space of dynamic systems*, Nauka publ., Moscow, 1988 (in Russian).

[66] F. L. Chernousko, *State estimation for dynamic systems*, CRC Press, Boca Raton, FL, 1994.

[67] F. Chorlton, *Ordinary differential and difference equations. Theory and applications*, D. van Nostrand, London, Toronto, Princeton, NJ, N.Y., 1965.

[68] J. R. Claerbout and F. Muir, "Robust modeling with erratic data", *Geophysics*, 1973, Vol. 38, pp. 826–844.

[69] B. Cloteaux, *On the Computational Power of Using Chemical Reactions*, Master Thesis, Department of Computer Science, University of Texas at El Paso, 1996.

[70] L. Collatz, "Aufgaben monotoner Art", *Arch. Math.*, 1952, Vol. 3, pp. 366–376.

[71] L. Collatz, *The Numerical Treatment of Differential Equations*, Springer-Verlag, 1960.

[72] G. E. Collins, "Quantifier elimination for the elementary theory of real closed fields by cylindrical algebraic decomposition", *Proc. Second GI Conf. Automata Theory and Formal Languages*, Springer-Verlag, Berlin, Lecture Notes in Computer Science, 1975, Vol. 33, pp. 134–183.

[73] G. E. Collins and H. Hong, "Partial Cylindrical Algebraic Decomposition for Quantifier Elimination", *Journal of Symbolic Computation*, 1991, Vol. 12, No. 3, pp. 299-328.

[74] D. E. Cooke, "An informal introduction to a high level language with applications to interval mathematics", *Reliable Computing*, 1995, Vol. 1, No. 1, pp. 65–76.

[75] D. E. Cooke, R. Duran, and A. Gates, "Bag Language Speeds Up Interval Computations", *Reliable Computing*, 1995, Supplement (Extended Abstracts of APIC'95: International Workshop on Applications of Interval Computations, El Paso, TX, Febr. 23–25, 1995), pp. 64–66.

[76] Th. H. Cormen, C. E. Leiserson, and R. L. Rivest, *Introduction to algorithms*, MIT Press, Cambridge, MA, and Mc-Graw Hill Co., N.Y., 1990.

[77] G. E. Coxson, "The P–matrix problem is co–NP–complete", *Mathematical Programming*, 1994, Vol. 64, pp. 173–178.

[78] G. E. Coxson, "Computing exact bounds on elements of an inverse matrix is NP-hard", *SCAN'95: IMACS/GAMM International Symposium on Scientific Computing, Computer Arithmetic, and Validated Numerics*, September 26–30, Wuppertal, Germany, Abstracts, p. 32 (the result is described in detail in 1995 Highes Aircraft Radar System Group Technical Report with the same title).

[79] G. E. Coxson and C. L. DeMarco, *Computing the real structured singular value is NP-hard*, Report ECE-92-4, Department of Electrical and Computer Engineering, University of Wisconsin, Madison, 1992.

[80] G. E. Coxson and C. L. DeMarco, "The computational complexity of approximating the minimal perturbation scaling to achieve instability in an interval matrix", *Mathematics of Control, Signals, and Systems*, 1994, Vol. 7, pp. 279–291.

[81] F. Cucker and P. Koiran, "Computing over the reals with addition and order: Higher complexity classes", *J. Complexity*, 1995, Vol. 11, No. 3, pp. 358–376.

[82] F. Cucker and F. Rosselló, "Recursiveness over the complex numbers is time-bounded", In: R. K. Shyamasundar (ed.), *Foundations of Software Technology and Theoretical Computer Science, Proceedings of the 13th Conference, Bombay, India, December 15–17, 1993*, Springer-Verlag, Berlin, 1993, pp. 260–267.

[83] E. Ya. Dantsin, "Parameters defining the time of tautology recognition by the splitting method", *Semiotics and Information Science*, VINITI, Moscow, 1979, Vol. 12, pp. 8–17 (in Russian).

[84] E. Ya. Dantsin, "Algorithmics of propositional satisfiability problems", *Problems of Cybernetics*, Moscow, 1987, Vol. 131, pp. 7–29 (in Russian); English translation in: V. Kreinovich and G. Mints (eds.), *Problems of reducing the exhaustive search*, American Mathematical Society, Providence, RI, 1997, pp. 5–22.

[85] J. H. Davenport and J. Heintz, "Real quantifier elimination is doubly exponential", *Journal of Symbolic Computations*, 1988, Vol. 5, No. 1/2, pp. 29–35.

[86] M. Davis, Yu. V. Matiyasevich, and J. Robinson, "Hilbert's tenth problem. Diophantine equations: positive aspects of a negative solution", In:

Mathematical developments arising from Hilbert's problems, Proceedings of Symposia in Pure Mathematics, Vol. 28, American Math. Society, Providence, RI, 1976, Part 2, pp. 323–378.

[87] G. J. Deboeck, K. Villaverde, and V. Kreinovich, "Interval Methods for Presenting Performance of Financial Trading Systems", *Reliable Computing*, 1995, Supplement (Extended Abstracts of APIC'95: International Workshop on Applications of Interval Computations, El Paso, TX, Febr. 23–25, 1995), pp. 67–70.

[88] J. J. D. Delgado-Romero, J. A. Rojas-Estrada, and J. M. Collado, "A simple test to determine the stability of an interval matrix", In *Proceedings of the American Control Conference 1994, Baltimore, Maryland*, 1994, pp. 642–643.

[89] J. W. Demmel, "The componentwise distance to the nearest singular matrix", *SIAM Journal on Matrix Analysis and Applications*, 1992, Vol. 13, pp. 10–19.

[90] D. Deutch, "Quantum theory, the Church-Turing principle, and the universal quantum computer", *Proc. Roy. Soc. (Lond.)*, 1985, Vol. A400, pp. 97–117.

[91] N. Dimitrova and S. Markov, "On the interval-arithmetic presentation of the range of a class of monotone functions of many variables", In: E. Kaucher, S. M. Markov, and G. Mayer (eds.), *Computer Arithmetic, Scientific Computation and Mathematical Modelling*, J. C. Baltzer Co., IMACS, 1991, pp. 213–228.

[92] V. G. Dmitriev, N. A. Zheludeva, and V. Kreinovich. "Applications of interval analysis methods to estimate algorithms errors in measuring systems," *Measurement, Control, Automatization*, 1985, No. 1(53), pp. 31–40 (in Russian).

[93] A. Dolzmann and T. Sturm, "REDLOG: Computer Algebra meets Computer Logic", *SIGSAM Bulletin*, 1997, Vol. 31, No. 2, pp. 2–9.

[94] A. Dolzmann, T. Sturm, and V. Weipfenning, "A new approach for automatic theorem proving in real geometry", *Journal of Automated Reasoning*, 1997 (to appear).

[95] D. I. Doser, K. D. Crain, M. R. Baker, Vladik Kreinovich, and Matthew C. Gerstenberger "Estimating uncertainties for geophysical tomography", *Reliable Computing*, 1998 (to appear; preliminary version appeared as University of Texas at El Paso, Department of Computer Science, Technical

Report UTEP-CS-95-22b, 1995; available by ftp from **cs.utep.edu**, login **anonymous**, file **pub/reports/tr95-22b.tex**).

[96] M. Dowd, *Forcing and the P hierarchy*, Technical Report LCSR-TR-35, Laboratory for Computer Science Research, Rutgers University, 1982.

[97] J. C. Doyle, "Analysis of feedback systems with structured uncertainties", *IEE Proceedings, Part D*, 1982, Vol. 129, pp. 242–250.

[98] J. Edmonds, "Systems of distinct representatives and linear algebra", *Journal of Research of the National Bureau of Standards (B)*, 1967, Vol. 71, pp. 241–245.

[99] T. Elguea and V. Kreinovich, "Mathematical programming with data perturbation and how to understand a visual scheme", *Abstracts of the Sixteenth Symposium on Mathematical Programming with Data Perturbation*, The George Washington University, Washington, D.C., 26–27 May 1994.

[100] P. van Emde Boas, "Machine models and simulations", In: J. van Leeuwen (ed.), *Handbook on Theoretical Computer Science*, Elsevier Science, N.Y., 1990, pp. 1–63.

[101] H. B. Enderton. *A mathematical introduction to logic.* Academic Press, N.Y., 1972.

[102] J. Eriksson and P. Lindström, "A parallel interval method implementation for global optimization using dynamic load balancing", *Reliable Computing*, 1995, Vol. 1, No. 1, pp. 65–76. pp. 77–92.

[103] R. L. Eubank, *Spline smoothing and nonparametric regression*, M. Dekker, N.Y., 1988.

[104] S. Fenner, "Notions of resource-bounded category and genericity", In: *Proceedings of the 6th IEEE Structure in Complexity Theory Conference*, 1991, pp. 196–212.

[105] R. Feynman, *The character of physical laws*, MIT Press, Cambridge, MA, 1965.

[106] R. P. Feynman, R. B. Leighton, and M. L. Sands, *The Feynman Lectures On Physics*, Addison-Wesley, Redwood City, CA, 1989.

[107] M. Fiedler and V. Pták, "On matrices with non–positive off–diagonal elements and positive principal minors", *Czechoslovak Mathematical Journal*, 1962, Vol. 12, pp. 382–400.

[108] A. F. Filippov, "Ellipsoidal estimates for a solution of a system of differential equations", *Interval Computations*, 1992, Vol. 2, No. 2(4), pp. 6–17.

[109] A. Finkelstein, O. Kosheleva, and V. Kreinovich, "Astrogeometry, error estimation, and other applications of set-valued analysis", *ACM SIGNUM Newsletter*, 1996, Vol. 31, No. 4, pp. 3–25.

[110] E. Fogel and Y. F. Huang, "On the value of information in system identification. Bounded noise case", *Automatica*, 1982, Vol. 18, No. 2, pp. 229–238.

[111] B. van Fraassen, *Laws and symmetry*, Claredon Press, Oxford, 1989.

[112] M. Friedman, M. Ming, and A. Kandel, "On the theory of typicality", *International Journal of Uncertainty, Fuzziness, and Knowledge-Based Systems*, 1995, Vol. 3, No. 2, pp. 127–142.

[113] L. O. Fuentes, *Applying uncertainty formalisms to well- defined problems: experimental and theoretical foundations*, Master Thesis, University of Texas at El Paso, Department of Computer Science, 1991.

[114] L. O. Fuentes and V. Ya. Kreinovich. "Simulation of Chemical Kinetics as a Promising Approach to Expert Systems", *Abstracts of the Southwestern Conference on Theoretical Chemistry*, The University of Texas at El Paso, November 1990, p. 33.

[115] A. A. Gaganov, *Computational complexity of the range of the polynomial in several variables*, Leningrad University, Math. Department, M.S. Thesis, 1981 (in Russian).

[116] A. A. Gaganov, "Computational complexity of the range of the polynomial in several variables", *Cybernetics*, 1985, pp. 418–421.

[117] R. Gandy, "Church's thesis and principles for mechanisms", In: J. Barwise, H. J. Keisler, and K. Kunen, *The Kleene Symposium*, North Holland, Amsterdam, 1980, pp. 123–148.

[118] R. Gandy, "The confluence of ideas in 1936", In: R. Herken (ed.) *The universal Turing machine: a half-century survey*, Kammerer & Unverzagt, Hamburg, 1988.

[119] F. R. Gantmacher, *The Theory of Matrices, vol. I*, Chelsea Publishing Company, N.Y., 1959.

[120] E. Gardeñes, H. Mielgo, and A. Trepat, "Modal intervals: reason and ground semantics", In: K. Nickel (ed.), *Interval Mathematics 1985*, Lecture Notes in Computer Science, Vol. 212, Springer-Verlag, Berlin, Heidelberg, 1986, pp. 27–35.

[121] M. E. Garey and D. S. Johnson, *Computers and intractability: a guide to the theory of NP-completeness*, Freeman, San Francisco, 1979.

[122] M. R. Garey, D. S. Johnson, and L. Stockmeyer, "Some simplified NP–complete graph problems", *Theoretical Computer Science*, 1976, Vol. 1, pp. 237–267.

[123] A. I. Gerasimov and V. Kreinovich, *Piecewise- fractionally-linear approximation*, Leningrad Polytechnical University and National Research Institute for Scientific and Technical Information (VINITI), 1988, 15 pp. (in Russian)

[124] A. I. Gerasimov and V. Kreinovich, *On the problem of optimal approximation choice for metrological characteristics*, Leningrad Polytechnical University and National Research Institute for Scientific and Technical Information (VINITI), 1988, 13 pp. (in Russian).

[125] R. Geroch and J. B. Hartle, "Computability and physical theories", *Foundations of Physics*, 1986, Vol. 16, p. 533.

[126] M. C. Gerstenberger, *Development of new techniques in crosswell seismic travel time tomography*, M.S. Thesis, Department of Geological Sciences, University of Texas at El Paso, El Paso, TX, July 1994.

[127] V. B. Glasko, *Inverse problems of mathematical physics,* American Institute of Physics, N. Y., 1984.

[128] J. Gleick, *Genius: the life and science of Richard Feynman*, Pantheon Books, N.Y., 1992.

[129] B. V. Gnedenko and A. N. Kolmogorov, *Limit distributions for sums of independent random variables*, Addison-Wesley, Cambridge, 1954.

[130] G. H. Golub and C. F. van Loan, *Matrix Computations*, The Johns Hopkins University Press, Baltimore, 1983.

[131] J. S. Gomberg, K. M. Shedlock, and S. W. Roecker, "The effect of S-wave arrival times on the accuracy of hypocenter estimation", *Bull. Seismol. Soc. Am.*, 1990, Vol. 80, pp. 1605–1628.

[132] J. R. Gott, "Closed timelike curves produced by pairs of moving cosmic strings: exact solution", Physical Review Letters, 1991, Vol. 66, pp. 1126–1129.

[133] E. Grädel and K. Meer, "Descriptive complexity theory over the real numbers", *Proceedings of the Symposium on Theory of Computing STOC'95*, Las Vegas, NV, 1995.

[134] D. Yu. Grigor'ev and N. N. Vorobjov (Jr.), "Solving systems of polynomial inequalities in subexponential time", *Journal of Symbolic Computation*, 1988, Vol. 5, No. 1/2, pp. 37–64.

[135] R. Gurevic, "Decidability of the equational theory of positive numbers with raising to power", *Siberian Mathematical Journal*, 1984, Vol. 25, No.2 (144), pp. 216–219.

[136] R. Gurevic, "Equational theory of positive numbers with exponentiation", *Proc. Amer. Math. Soc.*, 1985, Vol. 94, pp. 135–141.

[137] R. Gurevic, "Transcendental numbers and eventual dominance of exponential functions", *Bull. Lond. Math. Soc.*, 1986, Vol. 18, pp. 560–570.

[138] R. Gurevic, "Exponentiation of reals: Effects of base choice", *Proc. Amer. Math. Soc.*, 1987, Vol. 99, pp. 155–162.

[139] R. Gurevic, "Equational theory of positive numbers with exponentiation is not finitely axiomatizable", *Ann. Pure Appl. Logic*, 1990, Vol. 49, No. 1, pp. 1–30.

[140] R. Hammer, M. Hocks, U. Kulisch, and D. Ratz, *Numerical Toolbox for Verified Computing I,* Springer-Verlag, New York, 1993.

[141] G. W. Harrison, "Dynamic models with uncertain parameters", In: X. J. R. Avul (Editor), *Proceedings of the First International Conference on Mathematical Modeling, St. Louis*, University of Missouri-Rolla, 1977, pp. 295–304.

[142] S. Hecht, S. Schlaer, and M. H. Pirenne, "Energy, quanta, and vision", Journal of General Physiology, 1941, Vol. 25, pp. 891–940.

[143] G. Heindl, V. Kreinovich, and A. V. Lakeyev, "Solving Linear Interval Systems is NP-Hard Even If We Exclude Overflow and Underflow", *Reliable Computing*, 1998 (to appear; preliminary version appeared as University of Texas at El Paso, Department of Computer Science, Technical Report UTEP-CS-95-40b, 1995; available by ftp from **cs.utep.edu**, login **anonymous**, file **pub/reports/tr95-40b.tex**).

[144] G. Heindl, V. Kreinovich, and M. Rifqi, "In case of interval (or more general) uncertainty, no algorithm can choose the simplest representative", *Reliable Computing*, 1998 (to appear; preliminary version appeared as University of Texas at El Paso, Department of Computer Science, Technical Report UTEP-CS-96-27b, 1996; available by ftp from **cs.utep.edu**, login **anonymous**, file **pub/reports/tr96-27b.tex**).

[145] G. Heindl and E. Reinhart, "Adjustment by the principle of minimal maximum error", In: *Beiträge aus der Bundesrepublik Deutschland zur Vorlage bei der XVI Generalversammlung der Internationalen Union für Geodäsie und Geophysik, Grenoble 1975*, Veröffentlichungen der Deutschen Geodätischen Kommission, Reihe B, München, 1975 Vol. 213, pp. 33–43.

[146] G. Heindl and E. Reinhart, *Ausgleichung im Sinne minimaler Maximalfehler*, Veröffentlichungen der Deutschen Geodätischen Kommission bei der Bayerischen Akademie der Wissenschaften, Reihe A, München, 1976, Vol. 84.

[147] G. Heindl and E. Reinhart, "Experience with a non-statistical method of directing outliers", In: *Proceedings of the International Symposium on Geodetic Networks and Computations of the International Association of Geodesy, Volume V, Network Analysis Models*, Veröffentlichungen der Deutschen Geodätischen Kommission bei der Bayerischen Akademie der Wissenschaften, Reihe B, München, 1982, Vol. 258/V, pp. 19–28.

[148] J. Heintz, M.-F. Roy, and P. Solerno, "On the theoretical and practical complexity of the existential theory of reals", *The Computer Journal*, 1993, Vol. 36, No. 5.

[149] D. Hertz, "The extreme eigenvalues and stability of real symmetric interval matrices", *IEEE Transactions on Automatic Control*, 1992, Vol. 37, pp. 532–535.

[150] T. Henriksen and K. Madsen, "Parallel algorithms for global optimization", *Interval Computations*, 1992, No. 3, pp. 88–95.

[151] N. J. Higham, *Accuracy and Stability of Numerical Algorithms*, SIAM, Philadelphia, 1996.

[152] D. Hilbert, "Mathematical Problems" (lecture delivered before the International Congress of Mathematics in Paris in 1900), translated in *Bull. Amer. Math, Soc.*, 1902, Vol. 8, pp. 437–479; reprinted in *Mathematical developments arising from Hilbert's problems*, Proceedings of Symposia in Pure Mathematics, Vol. 28, American Math. Society, Providence, RI, 1976, Part 1, pp. 1–34.

[153] C. V. Hollot and A. C. Bartlett, *On the eigenvalues of interval matrices*, Technical Report CCS–87–104, University of Massachusetts, Amherst, 1987.

[154] H. Hong, *Improvements in CAD-based Quantifier Elimination*, Ph.D. Dissertation, The Ohio State University, 1990.

[155] H. Hong, G. E. Collins, J. R. Johnson, and M. J. Encarnacion, *QEPCAD interactive version 12*, 1993.

[156] H. Hong, R. Liska, and S. Steinberg, "Testing stability by quantifier elimination", *Journal of Symbolic Computations*, 1997 (to appear).

[157] R. Horst and P. M. Pardalos, *Handbook of global optimization*, Kluwer Academic Publ., Boston, MA, 1995.

[158] Chenyi Hu, A. Frolov, R. B. Kearfott, and Qing Yang, "A general iterative sparse linear solver and its parallelization for interval Newton methods", *Reliable Computing*, 1995, Vol. 1, No. 3, pp. 251–264.

[159] Chenyi Hu, J. Sheldon, R. B. Kearfott, and Qing Yang, "Optimizing INTBIS on the CRAY Y-MP", *Reliable Computing*, 1995, Vol. 1, No. 3, pp. 265–274.

[160] *Inverse problems,* SIAM–AMS Proceedings, Vol. 14. American Mathematical Society, Providence, RI, 1983.

[161] *Inverse problems,* Birkhauser Verlag, Basel, 1986.

[162] *Inverse problems,* Lecture Notes in Mathematics, Vol. 1225, Springer-Verlag, Berlin–Heidelberg, 1986.

[163] C. Jansson, "Interval linear systems with symmetric matrices, skew-symmetric matrices and dependencies in the right hand side," *Computing*, 1991, Vol. 46, pp. 265–274.

[164] C. Jansson, "Calculation of exact bounds for the solution set of linear interval systems", *Linear Algebra Appl.*, 1997, Vol. 251, pp. 321–340.

[165] D. S. Johnson, "The NP-completeness column", *Journal of Algorithms*, 1985, Vol. 6, pp. 291–305.

[166] W. Kahan, "Numerical linear algebra", *Canadian Mathematical Bulletin*, 1966, Vol. 9, pp. 757–801.

[167] P. Kahl, *Solving Narrow-Interval Linear Equation Systems Is NP-Hard*, Master Thesis, University of Texas at El Paso, Department of Computer Science, 1996.

[168] W. C. Karl, J. P. Greschak, and G. C. Verghese, "Comments on *A necessary and sufficient condition for the stability of interval matrices*," *International Journal of Control*, 1984, Vol. 39, pp. 849–851.

[169] N. Karmarkar, "A new polynomial-time algorithm for linear programming", *Combinatorica*, 1984, Vol. 4, pp. 373–396.

[170] R. M. Karp, "reducibility among combinatorial problems", In: R. E. Miller and J. W. Thatcher (eds.), *Complexity of Computer Computations*, Plenum Press, N.Y., 1972, pp. 85–103.

[171] R. B. Kearfott, "Some tests of generalized bisection", *ACM Trans. Math. Softw.*, 1987, Vol. 13, pp. 197–220.

[172] R. B. Kearfott, "Interval arithmetic techniques in the computational solution of nonlinear systems of equations: Introduction, examples, and comparisons", In: *Computational solution of nonlinear systems of equations, Proc. SIAM-AMS Summer Semin., Ft. Collins/CO (USA) 1988*, American Math. Society, Providece, RI, Lectures in Appl. Math. 1990, Vol. 26, pp. 337–357.

[173] R. B. Kearfott, "Interval Newton/generalized bisection when there are singularities near roots", *Ann. Oper. Res.*, 1990, Vol. 25, No. 1–4, pp. 181–196.

[174] R. B. Kearfott, "Preconditioners for the interval Gauss-Seidel method", *SIAM J. Numer. Anal.*, 1990, Vol. 27, No. 3, pp. 804–822.

[175] R. B. Kearfott, *Rigorous global search: continuous problems*, Kluwer, Dordrecht, 1996.

[176] R. B. Kearfott and V. Kreinovich (eds.), *Applications of Interval Computations*, Kluwer, Dordrecht, 1996.

[177] N. A. Khlebalin and Yu. I. Shokin, "Interval variant of modal control method", *Dokl. Akad. Nauk SSSR*, 1991, Vol. 316, pp. 846–850 (in Russian); English trans. in *Soviet. Phys. Dokl.*, 1991, Vol. 36.

[178] I. S. Kirillova, V. Ya. Kreinovich, and G. N. Solopchenko, "Distribution–independent estimators of error characteristics of measuring instruments", *Measuring Techniques*, 1989, Vol. 32, No. 7, pp. 621–627.

[179] U. Kohlenbach. *Theorie der Majorisierbaben ...*, Ph.D. Dissertation, Frankfurt am Main, 1990.

[180] U. Kohlenbach, "Effective moduli from ieffective uniqueness proofs. An unwinding of de La Vallée Poussin's proof for Chebycheff approximation", *Annals for Pure and Applied Logic*, 1993, Vol. 64, No. 1, pp. 27–94.

[181] E. Koltik, V. G. Dmitriev, N. A. Zheludeva, and V. Kreinovich. "An optimal method for estimating a random error component," *Investigations in Error Estimation*, Proceedings of the Mendeleev Metrological Institute, Leningrad, 1986, pp. 36–41 (in Russian).

[182] M. Koshelev, L. Longpré, and P. Taillibert, *Optimal Approximation of Quadratic Interval Functions*, University of Texas at El Paso, Department of Computer Science, Technical Report UTEP-CS-97-8, 1997; available from the Web at `http://cs.utep.edu/vladik/1997/tr97-8.tex`.

[183] M. Koshelev and P. Taillibert, "Optimal approximation of quadratic interval functions", *Proceedings of the NASA University Research Centers Conference*, Albuquerque, New Mexico, February 16–19, 1997, pp. 425–430.

[184] O. M. Kosheleva and V. Ya. Kreinovich. *On the algorithmic problems of a measurement process*, Research Reports in Philosophy of Physics, University of Toronto, Ontario, Canada, Dept. of Philosophy, No. 5, 1978, 63 pp.

[185] O. M. Kosheleva, V. Ya. Kreinovich, "What can physics give to constructive mathematics?", In: *Mathematical logic and mathematical linguistics*, Kalinin, 1981, pp. 117–128 (in Russian).

[186] O. Kosheleva and V. Kreinovich, "Error estimation for indirect measurements: Interval computation problem is (slightly) harder than a similar probabilistic computational problem", *Reliable Computing*, 1998 (to appear; preliminary version appeared as a Technical Report: *Université Paris VI et VII, Institut Blaise Pascal, Laboratoire Formes et Intelligence Artificielle LAFORIA*, Technical Report 96/24, September 1996).

[187] O. Kosheleva and V. Kreinovich, "Only Intervals Preserve the Invertibility of Arithmetic Operations", *Reliable Computing,* 1998 (to appear; preliminary version appeared as University of Texas at El Paso, Department of Computer Science, Technical Report UTEP-CS-95-55a, 1995; available by ftp from `cs.utep.edu`, login `anonymous`, file `pub/reports/tr95-55a.tex`).

[188] O. M. Kosheleva and S. V. Soloviev, "On the logic of using observable events in decision making", In: *Proceedings of the IX National Symposium on Cybernetics,* Moscow, 1981, pp. 49–51 (in Russian).

[189] V. Kozlenko and V. Kreinovich, "Optimization in case of uncertain criteria", *Proceedings of the IV USSR National Conference on Applications of*

Methods of Mathematical Logics, Part 1, Estonian Academy of Sciences, Institute of Cybernetics, Tallinn, 1986, pp. 126–128 (in Russian).

[190] V. Kreinovich, *Remarks on the Margins of Kushner's Book: On the Constructive Mathematical Analysis Restricted to Polynomial Time Algorithms*, Manuscript, Leningrad, 1973 (in Russian).

[191] V. Kreinovich, "What does the law of the excluded middle follow from?," *Proceedings of the Leningrad Mathematical Institute of the Academy of Sciences*, 1974, Vol. 40, pp.37-40 (in Russian), English translation: *Journal of Soviet Mathematics*, 1977, Vol. 8, No. 1, pp. 266–271.

[192] V. Kreinovich, *Complexity measures: computability and applications*, Master Thesis, Leningrad University, Department of Mathematics, Division of Mathematical Logic and Constructive Mathematics, 1974 (in Russian).

[193] V. Kreinovich, "Uniqueness implies algorithmic computability", *Proceedings of the 4th Student Mathematical Conference*, Leningrad University, Leningrad, 1975, pp. 19–21 (in Russian).

[194] V. Kreinovich, Reviewer's remarks in a review of D. S. Bridges, *Constrictive functional analysis*, Pitman, London, 1979; Zentralblatt für Mathematik, 1979, Vol. 401, pp. 22–24.

[195] V. Kreinovich, *Categories of space-time models*, Ph.D. dissertation, Novosibirsk, Soviet Academy of Sciences, Siberian Branch, Institute of Mathematics, 1979, (in Russian).

[196] V. Kreinovich, "Unsolvability of several algorithmically solvable analytical problems", *Abstracts Amer. Math. Soc.*, 1980, Vol. 1, No. 1, p. 174.

[197] V. Ya. Kreinovich, "A General Approach to Analysis of Uncertainty in Measurements", *Proceedings of the the 3-rd USSR National Symposium on Theoretical Metrology*, Leningrad, Mendeleev Metrology Institute (VNIIM), 1986, pp. 187–188 (in Russian).

[198] V. Kreinovich, "Semantics of S. Yu. Maslov's iterative method," *Problems of Cybernetics*, Moscow, 1987, Vol. 131, pp. 30–62 (in Russian); English translation in: V. Kreinovich and G. Mints (eds.), *Problems of reducing the exhaustive search*, American Mathematical Society, Providence, RI, 1997, pp. 23–51.

[199] V. Kreinovich, "Group-theoretic approach to intractable problems," *Proceedings of International Conference on Computer Logic COLOG-88, Tallinn, 1988*, Vol. 1, pp. 31–42.

[200] V. Ya. Kreinovich, *On the possibility of using physical processes when solving hard problems*, Leningrad Center for New Information Technology "Informatika", Technical Report, Leningrad, 1989 (in Russian).

[201] V. Ya. Kreinovich, *Philosophy of Optimism: Notes on the possibility of using algorithm theory when describing historical processes*, Leningrad Center for New Information Technology "Informatika", Technical Report, Leningrad, 1989 (in Russian).

[202] V. Kreinovich, "Group-theoretic approach to intractable problems," *Lecture Notes in Computer Science*, Springer-Verlag, Berlin, Vol. 417, 1990, pp. 112–121.

[203] V. Kreinovich, "Spacetime isomorphism problem is intractable (NP-hard)," *International Journal of Theoretical Physics*, 1991, Vol. 30, No. 9, pp. 1249–1257.

[204] V. Kreinovich, *Ellipsoid computations are intractable even for polynomials*, University of Texas at El Paso, Department of Computer Science, Technical Report UTEP-CS-93-53a, August 1993 (updated December 1993); available by ftp from `cs.utep.edu`, login **anonymous**, file `pub/reports/tr93-53a.tex`.

[205] V. Kreinovich, "Computers of generation Omega: non-traditional computational models", *Technical Program, Winter Meeting, Rio Grande Chapter of the ACM*, January 28, Carlsbad, NM (*Digitizer*, January 1994, p. 4).

[206] V. Kreinovich, "Error estimation for indirect measurements is exponentially hard," *Neural, Parallel, and Scientific Computations,* 1994, Vol. 2, No. 2, pp. 225–234.

[207] V. Kreinovich, "Dirty pages of logarithm tables, lifetime of the Universe, and computer representation of real numbers", *Technical Program, Winter Meeting, Rio Grande Chapter of the ACM*, December 2, Las Cruces, NM (*Digitizer*, September/October 1994, p. 3).

[208] V. Kreinovich, "Interval rational = algebraic", *ACM SIGNUM Newsletter*, 1995, Vol. 30, No. 4, pp. 2–13.

[209] V. Kreinovich (ed.), *Reliable Computing*, 1995, Supplement (Extended Abstracts of APIC'95: International Workshop on Applications of Interval Computations, El Paso, TX, Febr. 23–25, 1995).

[210] V. Kreinovich, "Why intervals? A simple limit theorem that is similar to limit theorems from statistics." *Reliable Computing,* 1995, Vol. 1, No. 1, pp. 33–40.

[211] V. Kreinovich, "S. Maslov's Iterative Method: 15 Years Later (Freedom of Choice, Neural Networks, Numerical Optimization, Uncertainty Reasoning, and Chemical Computing)", In: V. Kreinovich and G. Mints, *Problems of reducing the exhaustive search*, American Mathematical Society, Providence, RI, 1997, pp. 175–189.

[212] V. Ya. Kreinovich et al., *Theoretical foundations of estimating precision of software results in intellectual systems for control and measurement,* Soviet National Institute for Electrical Measuring Instruments, Technical Report No. 7550-8170-40, Leningrad, 1988 (in Russian).

[213] V. Kreinovich and A. Bernat, "Parallel algorithms for interval computations: an introduction", *Interval Computations*, 1994, No. 3, pp. 6–62.

[214] V. Kreinovich, A. Bernat, E. Villa, and Y. Mariscal, "Parallel computers estimate errors caused by imprecise data," *Proceedings of the Fourth ISMM (International Society on Mini and Micro Computers) International Conference on Parallel and Distributed Computing and Systems*, Washington, 1991, Vol. 1, pp. 386–390.

[215] V. Kreinovich, A. Bernat, E. Villa, and Y. Mariscal. "Parallel computers estimate errors caused by imprecise data," *Interval Computations*, 1991, No. 2, pp. 31–46.

[216] V. Kreinovich, A. Bernat, E. Villa, and Y. Mariscal, "Parallel computers estimate errors caused by imprecise data", *Technical Papers of the the Society of Mexican American Engineers and Scientists 1992 National Symposium*, San Antonio, Texas, April 1992, pp. 192–199.

[217] V. Kreinovich and L. O. Fuentes. "Simulation of chemical kinetics - a promising approach to inference engines," in: J. Liebowitz (ed.), *Proceedings of the World Congress on Expert Systems, Orlando, Florida, 1991*, Pergamon Press, N.Y., Vol. 3, pp. 1510–1517.

[218] V. Kreinovich and R. B. Kearfott "Computational complexity of optimization and nonlinear equations with interval data", *Abstracts of the Sixteenth Symposium on Mathematical Programming with Data Perturbation*, The George Washington University, Washington, D.C., 26–27 May 1994.

[219] V. Kreinovich and A. V. Lakeyev, " 'Interval Rational = Algebraic' Revisited: A More Computer Realistic Result", *ACM SIGNUM Newsletter*, 1996, Vol. 31, No. 1, pp. 14–17.

[220] V. Kreinovich and A. V. Lakeyev, "Linear Interval Equations: Computing Enclosures with Bounded Relative Or Absolute Overestimation is NP-Hard", *Reliable Computing*, 1996, Vol. 2, No. 4, pp. 341–350.

[221] V. Kreinovich, A. V. Lakeyev, and S. I. Noskov, "Optimal solution of interval linear systems is intractable (NP-hard)." *Interval Computations*, 1993, No. 1, pp. 6–14.

[222] V. Kreinovich, A. V. Lakeyev, and S. I. Noskov, "Approximate linear algebra is intractable", *Linear Algebra and its Applications,* 1996, Vol. 232, No. 1, pp. 45–54.

[223] V. Kreinovich, A. Lakeyev, and J. Rohn, "Computational Complexity of Interval Algebraic Problems: Some Are Feasible And Some Are Computationally Intractable: A Survey", *SCAN'95: IMACS/GAMM International Symposium on Scientific Computing, Computer Arithmetic, and Validated Numerics*, September 26–30, Wuppertal, Germany, Abstracts, p. 72.

[224] V. Kreinovich, A. Lakeyev, and J. Rohn, "Computational Complexity of Interval Algebraic Problems: Some Are Feasible And Some Are Computationally Intractable – A Survey", In: G. Alefeld, A. Frommer, and B. Lang (eds.), *Scientific Computing and Validated Numerics*, Akademie-Verlag, Berlin, 1996, pp. 293–306.

[225] V. Kreinovich, G. Mayer, and S. Starks, "On a Theoretical Justification of The Choice of Epsilon-Inflation in PASCAL-XSC", *Reliable Computing*, 1997, Vol. 3, No. 4 (to appear; preliminary version appeared as University of Texas at El Paso, Department of Computer Science, Technical Report UTEP-CS-96-22a, 1996; available by ftp from `cs.utep.edu`, login `anonymous`, file `pub/reports/tr96-22a.tex`).

[226] V. Kreinovich and G. Mints, *Problems of reducing the exhaustive search*, American Mathematical Society, Providence, RI, 1997.

[227] V. Kreinovich, V. M. Nesterov, and N. A. Zheludeva, "Interval Methods That Are Guaranteed to Underestimate (and the resulting new justification of Kaucher arithmetic)", *Reliable Computing*, 1996, Vol. 2, No. 2, pp. 119–124.

[228] V. Kreinovich and M. I. Pavlovich, "Error estimate of the result of indirect measurements by using a calculational experiment," *Izmeritelnaya*

Tekhnika, 1985, No. 3, pp. 11–13 (in Russian), English translation: *Measurement Techniques*, 1985, Vol. 28, No. 3, pp. 201–205.

[229] V. Ya. Kreinovich and M. I. Pavlovich, "Estimation of the error of the result of indirect measurements in a measuring and information system by computer simulation", In: *VII USSR National Conference on Information and Measurement Systems*, Vinnitsa, 1985, pp. 45–46 (in Russian).

[230] V. Kreinovich, B. Penn, and J. W. Harbaugh, "Computational Complexity of Interval Computations, of Taylor Methods, and of Their Fractal Extensions", *Abstracts of the SIAM Annual Meeting*, Stanford, CA, July 14–18, 1997.

[231] V. Kreinovich, C. Quintana, R. Lea, O. Fuentes, A. Lokshin, S. Kumar, I. Boricheva, and L. Reznik, "What non-linearity to choose? Mathematical foundations of fuzzy control," *Proceedings of the 1992 International Conference on Fuzzy Systems and Intelligent Control*, Louisville, KY, 1992, pp. 349–412.

[232] V. Kreinovich and D. Schirmer, "Towards a more realistic definition of feasibility", *SC-COSMIC, South Central Computational Sciences in Minority Institutions Consortium, First Student Conference in Computational Sciences*, October 21–22, 1995, Houston, TX, Abstracts, pp. 33–34.

[233] V. Kreinovich, N. Strickland, and O. Kosheleva, *What can we compute if we use computational devices based on non-Newtonian physical phenomena: Church's thesis revisited*, University of Texas at El Paso, Department of Computer Science, Technical Report UTEP-CS-93-54, September 1993; available by ftp from `cs.utep.edu`, login `anonymous`, file `pub/reports/tr93-54.tex`.

[234] V. Kreinovich and R. Trejo, "Optimal interval computation techniques: optimization of numerical methods in case of uncertainty", In: Marcilia A. Campos (ed.), *Abstracts of the II Workshop on Computer Arithmetic, Interval and Symbolic Computation (WAI'96)*, Recife, Pernambuco, Brazil, August 7-8, 1996, pp. 48–50.

[235] V. Kreinovich, A. Vazquez, and O. M. Kosheleva, "Prediction problem in quantum mechanics is intractable (NP-hard)," *International Journal of Theoretical Physics*, 1991, Vol. 30, No. 2, pp. 113–122.

[236] V. Kreinovich and K. Villaverde, *Towards modal interval analysis: how to compute maxima and minima of a function from approximate measurement results*, University of Texas at El Paso, Computer Science Department, Technical Report UTEP-CS-91-9, 1991.

[237] V. Kreinovich and K. Villaverde, "A Quadratic-Time Algorithm For Smoothing Interval Functions", *Reliable Computing,* 1996, Vol. 2, No. 3, pp. 255–264.

[238] G. Kreisel, "A notion of mechanistic theory", *Synthese*, 1974, Vol. 29, pp. 11–26.

[239] G. Kreisel, "A review of [330] and [331]", *Journal of Symbolic Logic*, 1982, Vol. 47, pp. 900–902.

[240] I. N. Krotkov, V. Kreinovich and V. D. Mazin, "Methodology of designing measuring systems, using fractionally linear transformations," *Measuring Systems. Theory and Applications. Proceedings of the Novosibirsk Electrical Engineering Institute*, 1986, pp. 5–14 (in Russian).

[241] I. N. Krotkov, V. Kreinovich and V. D. Mazin, "General form of measurement transformations which admit the computational methods of metrological analysis of measuring- testing and measuring-computing systems," *Izmeritelnaya Tekhnika*, 1987, No. 10, pp. 8–10 (in Russian); English translation: *Measurement Techniques*, 1987, Vol. 30, No. 10, pp. 936–939.

[242] J. Kuttler, "A fourth–order finite–difference approximation for the fixed membrane eigenproblem", *Mathematics of Computation*, 1971, Vol. 25, pp. 237–256.

[243] A. V. Lakeyev and V. Kreinovich, "If Input Intervals Are Small Enough, Then Interval Computations Are Almost Always Easy", *Reliable Computing*, 1995, Supplement (Extended Abstracts of APIC'95: International Workshop on Applications of Interval Computations, El Paso, TX, Febr. 23–25, 1995), pp. 134–139.

[244] A. V. Lakeyev and V. Kreinovich, "NP-hard classes of linear algebraic systems with uncertainties", *Reliable Computing*, 1997, Vol. 3, No. 1, pp. 51–81.

[245] A. V. Lakeyev and S. I. Noskov, "A description of the set of solutions of a linear equation with interval defined operator and right-hand side" *Russian Academy of Sciences, Doklady, Mathematics*, 1993, Vol. 47, No. 3, pp. 518–523.

[246] A. V. Lakeyev and S. I. Noskov, "On the solution set of a linear equation with the right-hand side and operator given by intervals", *Siberian Math. J.*, 1994, Vol. 35, No. 5, pp. 957-966.

[247] V. Lakshmikantham, S. Leela, *Differential and integral inequalities*, Academic Press, N.Y., 1969.

[248] M. M. Lavrentiev, V. G. Romanov, and S. P. Shishatskii, *Ill-posed problems of mathematical physics and analysis*, American Mathematical Society, Providence, RI, 1986.

[249] R. Lea, V. Kreinovich, and R. Trejo, "Optimal interval enclosures for fractionally-linear functions, and their application to intelligent control", *Reliable Computing*, 1996, Vol. 2, No. 3, pp. 265–286.

[250] A. Leclerc, "Parallel Interval Global Optimization and Its Implementation in C++", *Interval Computations*, 1993, No. 3, pp. 148–163.

[251] H. R. Lewis and C. H. Papadimitriou, *Elements of the Theory of Computation*, Prentice-Hall, Inc., New Jersey, 1981.

[252] O. Lhomme, "Consistency Techniques for Numeric CSPs," In: *Proceedings of IJCAI'93, the Thirteenth International Joint Conference on Artificial Intelligence*, 1993.

[253] O. Lhomme, A. Gottlieb, M. Rueher, and P. Taillibert, "Boosting the Interval Narrowing Algorithm," In: *Proceedings of JICSLP'96, Joint International Conference and Symposium on Logic Programming, Bonn, Germany, September 2–6*, 1996.

[254] M. Li and P. Vitányi, *An Introduction to Kolmogorov Complexity and its Applications*, Springer-Verlag, N.Y., 1997.

[255] Th. Lickteig and K. Meer, "A note on testing the resultant", *J. Complexity*, 1995, Vol. 11, No. 3, pp. 344–351.

[256] E. Loiez and P. Taillibert, "Analog Systems Diagnosis: Modeling with Temporal Bands," In: *Proceedings of CESA'96, IEEE International Conference on Computations in Engineering and Simulations Applications, Lille, France, July 9–12, 1996.*

[257] E. Loiez and P. Taillibert, "Polynomial Temporal Band Sequences for Analog Diagnosis", In: *Proceedings of the 7th International Workshop on Principles of Diagnosis DX'96, Val Morin, Quebec, Canada, October 13-16, 1996.*

[258] L. Longpré and M. Berz, "Interval and Complexity Workshops Back-to-Back with 1997 ACM Symposium on Theory of Computing (STOC'97)", *Reliable Computing*, 1997, Vol. 3, No. 4 (to appear).

[259] R. Loos and V. Weispfenning, "Applying linear quantifier elimination", *Computer Journal*, 1993, Vol. 36, No. 5, pp. 450–462.

[260] C. J. Lumsden and E. O. Wilson, *Genes, mind, and culture: the coevolutionary process*, Harvard University Press, Cambridge, MA, 1981.

[261] C. J. Lumsden and E. O. Wilson, *Promethean fire: reflections on the origin of mind*, Harvard University Press, Cambridge, MA, 1983.

[262] J. Lutz, "Category and measure in complexity classes", *SIAM Journal of Computing*, 1990, Vol. 19, pp. 1100–1131.

[263] W. Maass, "Recursively enumerable generic sets", *Journal of Symbolic Logic*, 1982, Vol. 47, pp. 809–823.

[264] K. B. MacDonald, ed., *Sociobiological perspectives on human development*, Springer-Verlag, N.Y., 1988.

[265] A. Macintyre and A. Wilkie, *On the decidability of the real exponential field*, manuscript, 1997.

[266] Maimonides, *The guide of the perplexed*, East and West Library, Horovitz Publ. Co., London, 1952.

[267] K. Makino and M. Berz, "Remainder differential algebras and their applications", In: M. Berz, C. Bischof, A. Griewank, and G. Corliss (Eds.), *Computational Differentiation: Techniques, Applications, and Tools*, SIAM, Philadelphia, 1996.

[268] K. Makino and M. Berz, "COSY INFINITY version 7" *Fourth Computational Accelerator Physics Conference*, AIP Conference Proceedings, 1996.

[269] K. Manders and L. Adleman, "NP-complete decision problems for binary quadratics", *J. Comput. Syst. Sci.*, 1978, Vol. 16, pp. 168–184.

[270] R. Mannshardt, "Enclosing a solution of an ordinary differential equation by sub- and superfunctions", In: C. Ulllrich (ed.), *Contributions to Computer Arithmetic and Self-Validating Numerical methods*, J. C. Baltzer AG, IMACS, 1990, pp. 319–328.

[271] M. Mansour, "Robust stability of interval matrices", *Proceedings of the 28th Conference on Decision and Control, Tampa, FL*, 1989, pp. 46–51.

[272] D. Marker, "Model theory and exponentiation", *Notices of the American Methematical Society*, 1996, Vol. 43, No. 7, pp. 753–759.

[273] S. M. Markov, "On the presentation of ranges of monotone functions using interval arithmetic", *Proceedings of the International Conference on Interval and Stochastic Methods in Science and Engineering INTERVAL'92*, Moscow, 1992, Vol. 2, pp. 66–74.

[274] J. C. Martin, *Introduction to languages and the theory of computation*, McGraw-Hill, N.Y., 1991.

[275] S. Yu. Maslov, *Theory of deductive systems and its applications.*, MIT Press, Cambridge, MA, 1987.

[276] Yu. V. Matiyasevich, "Enumerable sets are diophantine", *Soviet Math. Doklady*, 1970, Vol. 11, pp. 354–357.

[277] Yu. V. Matiyasevich and J. Robinson, "Reduction of an arbitrary Diophantine equation to one in 13 unknowns", *Acta Arithmetica*, 1974, Vol. 27, pp. 521–553.

[278] V. D. Mazin and V. Kreinovich, *An explanation of the universal character of the transformation functions* $y = (a\ln x + b)/(c\ln x + d)$, Leningrad Polytechnical Institute and National Research Institute for Scientific and Technical Information (VINITI), 1988, 6 pp. (in Russian).

[279] V. D. Mazin and V. Kreinovich, *An important property of fractional-linear transformation functions*, Leningrad Polytechnical Institute and National Research Institute for Scientific and Technical Information (VINITI), 1988, 13 pp. (in Russian).

[280] K. Meer, "On the complexity of quadratic programming in real number models of computation", *Theoretical Computer Science*, 1994, Vol. 133, pp. 85–94.

[281] K. Meer, "Real number computations: on the use of information", *J. Symbolic Computation*, 1994, Vol. 18, pp. 199-206.

[282] K. Meer, "On the relations between discrete and continuous complexity theory", *Math. Log. Quarterly*, 1995, Vol. 41, No. 2, pp. 281–286.

[283] K. Mehlhorn, "On the size of sets of computable functions", In: *Proc. 14th IEEE Symposium on Switching and Automata Theory*, 1973, pp. 190–196.

[284] G. Mints, *A short introduction to modal logic*, CSLI (Center for the Study of Language and Information), Stanford University, Stanford, CA, 1992.

[285] D. Misane and V. Kreinovich, "A new characterization of the set of all intervals, based on the necessity to check consistency easily", *Reliable Computing*, 1995, Vol. 1, No. 3, pp. 285–298.

[286] D. Misane and V. Kreinovich, "The necessity to check consistency explains the use of parallelepipeds in describing uncertainty", *Geombinatorics*, 1996, Vol. 5, No. 3, pp. 109–120.

[287] D. Morgenstein and V. Kreinovich, "Which algorithms are feasible and which are not depends on the geometry of space-time", *Geombinatorics*, 1995, Vol. 4, No. 3, pp. 80–97.

[288] R. E. Moore, *Automatic error analysis in digital computation*, Lockheed Missiles and Space Co. Technical Report LMSD-48421, Palo Alto, CA, 1959.

[289] R. E. Moore and C. T. Yang, *Interval analysis*, Lockheed Missiles and Space Co. Technical Report LMSD-285875, Palo Alto, CA, 1959.

[290] R. E. Moore, *Interval analysis*, Prentice Hall, Englewood Cliffs, NJ, 1966.

[291] R. E. Moore, *Methods and Applications of Interval Analysis*, SIAM, Philadelphia, 1979.

[292] H. Moravec, *Time travel and computing*, Carnegie-Mellon University, Computer Science Department, Preprint, 1991.

[293] D. Morgenstein and V. Kreinovich, "Which algorithms are feasible and which are not depends on the geometry of space-time", *Geombinatorics*, 1995, Vol. 4, No. 3, pp. 80–97.

[294] D. Morgenstein and Murphy, "Application of parallel interval techniques to geophysics", *Reliable Computing*, 1995, Supplement (Extended Abstracts of APIC'95: International Workshop on Applications of Interval Computations, El Paso, TX, Febr. 23–25, 1995), p. 155.

[295] M. S. Morris and K. S. Thorne, "Wormholes in spacetime and their use for interstellar travel: a tool for teaching general relativity", *American Journal of Physics*, 1988, Vol. 56, May, pp. 395–412.

[296] M. S. Morris, K. S. Thorne, and U. Yurtzever, "Wormholes, time machnies, and the weak energy condition", *Physical Review Letters*, 1988, Vol. 61, pp. 1446–1149.

[297] K. G. Murty, *Linear and Combinatorial Programming*, Wiley, N.Y., 1976.

[298] P. Nahin, *Time machines: time travel in physics, metaphysics, and science fiction*, American Institute of Physics, N.Y., 1993.

[299] D. Nemir, V. Kreinovich, and E. Gutierrez, "Applications of interval computations to earthquake-resistant engineering: how to compute derivatives of interval functions fast". *Reliable Computing*, 1995, Vol. 1, No. 2, pp. 141–172.

[300] A. Nemirovskii, "Several NP-hard problems arising in robust stability analysis", *Mathematics of Control, Signals, and Systems*, 1993, Vol. 6, pp. 99–105.

[301] S. Nesterov and V. Kreinovich, "The worse, the better: a survey of paradoxical computational complexity of interval computations", In: Marcilia A. Campos (ed.), *Abstracts of the II Workshop on Computer Arithmetic, Interval and Symbolic Computation (WAI'96)*, Recife, Pernambuco, Brazil, August 7-8, 1996, pp. 61A–63A.

[302] A. Neumaier, Letter to J. Rohn, March 1986.

[303] A. Neumaier, *Interval Methods for Systems of Equations*, Cambridge University Press, Cambridge, 1990.

[304] H. T. Nguyen and V. Kreinovich, "Nested Intervals and Sets: Concepts, Relations to Fuzzy Sets, and Applications", In: R. B. Kearfott et al (eds.), *Applications of Interval Computations*, Kluwer, Dordrecht, 1996, pp. 245–290.

[305] H. T. Nguyen and V. Kreinovich, *Applications of continuous mathematics to computer science*, Kluwer, Dordrecht, 1997 (to appear).

[306] H. T. Nguyen, V. Kreinovich, V. Nesterov, and M. Nakamura, "On hardware support for interval computations and for soft computing: a theorem", *IEEE Transactions on Fuzzy Systems*, 1997, Vol. 5, No. 1, pp. 108–127.

[307] H. T. Nguyen, V. Kreinovich, and Qiang Zuo, "Interval-valued degrees of belief: applications of interval computations to expert systems and intelligent control", *International Journal of Uncertainty, Fuzziness, and Knowledge-Based Systems (IJUFKS)*, 1997, Vol. 5, No. 3, pp. 317–358.

[308] M. Nogueira and A. Nandigam, *Why intervals? Because if we allow other sets, tractable problems become intractable*, University of Texas at El Paso, Department of Computer Science, Technical Report UTEP-CS-97-7, 1997; available from the Web at `http://cs.utep.edu/vladik/1997/tr97-7.tex`.

[309] J. P. Norton, "Identification and application of bounded parameter models", *Proceeding of the 7th IFAC Symposium on Identification and Parameter Estimation*, York, U.K., 1985.

[310] I. D. Novikov, "Analysis of the operation of a time machine", *Soviet Physics JETP*, 1989, Vol. 68, pp. 439–443.

[311] I. D. Novikov, "Time machines and self-consistent evolutions in problems with self-interaction", *Physical Review D*, 1992, Vol. 45, pp. 1989–1994.

[312] P. V. Novitskii and I. A. Zograph, *Estimating the measurement errors*, Energoatomizdat, Leningrad, 1991 (in Russian).

[313] D. Oelschlaegel and R. Schmietow, "Anwendung von M-Intervallmatrizen bei der Quadratischen Optimierung", *Wiss. Z. Tech. Hoschsh. Leuna-Merseburg*, 1980, Vol. 22, pp. 539–544.

[314] D. Oelschlaegel and H. Suesse, "Fehlerabschaetzung beim Verfahren von Wolfe zur Loesung Quadratischer Optimierungsprobleme mit Hilfe der Intervallarithmetik", *Math. Operationsforsch. Statist., Ser. Optimization*, 1978, Vol. 9, pp. 389–396.

[315] D. Oelschlaegel and H. Suesse, "Fehlerabschaetzung beim einem Speziellen Quadratischen Optimierungsproblem", *Z. Angew. Math. Mech.*, 1979, Vol. 59, pp. 482–483.

[316] D. Oelschlaegel and H. Suesse, "Einschliessungdes Minimalpunktes eines Quadratischen Optimierungsproblems mit Hilfe der Intervallarithmetik", *Wiss. Z. Tech. Hoschsh. Leuna-Merseburg*, 1980, Vol. 22, pp. 535–538.

[317] W. Oettli and W. Prager, "Compatibility of approximate solution of linear equations with given error bounds for coefficients and right–hand sides", *Numerische Mathematik*, 1964, Vol. 6, pp. 405–409.

[318] V. P. Orevkov, "New proof of the uniqueness theorem for constructive differentiable functions of a complex variable", *Zapiski Nauchn. Sem. LOMI*, 1974, Vol. 40, pp. 119–126 (in Russian); English translation in *J. Soviet Math.*, 1977, Vol. 8, pp. 329–334.

[319] A. Ori, "Rapidly moving cosmic strings and chronology protection", *Physical Reviews D*, 1991, Vol. 44, pp. 2214–2215.

[320] A. I. Orlov, "How often are the observations normal?", *Industrial Laboratory*, 1991, Vol. 57, No. 7, pp. 770–772.

[321] J. Oxtoby, *Measure and category*, Springer-Verlag, N.Y., 1980.

[322] V. Pan, "Solving a polynomial equation: some history and recent progress", *SIAM Review*, 1997, Vol. 39, No. 2, pp. 187–220.

[323] C. H. Papadimitriou, *Computational Complexity*, Addison Wesley, San Diego, 1994.

[324] P. M. Pardalos, *Complexity in Numerical Optimization*, World Scientific, Singapore, 1993.

[325] R. Penrose, *The emperor's new mind*, Oxford University Press, Oxford, N.Y., 1989.

[326] M. Pittarelli (ed.), "Anytime algorithms and deliberation scheduling", *ACM SIGART BUlletin*, 1996, Vol. 7, No. 2.

[327] B. Poizat, "$Q = NQ$?", *Journal of Symbolic Logic*, 1986, Vol. 51, pp. 22–32.

[328] S. Poljak and J. Rohn, *Radius of nonsingularity*, Research Report, KAM Series 88–117, Charles University, Prague, December 1988.

[329] S. Poljak and J. Rohn, "Checking robust nonsingularity is NP-hard", *Mathematics of Control, Signals, and Systems*, 1993, Vol. 6, pp. 1–9.

[330] M. B. Pour-El and I. Richards, "A computable ordinary differential equation which possesses no computable solutions", *Annals of Mathematical Logic*, 1979, Vol. 17, pp. 61–90.

[331] M. B. Pour-El and I. Richards, "The wave equation with computable initial data such that its unique solution is not computable", *Advances in Mathematics*, 1981, Vol. 39, pp. 215–139.

[332] M. B. Pour-El and I. Richards, *Computability in analysis and physics*, Springer-Verlag, N.Y., 1989.

[333] S. Rabinovich, *Measurement errors: theory and practice*, American Institute of Physics, N.Y., 1993.

[334] L. B. Rall, *Automatic differentiation: techniques and applications*, Lecture Notes in Computer Science, Springer-Verlag, Berlin, 1981, Vol. 120.

[335] L. B. Rall, "Mean value and Taylor forms in interval analysis", *SIAM J. Math. Anal.*, 1983, Vol. 2, pp. 223–238.

[336] H. Ratschek and J. Rokne, "Optimality of the Centered Form", In: K. Nickel (ed.), *Interval Mathematics 1980,* Acad. Press, N.Y., 1980, pp. 499–508.

[337] H. Ratschek and J. Rokne, *New Computer Methods for Global Optimization.* Ellis Horwood Limited, Chichester, England, 1988.

[338] H. Ratschek and J. Rokne, "Interval methods", in: R. Horst and P. M. Pardalos, *Handbook of global optimization*, Kluwer Academic Publ., Boston, MA, 1995, pp. 751–828.

[339] S. Reyes and M. Clarke, *Logic for computer science*, Addison-Wesley, 1990.

[340] J. Rohn, *Input-output planning with inexact data*, Freiburger Intervall-Berichte, 1978, Vol. 9.

[341] J. Rohn, "Input-output model with interval data", *Econometrica*, 1980, Vol. 48, No. 3, pp. 767–769.

[342] J. Rohn, *Proofs to "Solving interval linear systems"*, Freiburger Intervall–Berichte, 84/7, Freiburg University, Freiburg, 1984.

[343] J. Rohn, "Some Results on Interval Linear Systems", *Freiburger Intervall-Berichte 85/4*, Freiburg, 1985, pp. 93–116.

[344] J. Rohn, *Miscellaneous results on linear interval systems*, Freiburger Intervall–Berichte 85/9, Freiburg University, Freiburg, 1985.

[345] J. Rohn, "Inner solutions of linear interval systems", In: K. Nickel (ed.), *Interval Mathematics 1985*, Lecture Notes in Computer Science, Vol. 212, Springer-Verlag, Berlin, 1986, pp. 157–158.

[346] J. Rohn, "Inverse–positive interval matrices", *Zeitschrift für Angewandte Mathematik und Mechanik*, 1987, Vol. 67, pp. T492–T493.

[347] J. Rohn, "Systems of linear interval equations", *Linear Algebra and Its Applications*, 1989, Vol. 126, pp. 39–78.

[348] J. Rohn, "Interval matrices: Singularity and real eigenvalues", *SIAM Journal on Matrix Analysis and Applications*, 1993, Vol. 14, pp. 82–91.

[349] J. Rohn, "Inverse Interval Matrix", *SIAM J. Numer. Anal.*, 1993, Vol. 30, pp. 864–870.

[350] J. Rohn, "Positive definiteness and stability of interval matrices", *SIAM Journal on Matrix Analysis and Applications*, 1994, Vol. 15, pp. 175–184.

[351] J. Rohn, "Checking positive definiteness or stability of symmetric interval matrices is NP-hard", *Commentationes Mathematicae Universitatis Carolinae*, 1994, Vol. 35, pp. 795–797.

[352] J. Rohn, "Enclosing solutions of linear interval equations is NP-hard", *Computing*, 1994, Vol. 53, pp. 365–368.

[353] J. Rohn, *NP-hardness results for linear algebraic problems with interval data*, In J. P. Herzberger (ed.), *Topics in Validated Computations, Proceedings of the IMACS-GAMM International Workshop on Validated Computation, Oldenburg, Germany, 30 August – 3 September 1993*, Elzevier (North Holland), Amsterdam, 1994.

[354] J. Rohn, *Linear interval equations: computing sufficiently accurate enclosures is NP-hard*, Technical Report No. 621, Institute of Computer Science, Academy of Sciences of the Czech Republic, Prague 1995, 7 pp.

[355] J. Rohn, "Checking Bounds on Solutions of Linear Interval Equations is NP-Hard", *Linear Algebra and its Applications*, 1995, Vol. 223/224, pp. 589–596.

[356] J. Rohn, "Linear Interval Equations: Computing Enclosures with Bounded Relative Overestimation is NP-Hard", In: R. B. Kearfott and V. Kreinovich (eds.), *Applications of Interval Computations*, Kluwer, Boston, MA, 1996, pp. 81–89.

[357] J. Rohn, "An algorithm for checking stability of symmetric interval matrices", *IEEE Transactions on Automatic Control*, 1996, Vol. 41, pp. 133–136.

[358] J. Rohn, "Complexity of solving linear interval equations", *Zeitschrift für Angewandte Mathematik und Mechanik (ZAMM)*, 1996, Vol. 76, Supplement 3, pp. 271–274.

[359] J. Rohn, Unpublished results, 1996.

[360] J. Rohn and V. Kreinovich, "Computing exact componentwise bounds on solutions of linear systems with interval data is NP-hard," *SIAM Journal on Matrix Analysis and Applications (SIMAX)*, 1995, Vol. 16, pp. 415–420.

[361] J. Rohn and G. Rex, "Interval *P*–matrices", *SIAM Journal on Matrix Analysis and Appl.*, 1996, Vol. 17, No. 4, pp. 1020–1024.

[362] H. Rogers, Jr. *Theory of recursive functions and effective computability*, Mc-Graw Hill Co., N.Y., 1967.

[363] J. Rokne, "Optimal Computation of the Bernstein Algorithm for the Bound of an Interval Polynomial," *Computing*, 1982, Vol. 28, pp. 239–246.

[364] R. Rosen, "Church's thesis and its relation to the concept of realizability in biology and physics", *Bull. Math. Biophysics*, 1962, Vol. 24, pp. 375–393.

[365] R. Rosen, *Anticipatory systems*, Pergamon Press, N.Y., 1985.

[366] R. Rosen, In: R. Herken (ed.) *The universal Turing machine: a half-century survey*, Kammerer & Unverzagt, Hamburg, 1988, pp. 523–537.

[367] R. Rosen, *Life itself: a comprehensive inquiry into the nature, origin, and fabrication of life*, Columbia University Press, N.Y., 1991.

[368] R. Rosen, "What can we know?", In: J. L. Casti, A. Karlqvist (eds.), *Beyond belief: randomness, prediction, and explanation in science*, CRC Press, Boca Raton, FL, 1991, pp. 1–13.

[369] S. M. Rump, "Solving Algebraic Problems with High Accuracy", in: U. Kulisch and W. Miranker, eds., *A New Approach to Scientific Computation*, Academic Press, N.Y., 1983, pp. 51–120.

[370] S. M. Rump, "Verification methods for dense and sparse systems of equations", In: J. Herzberger, ed., *Topics in Validated Computations*, North–Holland, Amsterdam, 1994, pp. 63–135.

[371] S. M. Rump, "Almost sharp bounds for the componentwise distance to the nearest singular matrix", Submitted to *Linear and Multilinear Algebra*, 1996.

[372] S. M. Rump, "Bounds for the componentwise distance to the nearest singular matrix", *SIAM Journal on Matrix Analysis and Applications*, 1997, Vol. 18, No. 1, pp. 83–103.

[373] M. Ruse, *The Darwinian paradigm: essays on its history, philosophy, and religious implications*, Routledge, London, 1989.

[374] J. A. Scales, A. Gersztenkorn, and S. Treitel, "Fast l^p solutions of large, sparse, linear systems: Application to seismic travel time tomography", *J. Comput. Physics*, 1988, Vol. 75, pp. 314–333.

[375] M. J. Schaefer and T. Bubeck, "A parallel complex zero finder", *Reliable Computing*, 1995, Vol. 1, No. 3, pp. 317–324.

[376] T. J. Schaefer, "The complexity of satisfiability problems", *Proc. 10th Annual ACM Symposium on Theory of Computing*, Association for Computing Machinery, N.Y., 1978, pp. 216–226.

[377] U. Schendel, *Sparse matrices: numerical aspects with applications for scientists and engineers*, Ellis Horwood, Chichester, West Sussex, England, and Halsted Press, N.Y., 1989.

[378] D. Schirmer and V. Kreinovich, "Towards a More Realistic Definition of Feasibility", *Bulletin of the European Association for Theoretical Computer Science (EATCS)*, 1996, Vol. 90, pp. 151–153.

[379] G. Schmitgen, "Intervalanalytische Approximationsfragen", *Z. Angew. Math. Mech.*, 1971, Vol. 51, pp. T72–T73.

[380] J. R. Schoenfield, *Mathematical logic.* Addison-Wesley, 1967.

[381] C. A. Schnepper and M. A. Stadtherr, "Application of a Parallel Interval Newton/Generalized Bisection Algorithm to Equation-Based Chemical Process Flowsheeting", *Interval Computations*, 1993, No. 4, pp. 40–64.

[382] A. Schrijver, *Theory of Integer and Linear Programming*, Wiley, Chichester 1986.

[383] J. Schröder, *Operator Inequalities*, Academic Press, N.Y., 1980.

[384] M. J. Schulte and E. E. Swartzlander, Jr., "Parallel Hardware Designs for Correctly Rounded Elementary Functions", *Interval Computations*, 1993, No. 4, pp. 65–88.

[385] L. L. Schumaker, *Spline functions: basic theory*, Wiley, N.Y., 1981.

[386] F. C. Schweppe, "Recursive state estimation: unknown but bounded errors and system inputs", *IEEE Transactions on Automatic Control*, 1968, Vol. 13, p. 22.

[387] F. C. Schweppe, *Uncertain dynamic systems*, Prentice Hall, Englewood Cliffs, NJ, 1973.

[388] A. Seidenberg, "A new decision method for elementary algebra", *Annals of Math.*, 1954, Vol. 60, pp. 365–374.

[389] V. V. Shaidurov and S. P. Shary, *Solution of interval algebraic problem on tolerances*, Krasnoyarsk, Academy of Sciences Computing Center, Technical Report No. 5, 1988 (in Russian).

[390] S. P. Shary, "On computibility of linear tolerance problem", *Interval Computations*, 1991, No. 1, pp. 92–98 (in Russian).

[391] S. P. Shary, "A new class of algorithms for optimal solution of interval linear systems", *Interval Computations,*, 1992, No. 2(4), pp. 18–29.

[392] S. P. Shary, "On optimal solution of interval linear equations", *SIAM J. Numer. Anal.*, 1995, Vol. 32, pp. 610–630.

[393] S. P. Shary, "Algebraic approach to the interval linear static identification, tolerance, and control problems, or One more application of Kaucher arithmetic", *Reliable Computing*, 1996, Vol. 2, No. 1, pp. 3–34.

[394] G. L. Shevlyakov and N. O. Vil'chevskiy, "On the choice of an optimization criterion under uncertainty in interval computations - nonstochastic approach", *Reliable Computing*, 1995, Supplement (Extended Abstracts of APIC'95: International Workshop on Applications of Interval Computations, El Paso, TX, Febr. 23–25, 1995), p. 188–189.

[395] Z. C. Shi and W. B. Gao, "A necessary and sufficient condition for the positive–definiteness of interval symmetric matrices", *International Journal of Control*, 1986, Vol. 43, pp. 325–328.

[396] Yu. I. Shokin, "On interval problems, interval algorithms and their computational complexity", In: G. Alefeld, A. Frommer, and B. Lang (eds.), *Scientific Computing and Validated Numerics*, Akademie-Verlag, Berlin, 1996, pp. 314–328.

[397] I. E. Shvetsov and V. V. Telerman, "Intervals and multi-intervals in incompletely defined computational models", In: *Proceedings of the International Conference on Interval and Stochastic Methods in Science and Engineering INTERVAL'92*, Moscow, 1992, Vol. 1, pp. 201–203 (in Russian; English abstract Vol. 2, p. 100).

[398] S. Smale, "Some remarks on the foundations of numerical analysis", *SIAM Review*, 1990, Vol. 32, No. 2, pp. 211–220.

[399] S. Smith and V. Kreinovich, "In Case of Interval Uncertainty, Optimal Control is NP-Hard Even for Linear Plants, so Expert Knowledge is Needed", *Reliable Computing*, 1995, Supplement (Extended Abstracts of APIC'95: International Workshop on Applications of Interval Computations, El Paso, TX, Febr. 23–25, 1995), pp. 190–193.

[400] C. B. Soh, "Necessary and sufficient conditions for stability of symmetric interval matrices", *International Journal of Control*, 1990, Vol. 51, pp. 243–248.

[401] G. N. Solopchenko and V. Ya. Kreinovich, "Approximation of the uncertainty domain by ellipsoids when solving inverse problems in measurement techniques", *Proceedings of the IV USSR National Symposium on Methods of Identification Theory in the Problems of Measuring Techniques and Metrology*, Novosibirsk, 1985, pp. 75–76 (in Russian).

[402] S. T. Soltanov, "Asymptotic of the function of the outer estimation ellipsoid for a linear singularly perturbed controlled system", In: S. P. Shary and Yu. I. Shokin (editors), *Interval Analysis*, Krasnoyarsk, Academy of Sciences Computing Center, Technical Report No. 17, 1990, pp. 35–40 (in Russian).

[403] M. Stannett, "X-machines and the halting problem: building a super-Turing machine", *Formal Aspects of Computing*, 1990, Vol. 2, pp. 331–341.

[404] J. Stoer and R. Bulirsch, *Introduction to Numerical Analysis*, Springer–Verlag, Berlin, 1980.

[405] N. Strickland, *What can we compute if we use computational devices based on non-Newtonian physical phenomena: Church's thesis revisited*, Master Thesis, University of Texas at El Paso, Computer Science Department, 1994.

[406] P. Yu. Suvorov, "On the recognition of the tautological nature of propositional formulas", *Zap. Nauchn. Semin. Leningr. Otd. Mat. Inst. Steklov*, 1976, Vol. 60, pp. 197–206 (in Russian); Engl. translation in *J. Sov. Math.*, 1980, Vol. 14, 1556–1562.

[407] A. Tarantola, *Inverse problem theory: methods for data fitting and model parameter estimation*, Elsevier, Amsterdam, 1987.

[408] A. Tarski, *A decision method for elementary algebra and geometry*, 2nd ed., Berkeley and Los Angeles, 1951.

[409] K. S. Thorne, "Do the laws of physics permit closed timelike curves?", *Annals of the New York Academy of Sciences*, 1991, Vol. 631, pp. 182–193.

[410] A. N. Tikhonov, V. Y. Arsenin, *Solutions of ill-posed problems,* V. H. Winston & Sons, Washington, DC, 1977.

[411] F. J. Tipler, "Rotating cylinders and the possibility of global causality violation", *Physical Review D*, 1974, Vol. 9, pp. 2203–2206.

[412] F. J. Tipler, "Causality violation in asymptotically flat space-times", *Physical Review Letters*, 1976, Vol. 37, pp. 879–882.

[413] F. J. Tipler, "Singularities and causality violation", *Annals of Physics*, 1977, Vol. 108, pp. 1–36.

[414] B. Traylor and V. Kreinovich, "A bright side of NP-hardness of interval computations: interval heuristics applied to NP-problems", *Reliable Computing,* 1995, Vol. 1, No. 3, pp. 343–360.

[415] R. Trejo and A. I. Gerasimov, "Choosing interval functions to represent measurement inaccuracies: group-theoretic approach", *Reliable Computing,* 1995, Supplement (Extended Abstracts of APIC'95: International Workshop on Applications of Interval Computations, El Paso, TX, Febr. 23–25, 1995), pp. 207–210.

[416] G. Sh. Utyubaev, "On the ellipsoid method for a system of linear differential equations", In: S. P. Shary (editor), *Interval Analysis*, Krasnoyarsk, Academy of Sciences Computing Center, Technical Report No. 16, 1990, pp. 29–32 (in Russian).

[417] L. G. Valiant and V. V. Vazirani, "NP is as easy as detecting unique solutions", *Theoretical Computer Science*, 1986, Vol. 47, pp. 85–93.

[418] S. A. Vavasis, *Nonlinear optimization: complexity issues*, Oxford University Press, N.Y., 1991.

[419] E. Villa, A. Bernat, and V. Kreinovich "Estimating errors of indirect measurement on realistic parallel machines: routings on 2-D and 3-D meshes that are nearly optimal", *Interval Computations*, 1993, No. 4, pp. 154–175.

[420] K. Villaverde and V. Kreinovich. "A linear-time algorithm that locates local extrema of a function of one variable from interval measurement results," *Interval Computations*, 1993, No. 4, pp. 176–194.

[421] K. Villaverde and V. Kreinovich, "Parallel algorithm that locates local extrema of a function of one variable from interval measurement results", *Reliable Computing*, 1995, Supplement (Extended Abstracts of APIC'95: International Workshop on Applications of Interval Computations, El Paso, TX, Febr. 23–25, 1995), pp. 212–219.

[422] H. M. Wadsworth, Jr. (editor), *Handbook of statistical methods for engineers and scientists*, McGraw-Hill Publishing Co., N.Y., 1990.

[423] W. Walter, *Differential and integral inequalities*, Springer-Verlag, N.Y., 1970.

[424] K. Wang and A. Michel, "On sufficient conditions for the stability of interval matrices", *Systems and Control Letters*, 1993, Vol. 20, pp. 345–351.

[425] V. Weispfenning, "Parametric linear and quadratic optimization by elimination", *Journal of Symbolic Computation*, 1997 (to appear).

[426] V. Weispfenning, "Simulation and optimization by quantifier elimination", *Journal of Symbolic Computation*, 1997 (to appear).

[427] V. Weispfenning, "Quantifier elimination for real algebra – the quadratic case and beyond", *Appl. Alg. Engng. Commun. Comp.*, 1997, Vol. 8, No. 2, pp. 85–101.

[428] A. J. Wilkie, "Model completeness results for expansions of the ordered field of real numbers by restricted Pfaffian functions and the exponential function", *Journal of American Mathematical Society*, 1996, Vol. 9, No. 4, pp. 1051–1094.

[429] J. H. Wilkinson, *The Algebraic Eigenvalue Problem*, Oxford University Press, London, 1965.

[430] E. O. Wilson, *Sociobiology: the new synthesis*, Belknap Press, Cambridge, MA, 1975.

[431] E. O. Wilson, *On human nature*, Harvard University Press, Cambridge, MA, 1978.

[432] J. Wolff von Gudenberg, "Design of a parallel linear algebra library for verified computation", *Reliable Computing*, 1995, Vol. 1, No. 4, pp. 411–420.

[433] A. G. Yakovlev, "Machine arithmetic of multi-intervals", *Voprosy Kibernetiki (Problems of Cybernetics)*, 1987, Vol. 125, pp. 66–81 (in Russian).

[434] J. Yen and N. Pfluger, "Path planning and execution using fuzzy logic", In: *AIAA Guidance, Navigation and Control Conference*, New Orleans, LA, 1991, Vol. 3, pp. 1691–1698.

[435] J. Yen and N. Pfluger, "Designing an adaptive path execution system", In: *IEEE International Conference on Systems, Man and Cybernetics*, Charlottesville, VA, 1991.

[436] J. Yen, N. Pfluger, and R. Langari, "A defuzzification strategy for a fuzzy logic controller employing prohibitive information in command formulation", *Proceedings of IEEE International Conference on Fuzzy Systems, San Diego, CA, March 1992.*

INDEX